Innovative Materials and Processes for Removal of Biopersistent Pollutants

Innovative Materials and Processes for Removal of Biopersistent Pollutants

Editors

Andrea Petrella
Marco Race
Danilo Spasiano

MDPI

Basel • Beijing • Wuhan • Barcelona • Belgrade • Novi Sad • Cluj • Manchester

Editors
Andrea Petrella
Department of Civil, Environmental, Land, Building Engineering and Chemistry
Polytechnic University of Bari
Bari
Italy

Marco Race
Department of Civil and Mechanical Engineering
University of Cassino and Southern Lazio
Cassino
Italy

Danilo Spasiano
Department of Civil, Environmental, Land, Building Engineering and Chemistry
Polytechnic University of Bari
Bari
Italy

Editorial Office
MDPI
St. Alban-Anlage 66
4052 Basel, Switzerland

This is a reprint of articles from the Special Issue published online in the open access journal *Processes* (ISSN 2227-9717) (available at: www.mdpi.com/journal/processes/special_issues/biopersistent_removal).

For citation purposes, cite each article independently as indicated on the article page online and as indicated below:

Lastname, A.A.; Lastname, B.B. Article Title. *Journal Name* **Year**, *Volume Number*, Page Range.

ISBN 978-3-0365-8945-9 (Hbk)
ISBN 978-3-0365-8944-2 (PDF)
doi.org/10.3390/books978-3-0365-8944-2

Contents

About the Editors

Andrea Petrella

Andrea Petrella (Associate Professor) was born in Bari in 1976. He graduated with an M.Sc. in Chemistry from the University of Bari in July 2001 and received his Ph.D. in Chemical Sciences from the University of Bari in 2005. In 2007, he became the Assistant Professor in Materials Science and Technology at the Polytechnic University of Bari. Since March 2022, he has been an Associate Professor in Materials Science and Technology at the Polytechnic University of Bari. He is the author of more than 70 papers in refereed scientific journals (SCOPUS: h-index 26, citations 1748). His research fields are summarized as follows: use of recycling organic and inorganic materials in the building trade and/or in the removal of heavy metals present in wastewater; photocatalytic materials for the degradation of bio-persistent pollutants in water and wastewater; and nanocomposites for energy conversion and for novel optical devices.

Marco Race

Marco Race, an Associate Professor, was born in Napoli. He graduated with an MSc in Environmental Engineering at Universita degli Studi di Napoli Federico II in May 2012 and gained his Ph.D. in Environmental Systems Analysis at UNINA in 2016, according to the European Label. His main research fields concern the treatment of waste or wastewater, the remediation of soil and groundwater, novel contaminant (bio)monitoring and risk assessment approaches, and trace metals and organics in biogeochemical cycles. He is the author of more than 100 papers published in international journals, conference proceedings, and book chapters. He won the international award for soil reclamation.

Danilo Spasiano

Danilo Spasiano, Associate Professor, was born in Naples in 1984. He graduated with an MSc in Environmental Engineering at Universita degli Studi di Napoli Federico II in October 2009 and obtained his Ph.D. in Chemical Engineering at the same university in 2013. His main research fields concern the treatment of asbestos, containing wastes, and wastewater treatment by means of an advanced oxidation process. He has coauthored more than 60 papers published in international journals and indexed on Scopus.

Preface

Hazardous artificial contaminants are molecules of inorganic and organic nature (pharmaceuticals, food sources, heavy metals, dyes, personal care products, detergents, flame retardants, cosmetics, and pesticides) with potential toxicological effects on human health and the environment (air, water, and soil) due to their ubiquity at trace levels.

These products could be bio-persistent during conventional treatment processes; accordingly, the adoption of proper and innovative technologies is necessary for the removal of these hazardous, persistent chemicals before their release into the environment.

The aim of this reprint, entitled "Innovative Materials and Processes for the Removal of Biopersistent Pollutants", was to collect studies devoted to the recent progress and new perspectives in the treatment and removal of hazardous artificial contaminants in the air, soil, and water supply.

Andrea Petrella, Marco Race, and Danilo Spasiano
Editors

processes

MDPI

Editorial

Innovative Materials and Processes for Removal of Biopersistent Pollutants

Andrea Petrella [1,*], Marco Race [2] and Danilo Spasiano [1,*]

1 Dipartimento di Ingegneria Civile, Ambientale, Edile, Del Territorio e di Chimica, Politecnico di Bari, Via E. Orabona 4, 70125 Bari, Italy
2 Dipartimento di Ingegneria Civile e Meccanica, Università degli Studi di Cassino e del Lazio Meridionale, Via di Biasio 43, 03043 Cassino, Italy
* Correspondence: andrea.petrella@poliba.it (A.P.); danilo.spasiano@poliba.it (D.S.)

1. Introduction

The aim of this Special Issue "Innovative Materials and Processes for Removal of Biopersistent Pollutants" (https://www.mdpi.com/journal/processes/special_issues/biopersistent_removal; accessed on 12 January 2023) was to collect researches devoted to the recent progress and new perspectives in the processes of treatment and removal of hazardous artificial contaminants in air, soil, and water supply.

For this purpose, the Guest Editors share some comments. Fifteen papers were submitted, ten of which were published, and the publication times reflected those of the journal.

A range of different but complementary topics were addressed, from environmental regulation [1] to the development of applications that can be useful in supporting decisions [2–4]. Furthermore, many studies analyzed how the modification of some materials can be effective for the reduction in pollutant dispersions [5–9] and the application of an innovative plant configuration [10].

As evidence of the high interest in the topics covered, to date, all the papers have been cited in other works, reaching 18 citations [5], 6 citations [6,7], 5 citations [3,10], 4 citations [4], 2 citations [1,8,9], and 1 citation [2].

The Guest Editors want to thank all the authors for the appreciation received for this Special Issue, and believe that these topics can be a starting point for future research. They also thank the Editors in Chief for the opportunity to coordinate this SI.

check for updates

Citation: Petrella, A.; Race, M.; Spasiano, D. Innovative Materials and Processes for Removal of Biopersistent Pollutants. *Processes* **2023**, *11*, 336. https://doi.org/10.3390/pr11020336

Received: 11 January 2023
Accepted: 13 January 2023
Published: 20 January 2023

2. Innovative Materials for Removal of Biopersistent Pollutants

In the paper by Lin et al. [8], a mesoporous activated carbon (AC) was prepared from water caltrop husk at 750 °C for 90 min. This material could be used as an excellent adsorbent for the removal of methylene blue from the liquid phase due to its fast adsorption rate and maximal adsorption capacity (126.6 mg/g), and the process could be represented by a pseudo-second-order model.

In the paper by Loffredo et al. [9], a solid by-product named digestate, obtained through anaerobic digestion, was used as a biosorbent of organic and inorganic pollutants in wastewater treatment and soil remediation. The characterization of this material, the qualitative and quantitative aspects of the adsorption/desorption of pesticides and xenoestrogens, and data modeling were examined.

In the paper by Huong at al. [7], Fe-Cu materials were synthesized using the chemical plating method and tested for the removal of phenol from aqueous solution through internal microelectrolysis. Various parameters such as pH, time, Fe-Cu material weight, phenol concentration, and shaking speed were investigated. An evaluation of the optimal process was carried out in real coking wastewater from a coal factory, and resulted in treated wastewater with favorable water indicators.

In the study by Wang et al. [6], TiO_2 nanoparticles were synthesized, characterized, and combined with aged waste reactors to treat landfill leachate. The optimal process conditions were determined, such as the effects of the ultraviolet irradiation time, amount of the catalyst, and pH on the removal efficiency for COD and color in the leachate. The photocatalytic/biological combined treatment of landfill leachate was shown, together with excellent recyclability of the catalyst.

Zhang et al. [5] studied how to modify biochar to cover landfills and simultaneously oxidize methane, in order to reduce emissions into the atmosphere. The biochar was modified in order to increase its hydrophobicity and excellent results were achieved by coupling the silane agent KH-570.

3. Processes for Removal of Biopersistent Pollutants

Through legislation, it is possible to incentivize the progress of green industrial technology. For this reason, Shi et al. [1], took China as a case study and evaluated the effects of environmental regulation on human activities.

In the paper by Petrella et al. [10], an innovative unit was employed for the study of the UV/TiO_2 photo-catalytic degradation of biopersistent textile azo-dyes. The chemical, physical, and hydraulic/hydrodynamic parameters of the system influenced the degradation kinetics. A comparison of the removal efficiencies between dyes such as methyl red and methylene blue was carried out in consideration of the pH of the solution.

Innovative processes for the removal of heavy metals from aqueous solutions were also analyzed in consideration of the sanitary risks and environmental hazards of these toxic compounds [11,12]. In this context, parametric mathematical modelling techniques, such as response surface methodology (RSM) and artificial neural networks (ANNs), have been chosen as a tool for the optimization of operating conditions [13]. In the first paper by Fertu et al. [3], experimental laboratory data on the biosorption of Pb(II), Cd(II), and Zn(II) from aqueous media using soybean and soybean waste biomasses were exploited through modeling and optimization. For this purpose, RSM was used as a model, followed by optimization based on numerical methods. The solutions confirmed the efficiency of the sorbents in the removal of heavy metals and the results were validated experimentally.

In the second paper by Fertu et al. [4], the results of the previous research based on heavy metal retention in soybean and soybean waste biomasses were capitalized. The data were processed by applying a methodology based on ANNs and evolutionary algorithms (EAs), the latter represented by the differential evolution (DE) algorithm. A simultaneous training and determination of the topology was performed, and the hSADE-NN hybrid algorithm was applied to obtain optimal models for the heavy metal retention process.

Finally, a platform to support the drafting of strategic plans aimed at safeguarding water resources was created by Liang et al. [2]. This tool can be used to prevent water pollution and manage emergencies.

Acknowledgments: The guest editors would like to thank all the authors and the reviewers. Special acknowledgments to all the staff of the *Processes* Editorial Office for their great support during the preparation of this Special Issue.

Conflicts of Interest: The authors declare no conflict of interest.

References

1. Shi, J.; Yu, Y. To Advance Industrial Green Technology via Environmental Governance—Evidence from China's Industrial Sector. *Processes* **2021**, *9*, 1797. [CrossRef]
2. Liang, J.; Xie, J.; Wang, X.; Wang, R.; Jin, T.; Wang, S. Research on the Construction and Application Mode of Digital Plans for Sudden Water Pollution Events. *Processes* **2022**, *10*, 833. [CrossRef]
3. Fertu, D.I.; Bulgariu, L.; Gavrilescu, M. Modeling and Optimization of Heavy Metals Biosorption by Low-Cost Sorbents Using Response Surface Methodology. *Processes* **2022**, *10*, 523. [CrossRef]
4. Fertu, D.I.; Dragoi, E.N.; Bulgariu, L.; Curteanu, S.; Gavrilescu, M. Modeling the Biosorption Process of Heavy Metal Ions on Soybean-Based Low-Cost Biosorbents Using Artificial Neural Networks. *Processes* **2022**, *10*, 603. [CrossRef]

5. Zhang, M.; Zhu, H.; Xi, B.; Tian, Y.; Sun, X.; Zhang, H.; Wu, B. Surface Hydrophobic Modification of Biochar by Silane Coupling Agent KH-570. *Processes* **2022**, *10*, 301. [CrossRef]
6. Wang, C.; Sun, X.; Shan, H.; Zhang, H.; Xi, B. Degradation of Landfill Leachate Using UV-TiO2 Photocatalysis Combination with Aged Waste Reactors. *Processes* **2021**, *9*, 946. [CrossRef]
7. Huong, D.T.; Van Tu, N.; Anh, D.T.T.; Tien, N.A.; Ngan, T.T.K.; Van Tan, L. Removal of Phenol from Aqueous Solution Using Internal Microelectrolysis with Fe-Cu: Optimization and Application on Real Coking Wastewater. *Processes* **2021**, *9*, 720. [CrossRef]
8. Lin, Y.-Q.; Tsai, W.-T. Liquid-Phase Removal of Methylene Blue as Organic Pollutant by Mesoporous Activated Carbon Prepared from Water Caltrop Husk Using Carbon Dioxide Activation. *Processes* **2021**, *9*, 238. [CrossRef]
9. Loffredo, E.; Carnimeo, C.; Silletti, R.; Summo, C. Use of the Solid By-Product of Anaerobic Digestion of Biomass to Remove Anthropogenic Organic Pollutants with Endocrine Disruptive Activity. *Processes* **2021**, *9*, 2018. [CrossRef]
10. Petrella, A.; Spasiano, D.; Cosma, P.; Rizzi, V.; Race, M.; Mascolo, M.C.; Ranieri, E. Methyl Orange Photo-degradation by Tio2 in a Pilot Unit under Different Chemical, Physical, and Hydraulic Conditions. *Processes* **2021**, *9*, 205. [CrossRef]
11. Petrella, A.; Petruzzelli, V.; Basile, T.; Petrella, M.; Boghetich, G.; Petruzzelli, D. Recycled Porous Glass from Municipal/Industrial Solid Wastes Sorting Operations as a Lead Ion Sorbent from Wastewaters. *React. Funct. Polym.* **2010**, *70*, 203–209. [CrossRef]
12. Petrella, A.; Petrella, M.; Boghetich, G.; Basile, T.; Petruzzelli, V.; Petruzzelli, D. Heavy Metals Retention on Recycled Waste Glass from Solid Wastes Sorting Operations: A Comparative Study among Different Metal Species. *Ind. Eng. Chem. Res.* **2012**, *51*, 119–125. [CrossRef]
13. Policastro, G.; Luongo, V.; Frunzo, L.; Fabbricino, M. A Comprehensive Review of Mathematical Models of Photo Fermentation. *Crit. Rev. Biotechnol.* **2021**, *41*, 628–648. [CrossRef] [PubMed]

processes

Article

Research on the Construction and Application Mode of Digital Plans for Sudden Water Pollution Events

Jichao Liang [1], Jiancang Xie [1,*], Xue Wang [1,2], Rui Wang [1], Tao Jin [1] and Shaojiu Wang [1]

1 State Key Laboratory of Eco-Hydraulics in Northwest Arid Region of China, Xi'an University of Technology, Xi'an 710048, China; xutljchao@163.com (J.L.); doraemon2129@126.com (X.W.); wr92333@163.com (R.W.); jintao20200515@163.com (T.J.); shaojiuwang01@163.com (S.W.)
2 Coordination Center of Water Resources Development Survey and Hanjiang to Weihe River Project of Shaanxi Province, Xi'an 710004, China
* Correspondence: jcxie@mail.xaut.edu.cn; Tel.: +86-18792713251

Abstract: Water pollution is an important aspect of a national water treatment. Sudden water safety incidents are random and destructive, often bringing about huge losses of life and property. Due to the uncertainty of sudden water pollution, it is difficult to respond in a timely and rapid manner. Emergency personnel must deal with emergencies quickly and effectively to reduce the harm caused by these emergencies. The randomness and uncertainty of sudden water pollution events make emergency work more complicated; it is difficult for current emergency plans to play guiding roles in complex responses. The decision-making and use of traditional water safety procedures largely depend on the experiences of command personnel, as well as on the emergency plan, which often has poor applicability. This can result in ineffective implementation of emergency actions and use of resources stemming from the high subjectivity and low efficiency of emergency plans. In this paper, we summarize previous research on digital planning and platform component technology exploration in order to evaluate the use of sudden water safety emergency procedures. We first analyze the main problems in the construction and use of emergency plans (e.g., the lack of experience and adaptability). Secondly, based on the decision-making support platform, a digital emergency plan database for water pollution emergencies was established by using component technology and knowledge map technology. In doing so, the decision support platform could enable the rapid construction of digital plans that improve application efficiency in an actual response scenario. Finally, through the system example, this system model can be quickly matched from the plan database to the emergency plan that meets the current scenario. It is a recommended model used to provide rapid and effective assistance for emergency management and improve emergency efficiency.

Keywords: sudden water pollution events; digital emergency plan; emergency response; emergency plan management; knowledge visualization; decision support system

check for updates

Citation: Liang, J.; Xie, J.; Wang, X.; Wang, R.; Jin, T.; Wang, S. Research on the Construction and Application Mode of Digital Plans for Sudden Water Pollution Events. *Processes* **2022**, *10*, 833. https://doi.org/10.3390/pr10050833

Academic Editors: Andrea Petrella, Marco Race and Danilo Spasiano

Received: 11 March 2022
Accepted: 21 April 2022
Published: 23 April 2022

1. Introduction

Due to natural disasters, manmade accidents, etc., sudden water pollution incidents frequently occur. These pollutants enter a water body in a short period of time, causing pollution and endangering the normal social order, economic activities, and aquatic ecosystems. The occurrences of sudden water pollution incidents are beyond the scope of normal human disposal; they are generally instantaneous occurrences, with complex performances, destructive consequences, with an urgent need of disposal, and surrounded by uncertainty. It is precisely because of these characteristics that sudden water pollution events are often beyond the scope of human disposal; they seriously endanger the normal social order, economic activities, and water ecological environment [1].

Emergency plans are pre-prepared action arrangements for possible emergencies. The specific work of each emergency phase can be quickly identified in the disposal process through the plan. An emergency plan is a principled plan for emergency response, which plays an operational guiding role in emergency response and in the full disposal of events. Although emergency plan systems have been formed, at present, in response to sudden water pollution events, their guiding roles have not been effectively utilized. Until now, even if emergency measures have (basically) been formed, their value in solving sudden water pollution events has not yet been wielded. For one thing, a lack of concrete precautionary measures, such as experience assistance, plays a pivotal role in successfully solving a real emergency. There is a lack of (specific) directions on how to respond to sudden water pollution incidents, and there are no specific expressions of incident classification, division of responsibility, disposal operations, etc., regarding the occurrence of the incident. Moreover, the application mode is not flexible enough. Most existing emergency response plans are documents in written text, which can neither achieve cross-combinations between available plans nor adapt to changes (regarding time and flexibility) in emergency processes [2]. Therefore, based on modern information technology, one important way to solve current emergencies is to make better use of precautionary measures and to support emergency solutions.

Scientists in European countries have set up corresponding departments to respond to sudden water pollution events. Robert Health proposed an emergency framework for emergencies from a macro perspective, i.e., weakening them from different sources, locations, scopes, etc., around the root causes of disasters, effectively reducing the negative impact of sudden disasters [3]. Most scholars focus on micro-technical processing, such as Deroux, who built a monitoring database for reservoir water quality based on the laws of space–time evolution of organic compounds in surface water [4]. Shafiee Mohammadmehdi analyzed the impact of sudden water pollution events on human water use and realized a real-time simulation by establishing a dynamic transport model of pollutants [5]. Chinese scholarly research on emergency management regarding sudden water pollution events mostly focus on emergency early warning and risk decision-making; i.e., Yu Zhaohui discussed techniques surrounding the early warning of sudden water pollution events [6]. Guo Yuan introduced methods to deal with sudden water pollution in reservoirs [7].

With the development of information technology, information research aimed at emergencies is increasing. The German Emergency Planning Information System (deNIS) was set up to support the jobs of victim managers [8]; the deNISII was set up to provide information services for securing emergencies, to estimate current situations, to confront problems of a disaster, to analyze securing resorts, etc. Integrated emergency management (IEM) boasts such functions as estimating the optional risks of an accident, recognizing the accident, making/evaluating precautionary measures, urgent deployment, etc. [9]. Ohio's Oil and Gas Field Emergency Response System in America provides leaders and operational staff facing emergencies with real-time contingent information; analyzing functions and information could help support the deployment of emergencies in the oil and gas field and in the formulation of solving an emergency [10]. The e-FEMA strategy from the Federal Emergency Management Agency (FEMA) includes an information hierarchical structure for emergencies [11]. Various resources in information systems could be updated, promoting information sharing among the systems, providing a decision-making process to handle emergencies with real-time urgent information and technological support.

Timperio G, from the perspective of operating precautionary measures and real-time information collection for accidents, focused on vague and dispersive information and uncertain dependent relations among urgent activities, setting up a dynamic and static information model to handle urgent activities based on certain task frameworks [12]. Marco Scaioni, considering the necessity to search for appropriate and available data information in different geographical databases during accidents, exploited an application EPM to handle and control emergencies based on GIS [13]. Mark Hoogendoorn put forward a formalized framework based on the urgent analysis and construction of a precautionary measure, applying formalized temporal trace language (TTL) to build a model and expand upon previous research projects [14]. Gheorghe Tecuci put forward an emergent application,

Disciple-VPT (virtual planning team), based on the American plan of responding to an emergency, applied mainly to train and simulate maneuvers of precautionary measures [15].

Be analyzing the research, we found that domestic and overseas institutions have set up (relatively fulfilling) emergency systems. However, the ability to collect information and handle emergencies need to be explored [16–19]. Research studies on simulated maneuvers that are suitable for sudden water pollution incidents are relatively insufficient. As time progresses, technology continues to expand; however, we still do not have systematic achievements or complete theories in the current research that could solve sudden water pollution emergencies. Therefore, in an emergency that cannot be prevented, there is a need to respond quickly to the emergency; that is, when emergencies occur, managers should have ways to deal with them. Compared with previous studies, the applicability of an existing emergency plan system is very poor, because the content of the plan and the emergency do not correspond, which makes it difficult for managers to perform their work according to the plan [20–24].

In current digitalized precautionary measures, back-stage database restock and front-stage user check are mostly used, only achieving inquiries, glances, and modifications of textual precautionary measures; organic cooperation among the available multi-precautionary measures cannot be complete without links among the content, law, system, and real case; there is a lack of adaptability and low informatizing levels.

In order to reduce the influence from sudden disasters, to a large extent, digitalized precautionary measures should be built, aimed at handling cases in suitable ways, such as setting a 'case base' to provide complete assistance, analyzing current precautionary measures and case characteristics to build a precautionary mode (to handle water pollution), realizing the digitalized management and flexible applications by dividing and combining each opinion about a precautionary measure, and using visible forms to deploy logical arrangements, according to data, information, the model, and the method organized by the procedure.

In view of the above problems in the response to sudden water pollution incidents, the purpose of this paper was to create a system that could quickly push out emergency plans for sudden water pollution events. In order to achieve the purpose of rapid response and rapid matching, it must be done with the help of information technology. SOA and Java EE architecture were used to build a visualization platform; the logical relationship between businesses is expressed by knowledge map technology, and the business function is realized by component technology. This information mode can enable managers to quickly form effective response plans for different kinds of emergencies. Finally, regarding the system, we provide an example of how to quickly match the plan and apply it. We believe this approach can draw on many areas of emergency response to help emergency managers.

2. Materials and Methods

2.1. Data Collection

Data from the emergency plan were downloaded from the official websites of city governments (Bureau of Emergency Management of Tianjin, Available online: http://yjgl.tj.gov.cn/, accessed on 2 December 2013) and counties directly under the central government (Ministry of Emergency Management of the People's Republic of China, Available online: https://www.mem.gov.cn/, accessed on 1 November 2018). The various levels of water pollution event profiles are stored in a local database so that they can be invoked later in component-based development.

2.2. Support Platform

SOA is a software architecture designed with service orientation at its core, enabling the separation of business and technology, which communicates through interfaces between different services without involving the underlying programming interface and communication patterns by building a service architecture that is coarse-grained, loosely coupled, with location and a transparent transmission protocol. By combining the different services of the application with good interfaces and standards, the service components scattered in the distributed environment are integrated into a new whole process, and the user is provided with services in the form of components to solve the heterogeneous problems faced in the distributed application system based on the components. Using SOA and Java EE architectures, a comprehensive integrated support platform for knowledge visualization has been developed [25]. Users can build a variety of theme application services on the platform. The platform is the carrier, and the knowledge map is the tool to describe the logical relationship. Business process relationships can be expressed by drawing knowledge maps, and components can be customized to enable business functions. The platform provides a business environment, which provides a strong support and guarantee for the business environment regarding emergency management of sudden water pollution events [26].

2.3. The Assembly of Precautionary Measures Base

In the process of solving an emergency, besides precautionary measures being used in decision assistance, we usually depend on the solving experiences of decision makers themselves. Therefore, analyzing real cases that took place before is an effective way for a decision maker to improve his/her experience in solving a similar case. The combination of precautionary measures and real cases could improve the operating efficiencies in solving emergencies using these precautionary measures and reduce unreasonable behaviors that lack overall viewpoints in solving emergencies.

A precautionary measure 'base' for an emergency should be built. Files on precautionary measures, laws, regulations, and regulating systems to solve water pollution are massive, because state-owned and local enterprises all have corresponding precautionary measures for emergencies. According to different characteristics designed on the bases of level and district divides, precautionary measures are classified and managed, providing application with foundation. The classification tables of emergency response plans at different levels are shown in Table 1.

Table 1. Part of the pre-plan hierarchical management table.

Code	Grade	Administrative Divisions	Type	Plan
1	National	National	emergency plan	National emergency plan for water pollution emergencies
2	Municipal	City	emergency plan	Emergency Plan for Sudden Water Pollution Incidents in Beijing
3	Municipal	City	emergency plan	Emergency Plan for Water Pollution Incidents in Tianjin
4	District and County	District	emergency plan	Emergency Plan for Sudden Water Pollution Incident in Changping District
5	Company	/	emergency plan	Emergency Plan for Water Pollution Incidents of Water Supply Company
6	Company	/	Regulations	Guiyang Railway Investment Corporation's emergency handling system

A real case base should be built. Normal cases could be handled by following stages. Firstly, when an accident occurs, an abnormal 'inspecting' alarm or public report will be received. Previous management will be 'applied', according to the relative situation. The professional team will have discussions while the operational staff and supplies are sent to the emergency spot. The discussed managing project will be sent to the spot. After the emergency is under control, remedial work will proceed, and the leadership and professionals will have to make conclusions and estimations of the process of handling the emergency. Analyzing previous water disaster cases and simulated maneuver cases, referring to the process–managing mode, sequenced according to time, the normal procedure of a case is usually divided into three stages—preparation before the case, management during the case, and management after the case. Extracting the key points from each stage, we can divide the main work of each stage in detail. According to the divided models, information of a case is written into a standard precautionary model, and the case will be a precautionary measure, from which we can search details to solve an emergency and provide experience for a decision, as shown in Figure 1.

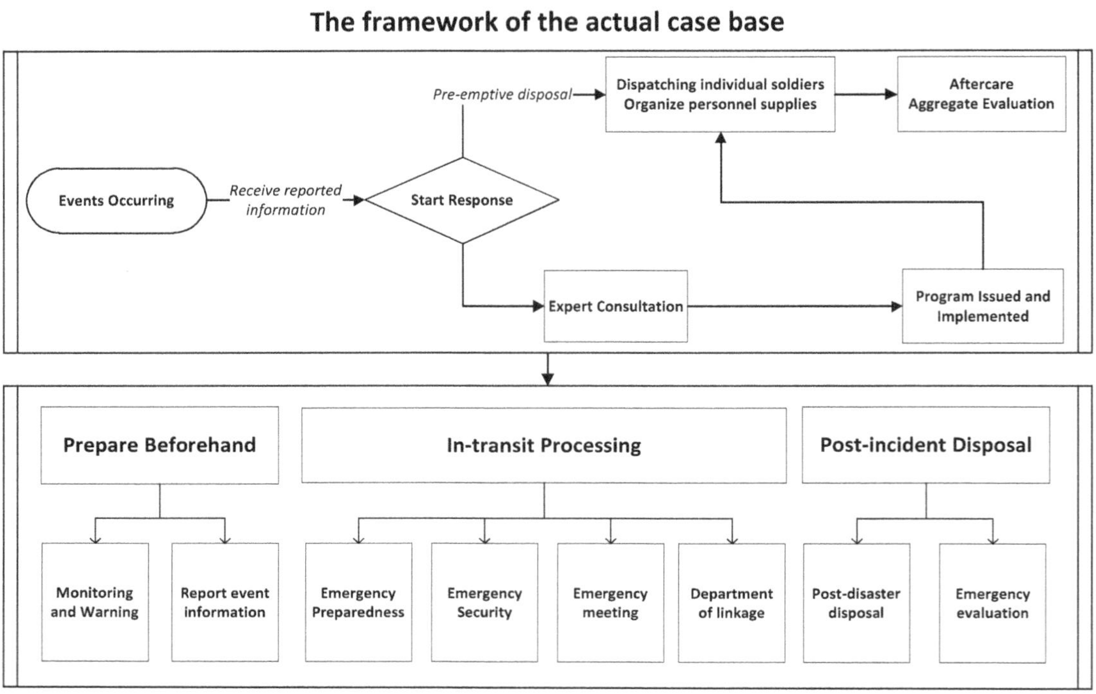

Figure 1. The framework of the actual case base.

2.4. Construction Method

Regarding the construction of traditional precautionary cases—there is more of an emphasis on showing the content, but not on the connection between precautionary measures and real cases. In the precautionary measures, there is no targeted or specific case/detailed solving method. Through contractually dismantling precautionary measures, the working content of an emergency is extracted and filed in a precautionary database by building a knowledge graph. In the application, suitable entries are inquired about and extracted constantly according to real situations, and digital precautional measures are built quickly to adapt to current emergencies on the basis of describing the procedure of precautionary measures, as shown in Figure 2.

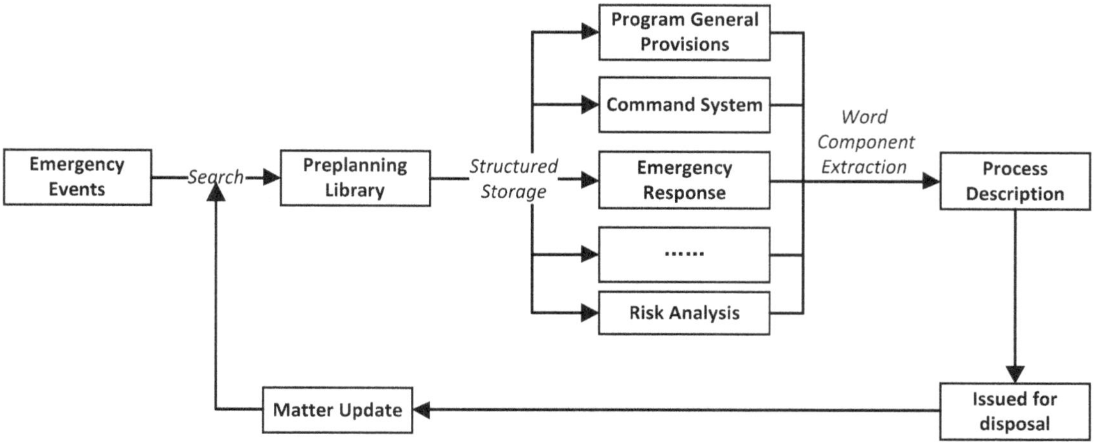

Figure 2. Digital plan construction technical frame diagram.

(1) Structural dismantle. Horizontal dismantle means that, based on the chapter and concrete content of precautionary measures, precautionary measures are dismantled structurally according to different level. Working contents to handle an emergency, after being analyzed and extracted stage-by-stage, are packaged and stored separately, as shown in Figure 3.

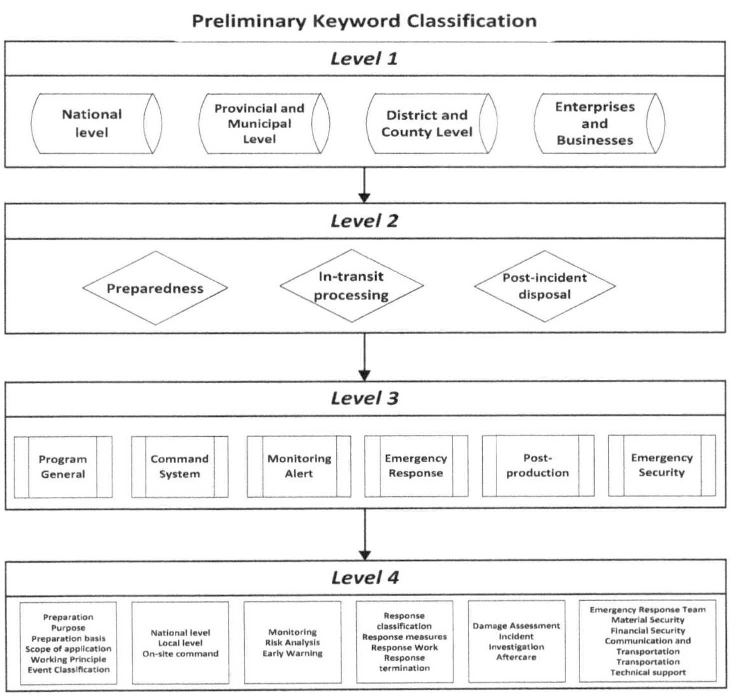

Figure 3. Pre-plan structure classification table.

Vertical dismantle means utilizing special keywords to separate these cases according to the causes or handling methods. Moreover, those cases are packaged and stored according to the causes and handling methods. The specific classification is shown in Table 2.

Table 2. Keywords of emergency water events.

Classification Basis	Factor 1	Factor 2	Factor 3
Cause of Accident	Pollution caused by traffic accidents	Oil pollution	Petroleum substances
	Pollution caused by production accidents	Heavy metal pollution	zinc, etc.
	Pollution caused by natural changes	Aquatic Biological Pollution	Algae
	Pollution caused by wastewater discharge	Chemical pollution	Mercury, etc.
	Pollution caused by natural disasters	Construction waste pollution	stone
	Pollution caused by manmade damage		
Disposal Methods	Engineering measures	Emergency Dispatch	
		Adsorption interception	
	Non-engineering measures	Catalytic oxidation	
		Biodegradation	

All precautionary measures or cases, matched according to specific keywords inquired through a fuzzy query function, were filtrated. Moreover, part modules of those precautionary entries or handling methods suitable to the current real situation were extracted, so that quick construction responding to previous precautionary measures and the project to adapt change in the duration were achieved.

(2) Construction of precautionary measure modules. With the help of technologies, such as module technology and Web Service, constructions, such as J2EE and SOA, and procedures, such as business extraction and classification, module dismantling, exploitation, registration and publishment, working content, case information, and handling methods dismantled in precautionary measures, cases are exploited and filed into imputing and outputting standard modules [27]. Building a business application module base could support the management and the adaptable application of digital precautionary measures. The planned component development process is as follows:

Step 1: build a new project on the file menu and name it. Compile new Java code that a program needs in the "*scr*" file package in the project file.

Step 2: edit the interface accustomed by a user and the Jelly file corresponding to the main Java program.

Step 3: according to the edited Jelly file, make sure the "*LOVResponse*" receptor will be used in the Jelly file, including the methods of "*getLOV*()" and "*getLOVSchema*()".

Step 4: make sure the "*ActionCode*()" corresponds to each main program in "*ActionResponse*" and complete the registration in the "*ActionRegistray*()" method.

Step 5: in the "*ActionHelper*()" method of "*ActionResponse*", return to step 2 to make the Jelly file.

Step 6: compile input "*SCHEMA*" and return the "*ActionInputSchema*()" method in "*ActionResponse*".

Step 7: compile the "*execute*()" method in "*ActionResponse*", achieve the case detecting and searching function and return with the "*XML*" form.

Step 8: output the result with the "*SCHEMA*" form.

Step 9: return the module name through the "*ActionName*()" method in "*ActionResponse*". The flowchart is shown in Figure 4.

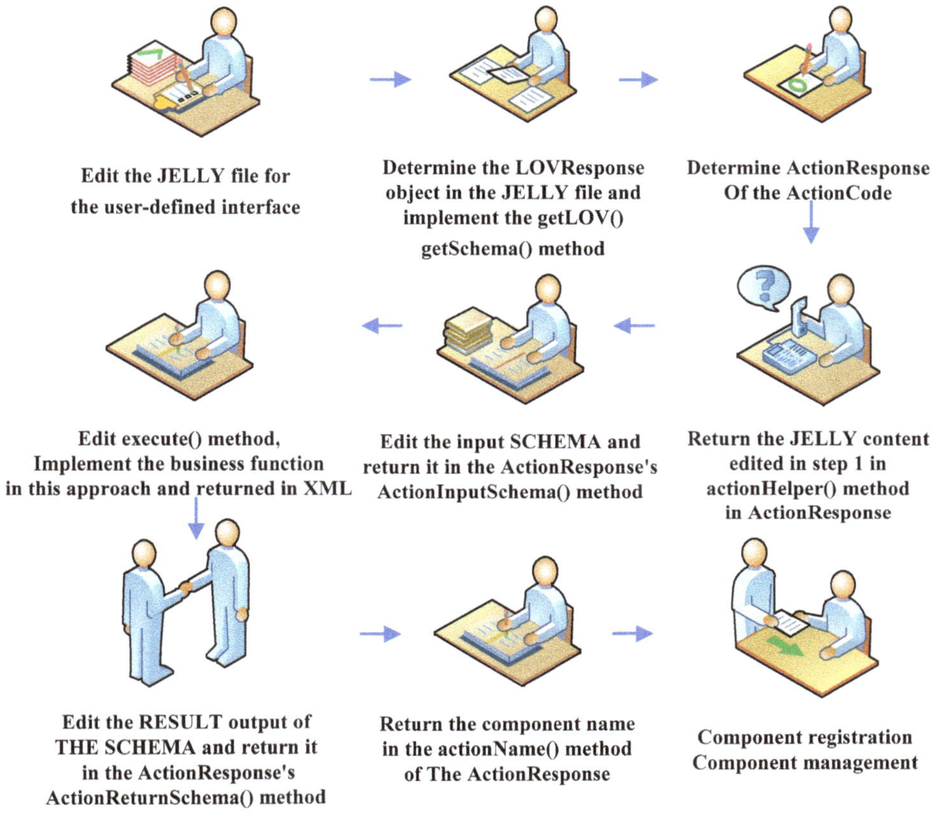

Figure 4. Development process diagram.

After the component is developed, the encapsulation process is as follows.

After the module is exploited, Web Service technology will be needed to package the module into standard form in accordance with Web Service. Then, the module should be registered in UDDI and deployed to the server that is needed.

Step 1: set up the exploiting environment. Choose the Tomcat applicating server and configurate corresponding environment variants. Set up Axis (based on Java language and published in the web application form.)

Step 2: package module. With help from the exploiting application, package the module and build a new file with ".aar" as the name suffix.

Step 3: upload the module. Uploading of the module is relatively easy. If the Tomcat servicer is built in a local computer, open the starup.bat in the Tomcat file and enter http://localhost:8080/axis2-1.1.1. Then, upload the packaged file after user log in.

Step 4: publish the server. The website address is http://localhost:8080/uddibrowser. After one enters, logs in, and registers, fill in the entrance URL that a service needs, and then publish it on the internet. All services could be inquired in uddibrowser. Part of the code for the development process is shown in Figure 5.

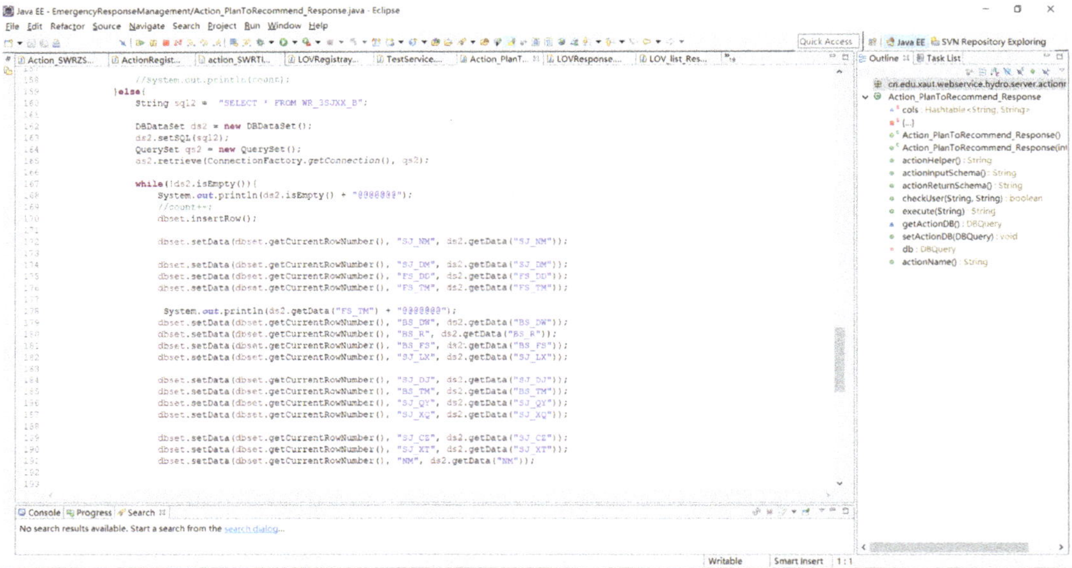

Figure 5. Part of the code for the development process.

2.5. Streamlined Description

The streamlined description combined with graphs can simulate people's minds, changing random and illogical information into highly organized graphs and expressing them. That is what is called a streamlined description. Based on the thinking of service combination, the assembling technology is applied to package modules and compile business procedures. Based on the visible platform in a graphical way, the business procedure is shown to improve the flexibility of business procedure. Moreover, the visible effect, based on the research and achievement of knowledge graphs, can help us understand the streamline deduction toward an emergency. The handling procedure of an emergency could be recorded in the figure.

Regarding the framework and chapter content of such precautionary measures of water pollution, in combination with the overall framework of the emergency handling platform made by the nation to solve an emergency, the standard handling procedure of an emergency should be built, including pre-emergency inspection guarantees, emergency discussions and cooperation, post-emergency management estimations, and other procedural modules. The procedure framework is shown in Figure 6.

Regarding precautionary measures, a precautionary measure module that is exploited should be filled with relative concrete content (of precautionary measures) with the help from module technology, to achieve the digital precautionary measures, to handling procedures and business logic, as well as the exhibition of related information, models, and methods. Based on the precautionary measures, business applications and the deduction of modularized precautionary measure procedures of each stage, direct visibility of precautionary measures could be completed.

Data transfer and change between knowledge pictures are usually stored in XML form. When used, the reflection relation between the information of the conception, link, and source in a knowledge picture and the XML file should be built in advance; the data flowing direction and module information should be described and the relationship between the file information and knowledge picture should be analyzed, so that the user can achieve flexible customization of system construction, modification, and business applications by using the working procedure in the application system according to XML, modifying the module of the knowledge picture. The business module in a knowledge picture features visible characteristics, which provides user with various business activity situations and the data flowing exhibition project, achieving knowledge visibility and showing users direct decision-making processes and outcome exhibitions. The knowledge picture could effectively describe overt knowledge in normalized and systemized ways, as well as show the description of covert knowledge and the procedure of transferring overt knowledge into covert knowledge.

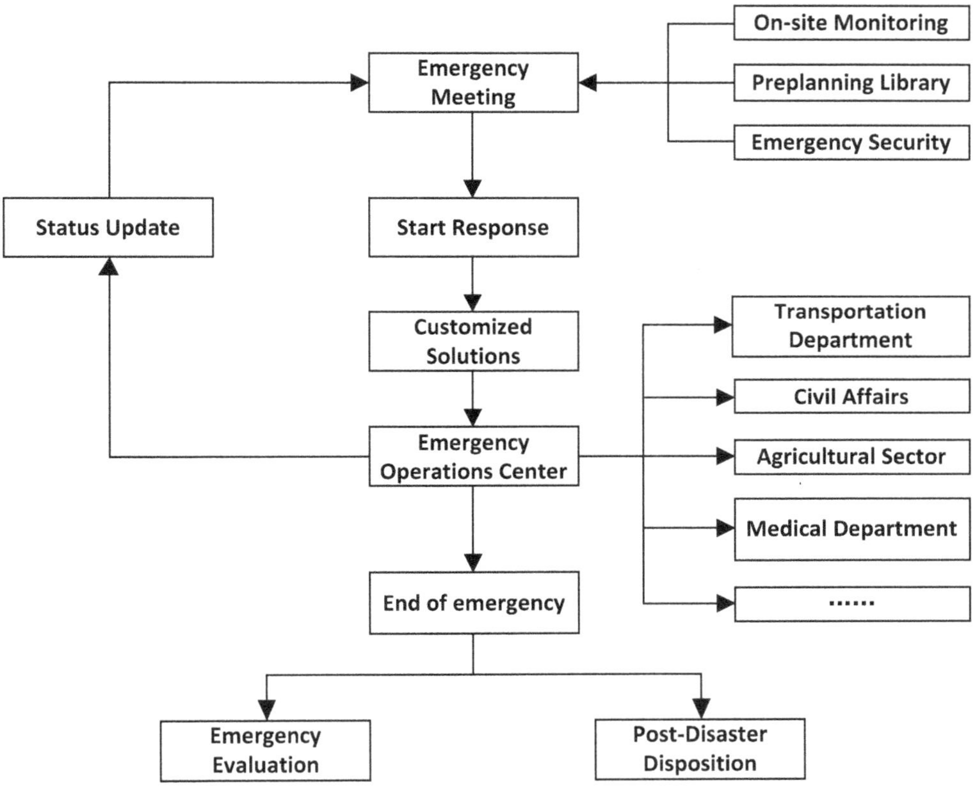

Figure 6. Modular standard disposal process.

3. Adaptive Application Models

Application is the criterion used for judging the value of a precautionary measure, and dynamic adaptation and operability are important manifestations of the application value of digital precautionary measure. Relying on the form of the process knowledge diagram to show the overall process of disposal, and using node components, the usability of the precautionary measure is realized; in response, by adding and deleting components to the corresponding nodes on the framework diagram of the precautionary measure, the rapid precautionary measure is realized. In the response, the precautionary measure can be built quickly and adapted to the development of the situation by adding and deleting components to the corresponding nodes on the precautionary measure framework. The textual emergency precautionary measure is transformed into a practical emergency response plan that meets the current situation and guides the response and emergency management of water pollution incidents. The framework is shown in Figure 7.

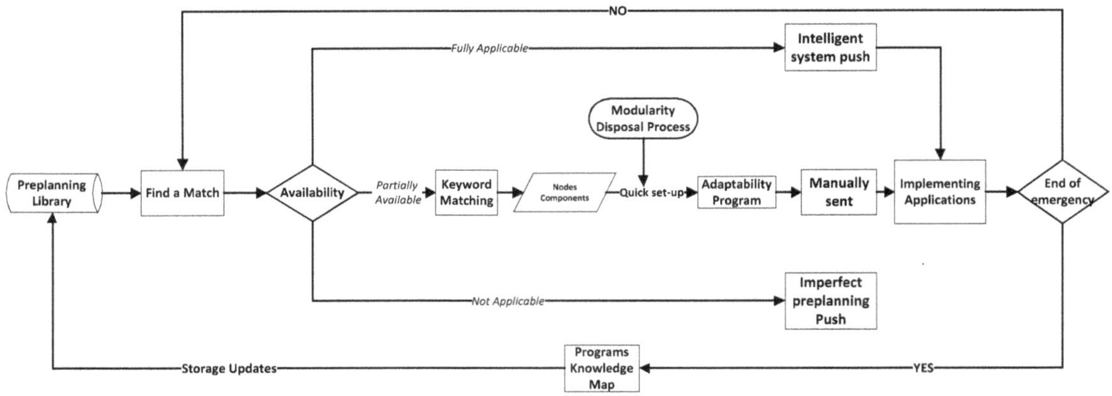

Figure 7. Modular standard disposal process.

3.1. Three Application Modes

(1) The system intelligently pushes; that is, the digital precautionary measure management platform can intelligently generate precautionary measures to push to users according to the current information, to realize the rapid application of the precautionary measure. According to the current input information, the system matches the extracted keywords with the phrases in the precautionary measure library and pushes the application after a small-scale adaptation of the same instances that have occurred before.

The system intelligent push of precautionary measures mainly responds to events that already have perfect disposal experiences. If an emergency event that has already occurred recurs, the system pushes out existing emergency response experience. For this type of emergency, the precautionary measure itself has the highest degree of "adaptability" because the precautionary measure library contains the previous application of the precautionary measure and disposal cases of the event.

(2) Human involvement in pushing. Due to the complexity of emergency response, in most cases, the precautionary measure cannot directly fit the process of handling the situation. In order to deal with the poor "adaptability" of this kind of precautionary measure, human participation is needed to modify the precautionary measure and push it to the application. First, according to the keywords entered by the front-end, the system precautionary measure library is matched. Then, according to the development and changes of the situation, the corresponding node components are modified in the matched precautionary measure modular disposition process to realize the rapid formation of the response plan. Then, according to the feedback information after the plan is issued and executed, real-time modification is carried out, and the response plan is provided continuously until the end of the emergency.

(3) Imperfect intelligence precautionary measure push. There is a situation in the emergency response; that is, "a new event that has not happened at all, without any similar and available precautionary measure". For this type of emergency, since it is impossible to match the available precautionary measure, the system can push out the modular disposal process framework of the digital precautionary measure, and according to the specific situations of real events, use the components in the framework to quickly build and issue the plan for execution. Since the system pushes out the precautionary measure process framework, the content of the plan needs to be prepared by the decision maker according to the actual situation, so this application mode is an imperfect intelligence 'pushing' precautionary measure.

3.2. Precautionary Measure Matching Acquisition

The precautionary measure is obtained using case-based reasoning techniques. Case-based reasoning (CBR) is a newly emerged method for problem solving and knowledge inference in artificial intelligence. The basic process: when a new problem arises, the system searches and retrieves the original instance database according to its relevant features to find the candidate instance with the most similar features to the new problem, and then reuses this candidate instance. If the solution of the candidate problem is not satisfactory, it can be modified to fit the problem to be solved, and finally the modified example is stored in the example base as a new example to be used as a new reference to solve the problem when a similar problem is encountered in the future. The reasoning process of CBR is as follows: (1) it retrieves the relevant examples similar to the new problem from the example base according to the characteristics of the new problem; (2) it retrieves the relevant examples from the retrieved, selecting the most similar examples or cases from the retrieved examples or by combining multiple examples to form a solution to the new problem; (3) it makes a revised solution to the new problem, verifying the obtained solution; (4) it stores the solved new problem as a case in the example base for future use. By extracting the characteristic keywords of the sudden water pollution event, the relevant process-oriented precautionary measure can be retrieved from the precautionary measure library, and after similarity selection, the precautionary measure with the highest applicability to the current sudden event is obtained as the emergency management solution [28]. The flow chart is shown in Figure 8.

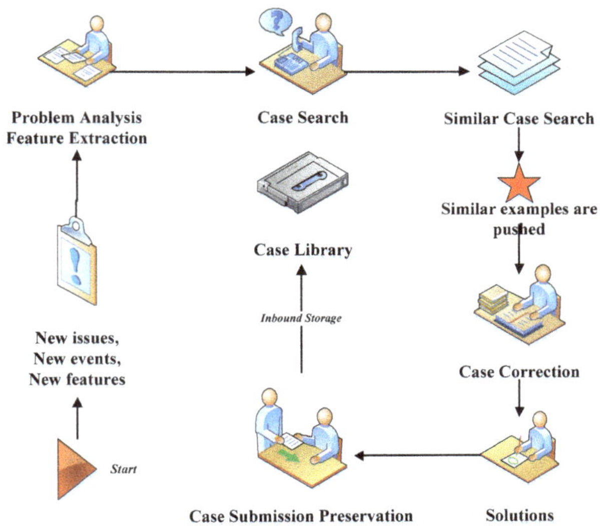

Figure 8. CBR workflow.

3.3. Adaptive Modifications of Precautionary Measure

The precautionary measures acquired through matching are more or less modified and newly created to achieve interconnection with emergencies due to differences in the degree of adaptation. Modifying the emergency precaution measure to adapt to the current emergency management requirements is valuable when implementing emergency solutions.

By adding and changing node components on the knowledge graph of the process-oriented precautionary measure, adaptive changes of the precautionary measure with the development of events are realized. Based on matching the precautionary measure, new components are created or applicable components are searched for, quick responses, according to specific events; applications are continuously issued, and then changes continue to be made according to the feedback information of implementation, which is continuously applied, accumulated, and improved.

Each disposal over the event processes precautionary measures stored in the database. With the continuous responses, the program knowledge map and components are constantly enriched to expand the existing emergency solution database. When a similar event occurs, the knowledge map of the scenario is acquired, modified, adapted to the real situation, and stored while being used, so that it is continuously accumulated, adapted, and developed. The open and growable precautionary measure and scheme involve the process of knowledge inheritance, accumulation, and expansion, in which the combination of multiple topics and the growth of the knowledge system, from a simple description and seminar process, gradually approach the complexity of the problem, realizing the human–machine combination and scientific decision-making.

The visual description is divided into three steps: first, the subject of the visual description as well as the information to be described under the subject are made clear; second, the appropriate description method according to the actual situation on the basis of the subject and information are chosen; finally, the visual description knowledge diagram is drawn on the comprehensive integration platform. The visual description process is shown in Figure 9.

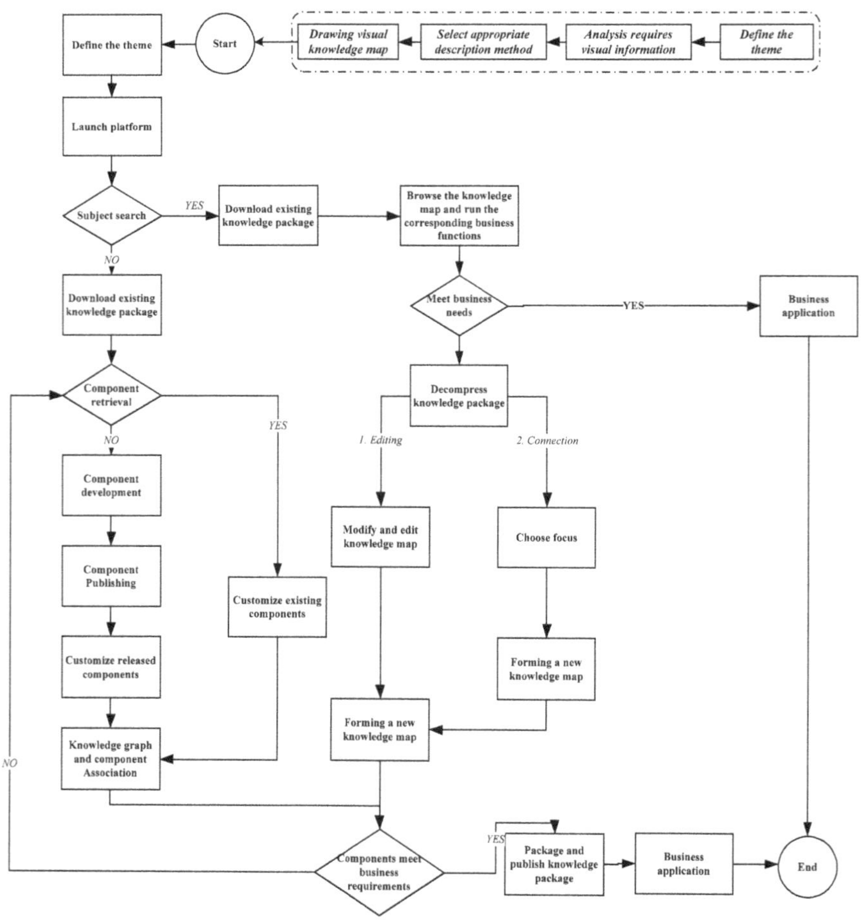

Figure 9. Visual description process.

Through visual expression, we built a digital pre-planning system of the platform. The visual part of the main interface is the processing process for emergency management, and the function is implemented through custom components under each node. The main system interface is shown in Figure 10.

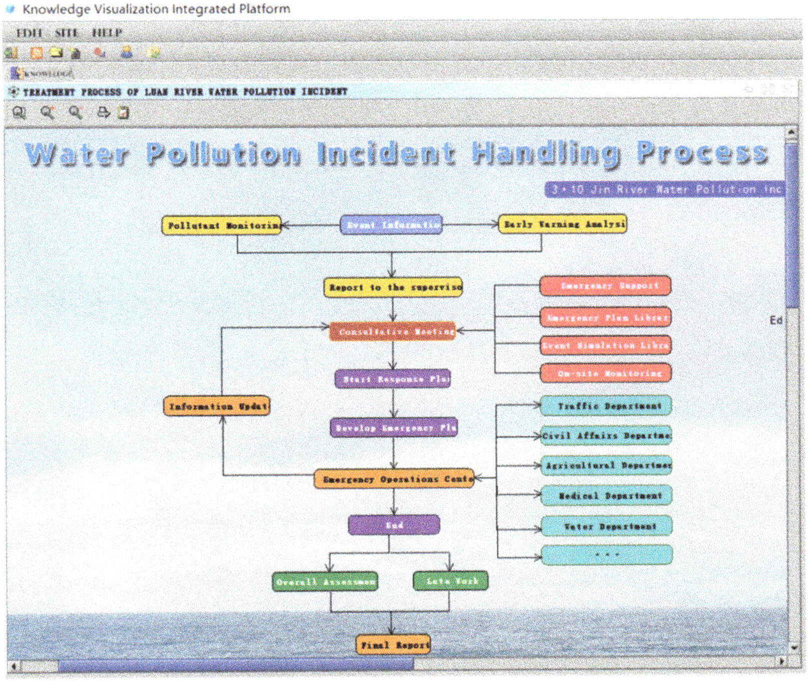

Figure 10. The main interface of the system.

4. System Application

The digital precautionary measure is based on a water pollution event in the Luanjiang–Tianjin river. The Luanjiang–Tianjin project is an urban water supply project that brings the Luan River from Hebei province to Tianjin, across the river basin. With the Panjiakou Reservoir as the water source, the project alleviated the water supply difficulties in Tianjin, improved water quality, reduced the intensity of groundwater extraction, and stabilized the ground subsidence in Tianjin city. The project starts from Panjiakou Reservoir and Daheiting Reservoir, and enters the Yuqiao Reservoir in Jixian County, Tianjin, through the main water transmission canal through Qianxi and Zunhua, with a total length of 234 km. Due to the wide range of the scope involved, the pollution situation is often complex and changeable, including reservoir aquaculture pollution, mineral sand pollution, domestic pollution, agricultural pollution, etc.

The water quality of the main water source in the bridge reservoir is the national surface water class IV; the total nitrogen seriously exceeds the standard. The total nitrogen value of the bridge reservoir in 2013 was 5.01 to 4.44 mg/L, which was higher than the national surface water class III; 1.0 mg/L standard. The total nitrogen and iron of the water entering the reservoir exceed the standard and cannot meet the three standards of surface water and drinking water requirements. Through the integration of data and historical experience, the digital plan has established a response mechanism model of timely response and flexible adaptation, and improved the supervision and emergency response capabilities of water environmental protection management, from the aspects of organizational coordination and organizational action. Through the modular organizational platform, the water management responsibilities among upstream and downstream, different levels, and different departments, are coordinated, and the "written plan" is further effectively transformed into a "disposal action", the emergency disposal task is implemented, and the quality of emergency management is improved. At the same time, in the event of a major environmental pollution incident, it can also cause significant harm. The river, emergency materials, text precautionary measures, other related objects, and water pollution emergency processes involved in the area were analyzed, extracted, and designed, and then a digital precautionary measure for the area was designed and implemented in accordance with the digital approach of this paper.

In the emergency response, we extracted keywords based on the front-end event information, and searched for similar precautionary measures in the precautionary measure database, such as "water pollution", "Tianjin", and other event features and keywords. The precautionary measures library automatically gives the required structured words. The query results are shown in Figure 11.

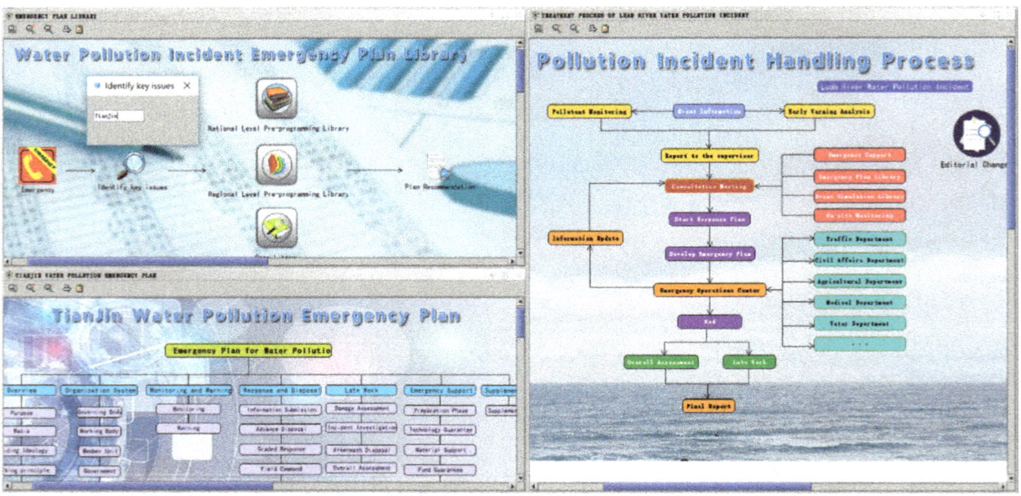

Figure 11. Pre-plan query matching results.

According to the event situation, the word components of the precautionary measure or the disposal components in the case are quickly added to the modular disposal process; the required digital precautionary measure is quickly assembled and built by combining to dynamically adapt to different events. Finally, the precautionary measure is transformed into an applicable and reliable knowledge graph scheme by combining with emergency resources. The result of the digital plan generation is shown in Figure 12.

As the situation progresses, the knowledge graph is continuously improved by adding and deleting components until the emergency situation is over. Finally, the completed knowledge map is uploaded into the example case library to reserve as experience for the next emergency.

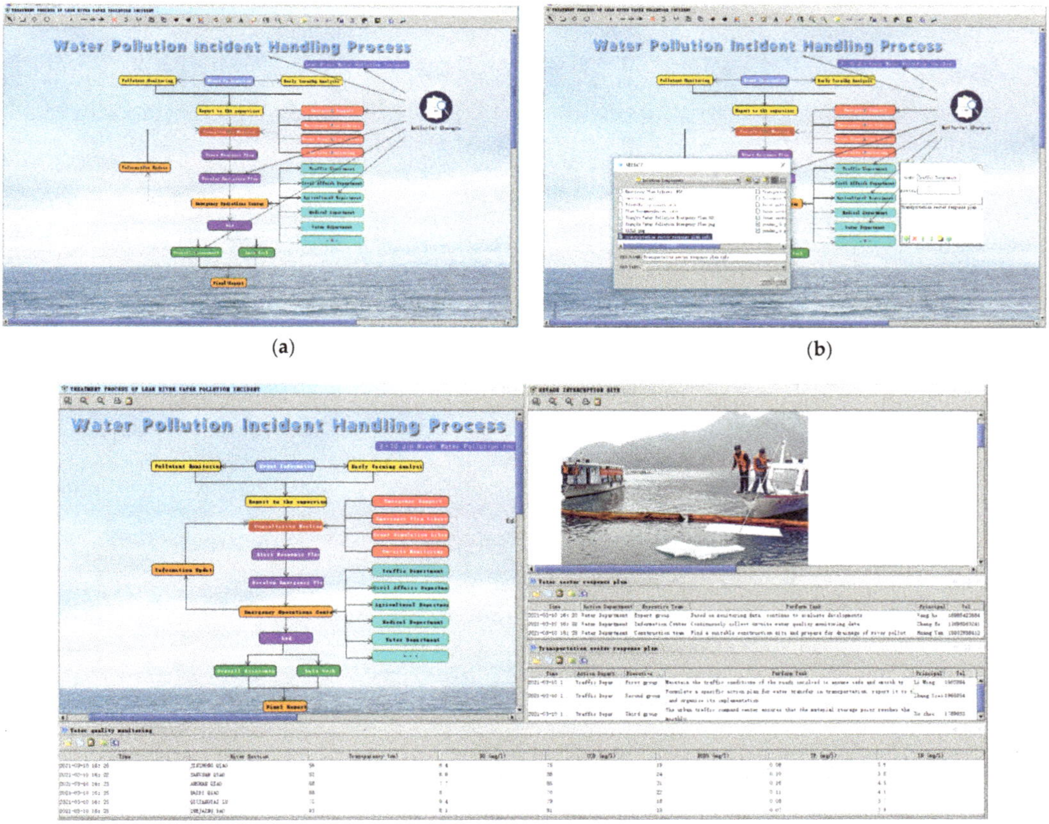

(**a**) (**b**)

(**c**)

Figure 12. The result of the digital plan generation. (**a**) Emergency process knowledge drawing. (**b**) Add components and modify. (**c**) Save and generate a new plan.

5. Discussion

The contingency plan for traditional water pollution incidents is based on specific pollutants and events (Figure 2). This approach is often aimed at specific pollutants and pollution situations, with clear directionality. Once the event changes, such as the appearance of mixed pollutants that need to adjust the plan, it is difficult to continue to use the original plan at this time [29–31]. In response to this phenomenon, it is necessary to propose a solution based on the existing plan and quickly generate a new action plan.

The construction framework of the case library mentioned in this article is built on the basis of the incident disposal process. The case database can be applied to the entire process of event response; that is, clicking on the corresponding nodes of different links on the system can display the corresponding event information and the current proposed action plan (Figures 11 and 12).

Rapid response and adaptability are reflected in the concept of an "emergency plan database", where the knowledge graph can be edited, corresponding to changes in emergency processes. Similarly, the functions of each node can be changed and customized, corresponding to the creation of new components through programming. Each time a knowledge graph is saved, a new course of action is generated and stored in the database

(Figures 4, 5, 8 and 9). In this way, the more scenarios that accumulate in the database, the higher the value of the scenarios that can be referred to in the future.

In this article, we sought to establish a digital plan library and propose a model for the application of plans that are suitable for the whole process of emergency management. The matching accuracy and rapid response of the system will be studied in future work.

6. Conclusions

Due to the rapid development of the economy, environmental pollution accidents are not uncommon, particularly sudden water pollution incidents. There are many problems in the current response process; because there is no systematic solution, blind responses to these events often miss their opportunities. In view of the complexity of emergency disposal work and the characteristics of the dynamic development of emergencies, based on summarizing and integrating conventional handling methods for emergencies, we conducted a visual emergency response plan for sudden water pollution incidents based on the emergency management platform, and realized the application value of the scheme in water pollution emergency response work through simulation of actual cases.

We explored and designed a digital emergency precautionary measure for water pollution emergencies. First, it analyzes the problems related to the construction and application of the traditional emergency precaution measure, and offers a process-oriented construction method for the emergency precaution measure of water pollution emergencies by splitting the principles of the preparation of the emergency precaution measure and the event disposal process. Second, in this paper, SOA and Java EE architecture were used to develop a pre-push system for sudden water pollution events, which could realize three levels of digital planning push. For events that have occurred before, the system can quickly push the existing disposal experience; for similar events, the system can use fuzzy query, or the CBR model can be artificially modified to participate in the plan for events that have not occurred; the system can push the plan template, the user, according to the development of the current event, quickly, to form a new plan. This method can support the rapid construction and dynamic changes of precautionary measures, and improve the practicality of precautionary measures in practical applications.

In the next step, we will continue to improve the digital precautionary measure system and further combine with GIS, simulation, the mock-up system, online evaluation, etc., to realize artificial intelligence methods, such as pollution state reasoning and comprehensive evaluation of disposal, to make decision support for the whole process of emergency response.

7. Patents

We applied for a Chinese computer software copyright using the relevant results of this article; the registration number is 2018SR788164.

Author Contributions: Conceptualization, J.L. and J.X.; methodology, J.L.; software, J.L.; validation, X.W., T.J. and S.W.; formal analysis, J.L.; investigation, J.L.; resources, X.W.; data curation, S.W.; writing—original draft preparation, R.W.; writing—review and editing, J.L.; visualization, T.J.; supervision, J.X.; project administration, J.X.; funding acquisition, J.X. All authors have read and agreed to the published version of the manuscript.

Funding: This research was funded by The National Key Research and Development Program of China, grant no. 2016YFC0401409, the Natural Science Basic Research Program of Shaanxi province, grant no. 2019JLZ-16, and the Science and Technology Program of Shaanxi Province, grant no. 2020slkj-16.

Institutional Review Board Statement: Not applicable.

Informed Consent Statement: Not applicable.

Data Availability Statement: The knowledge map of this paper and the software package can be downloaded from our website: Cloud service chain. Available online: http://www.yunqishui.com (accessed on 2 December 2018).

Acknowledgments: The authors thank the editor and the anonymous reviewers for their constructive comments and suggestions.

Conflicts of Interest: The authors declare no conflict of interest. The funders had no role in the design of the study; in the collection, analyses, or interpretation of data; in the writing of the manuscript, or in the decision to publish the results.

References

1. Claborn, D.M. Review of Security and Emergency Planning for Water and Wastewater Utilities. *J. Homel. Secur. Emerg. Manag.* **2011**, *8*. [CrossRef]
2. Jiang, R.; Han, H.; Xie, J.; Li, F. The new response mode of urban storm flood under changing environment. *J. Catastrophol.* **2017**, *32*, 12–17. [CrossRef]
3. Robert, H.; Wang, C.; Song, B.; Jin, E. Translation. In *Crisis Management*; CITIC Press: Beijing, China, 2004; pp. 35–36.
4. Deroux, J.M.; Gonzalez, C.; Cloirec, P.L.; Roumagnac, A. Long-term extractable compounds screening in surface water to prevent accidental organic pollution. *Sci. Total Environ.* **1997**, *203*, 261–274. [CrossRef]
5. Shafiee, M.; Zechman, E.M. An Agent-Based Modeling Approach to Evaluate Protective Action Strategies in a Water Distribution Contamination Event. *Am. Soc. Civ. Eng.* **2011**, *414*, 276–282. [CrossRef]
6. Yu, Z.; Xia, J.; Ren, H. Water Pollution Accident Emergency Response and Early-Warning Model in Gansu-Ningxia-Inner Mongolia section of the Yellow River. *Yellow River* **2014**, *36*, 37–40.
7. Liu, D.; Liu, M. Research progress on risk analysis and emergency management of sudden water pollution accidents. *China Public Saf. Acad. Ed.* **2009**, *1*, 167–170. [CrossRef]
8. Shao, Y.; Ingo, B. Crisis prevention information system in Germany. *Inf. Constr.* **2005**, *8*, 46–48.
9. Li, X. The characteristics and implications of British emergency management. *Adm. Manag. Reform* **2010**, *3*, 54–59.
10. Fan, W.; Yuan, H. Current Situation Analysis and Countermeasures of Emergency Platform Construction in China. *Inf. Constr.* **2006**, *9*, 14–17.
11. Deng, S. American Emergency Management System and Its Implications. *J. Chin. Acad. Gov.* **2008**, 102–104. [CrossRef]
12. Timperio, G.; Tiwari, S.; Lee, C.K.; Samvedi, A.; de Souza, R. Integrated Decision Support Framework for Enhancing Disaster Preparedness: A Pilot application in Indonesia. *Int. J. Disaster Risk Reduct.* **2020**, *51*, 101773. [CrossRef]
13. Scaioni, M.; Alba, M.; Rota, R.; Caragliano, S. A GIS-Based SW Prototype for Emergency Preparedness Management Applied to a Real Case Study. In *Proceedings of the International Conference on Computational Science & Its Applications, Seoul, Korea, 29 June–2 July 2009*; Springer: Cham, Switzerland, 2009.
14. Hoogendoorn, M.; Jonker, C.M.; Popova, V.; Sharpanskykh, A.; Xu, L. Formal modelling and compering of disaster plans. In Proceedings of the 2nd International ISCRAM Conference, Brussels, Belgium, 18–20 April 2005; pp. 98–107.
15. Tecuci, G.; Boicu, M.; Hajduk, T.; Marcu, D.; Barbulescu, M.; Boicu, C.; Le, V. A Tool for Training and Assistance in Emergency Response Planning. In *Proceedings of the Hawaii International Conference on System Sciences, Waikoloa, HI, USA, 3–6 January 2007*; IEEE Computer Society: Washington, DC, USA, 2007.
16. Cahn, M. Managing the human impact on the natural environment: Patterns and processes. *Appl. Geogr.* **1993**, *13*, 285. [CrossRef]
17. Fu, L.; Wang, J.; Peng, Z.H. Simulation of Wastewater Discharge in a Coastal Plain River Network Considering Different Boundary Conditions. *J. Coast. Res.* **2020**, *104* (Suppl. S1), 529–534. [CrossRef]
18. Ding, H.; Tong, J.; Wang, Y.; Zhang, L. Development of emergency planning zone for high temperature gas-cooled reactor. *Ann. Nucl. Energy* **2018**, *111*, 347–353. [CrossRef]
19. Li, W.; Zheng, Y.; Chao, Y.E.; Li, H. Emergency Plan for Water Supply in Consecutive Droughts and Sustainable Water Resources Management in Beijing. *Acta Geol. Sin.-Engl. Ed.* **2018**, *92*, 14. [CrossRef]
20. Long, Y.; Yang, Y.; Lei, X.; Tian, Y.; Li, Y. Integrated Assessment Method of Emergency Plan for Sudden Water Pollution Accidents Based on Improved TOPSIS, Shannon Entropy and a Coordinated Development Degree Model. *Sustainability* **2019**, *11*, 510. [CrossRef]
21. Duan, W.; He, B. Emergency Response System for Pollution Accidents in Chemical Industrial Parks, China. *Int. J. Environ. Res. Public Health* **2015**, *12*, 7868–7885. [CrossRef]
22. Long, Y.; Xu, G.; Ma, C.; Chen, L. Emergency control system based on the analytical hierarchy process and coordinated development degree model for sudden water pollution accidents in the Middle Route of the South-to-North Water Transfer Project in China. *Environ. Sci. Pollut. Res.* **2016**, *23*, 12332–12342. [CrossRef]
23. Zhang, X.J.; Ding, K.S.; Hao, M.R.; Long, M. Emergency water supply and water resources management in Tongzhou District, Nantong City, China. *Int. J. Environ. Stud.* **2017**, *74*, 412–427. [CrossRef]
24. Ni, J.; Zhang, C.; Ren, L.; Yang, S.X. Abrupt Event Monitoring for Water Environment System Based on KPCA and SVM. *IEEE Trans. Instrum. Meas.* **2012**, *61*, 980–989. [CrossRef]
25. Xie, J.; Luo, J. Integrated Service Platform for the Information Explosion Process in Water Resources Industry and its Application Pattern. *Water Resour. Informatiz.* **2010**, *5*, 18–23. [CrossRef]
26. Zhang, X.; Luo, J.; Zhang, X.; Xie, J. Dynamic Simulation of River Water Environmental Capacity Based on the Subsection Summation Model. *Water Environ. Res.* **2019**, *92*, 278–290. [CrossRef] [PubMed]

27. He, X.; Luo, J.; Xie, J. Research and implementation of the strictest water resources assessment management system in Shaanxi Province. *J. Water Resour. Water Eng.* **2016**, *27*, 55–60. [CrossRef]
28. Luo, J.G.; Xie, J.C.; Wang, N. A CBR-based flood scheduling model for reservoirs. *Syst. Eng. Theory Pract.* **2008**, *28*, 122–130. [CrossRef]
29. Sun, B.; Ma, W. An approach to evaluation of emergency plans for unconventional emergency events based on soft fuzzy rough set. *Kybernetes* **2016**, *45*, 461–473.
30. Aureli, F.; Prost, F.; Vacondio, R.; Dazzi, S.; Ferrari, A. A GPU-accelerated Shallow-Water Scheme for Surface Runoff Simulations. *Water* **2020**, *12*, 637. [CrossRef]
31. Wu, X.; Zhang, J.; Chen, Y.; Tang, Z. A System Methodology and Its Application in Developing an Emergency Response Plan for Natural Gas Leakage on an Offshore Platform. *J. Coast. Res.* **2020**, *111* (Suppl. S1), 89–92. [CrossRef]

processes
MDPI

Article

Modeling the Biosorption Process of Heavy Metal Ions on Soybean-Based Low-Cost Biosorbents Using Artificial Neural Networks

Daniela Ionela Fertu [1,2], Elena Niculina Dragoi [3,*], Laura Bulgariu [1,*], Silvia Curteanu [3] and Maria Gavrilescu [1,4,*]

[1] Department of Environmental Engineering and Management, "Cristofor Simionescu" Faculty of Chemical Engineering and Environmental Protection, "Gheorghe Asachi" Technical University of Iasi, 73 D. Mangeron Blvd., 700050 Iasi, Romania; danafertu2004@yahoo.com
[2] Department of Pharmaceutical Sciences, Faculty of Medicine and Pharmacy, "Dunarea de Jos" University of Galati, 800002 Galati, Romania
[3] Department of Chemical Engineering, "Cristofor Simionescu" Faculty of Chemical Engineering and Environmental Protection, "Gheorghe Asachi" Technical University of Iasi, 73 D. Mangeron Blvd., 700050 Iasi, Romania; scurtean@tuiasi.ro
[4] Academy of Romanian Scientists, 3 Ilfov Street, Sector 5, 50044 Bucharest, Romania
* Correspondence: elena.dragoi@tuiasi.ro (E.N.D.); lbulg@tuiasi.ro (L.B.); mgav@tuiasi.ro (M.G.)

Abstract: Pollution of the environment with heavy metals requires finding solutions to eliminate them from aqueous flows. The current trends aim at exploiting the advantages of the adsorption operation, by using some low-cost sorbents from agricultural waste biomass, and with good retention capacity of some heavy metal ions. In this context, it is important to provide tools that allow the modeling and optimization of the process, in order to transpose the process to a higher operating scale of the biosorption process. This paper capitalizes on the results of previous research on the biosorption of heavy metal ions, namely Pb(II), Cd(II), and Zn(II) on soybean biomass and soybean waste biomass resulting from biofuels extraction process. The data were processed by applying a methodology based on Artificial Neural Networks (ANNs) and evolutionary algorithms (EAs) capable of evolving ANN parameters. EAs are represented in this paper by the Differential Evolution (DE) algorithm, and a simultaneous training and determination of the topology is performed. The resulting hybrid algorithm, hSADE-NN was applied to obtain optimal models for the biosorption process. The expected response of the system addresses biosorption capacity of the biosorbent (q, mg/g), the biosorption efficiency (E, %), as functions of input parameters: pH, biosorbent dose (DS, mg/g), the initial concentration of metal in the solution (c_0, mg/L), contact time (t_c, h), and temperature (T, °C). Models were developed for the two output variables, for each metal ion, finding a high degree of accuracy. Furthermore, the combinations of input parameters were found which can lead to an optimal output in terms of biosorption capacity and biosorption efficiency.

Keywords: Artificial Neural Networks; biosorption; Differential Evolution; heavy metals; optimization; soybean waste

Citation: Fertu, D.I.; Dragoi, E.N.; Bulgariu, L.; Curteanu, S.; Gavrilescu, M. Modeling the Biosorption Process of Heavy Metal Ions on Soybean-Based Low-Cost Biosorbents Using Artificial Neural Networks. *Processes* **2022**, *10*, 603. https://doi.org/10.3390/pr10030603

Academic Editors: Andrea Petrella, Marco Race and Danilo Spasiano

Received: 15 February 2022
Accepted: 17 March 2022
Published: 20 March 2022

1. Introduction

Pollution of the environment with heavy metals has been and remains a problem, since they are toxic, non-biodegradable, persistent in the environment, and have the ability to bioaccumulate in the food chain, disrupting it and posing risks to human health. Their presence in the environment is the consequence of anthropogenic activities such as mining, steel, metallurgy, the metal coating industry, pesticide industry, chemical fertilizers, animal skin processing, etc. [1–4].

The presence of heavy metals in the environment generates a significant pressure on it, because a large part of these pollutants (e.g., arsenic (As), chromium (CrVI), cadmium

(Cd), mercury (Hg), lead (Pb), etc.) are very toxic and dangerous for any form of life, even at low concentrations. Therefore, heavy metals can generate serious impacts and risks in the environment, and, furthermore, for human health, because they do not degrade and are able to accumulate in the food chain [5–7]. That is why the elimination of heavy metals from the environment has long been the concern of many researchers, as well as environmental managers, who developed technologies for the prevention and control of heavy metal pollution. The technologies for the elimination of these pollutants are based on physical, chemical/electrochemical, biological, single or combined processes, such as: chemical precipitation, flotation, ozonation, (electro)coagulation, ion exchange, membrane processes, etc. The related literature mentions and demonstrates the advantages, but also the disadvantages of these technologies, mainly related to the fairly low efficiency of metal removal, the presence of some waste (sludge), and relatively high operating costs [8–10].

Adsorption is one of the mass transfer operations often agreed upon by operators interested in the removal of heavy metals from polluted aqueous effluents. It has been found that adsorption can treat effluents with low concentrations of heavy metals (1–100 mg/L), it is less expensive, and the adsorbent matrices can be easily regenerated. One of the most effective adsorbent materials, often used as a reference in all adsorption studies, is activated carbon, but it has the disadvantage of high production costs, since the carbonization process involves high energy consumption and working at high temperatures. To overcome this disadvantage, solutions were sought for the use of other categories of effective adsorbents, but cheaper than activated carbon, the so-called low-cost sorbents.

Biomass waste from industry and agriculture is proving to be a viable alternative for removing heavy metals from polluted effluents because they have functional groups on their porous surface capable of retaining metals by predominantly ionic bonds, but also hydrogen bonding and the coordination bond in metal ion complexation [11–13]. Some studies have used agricultural waste (fruit kernels, wood waste, plant waste, shells or shells from some vegetables or seeds) due to low procurement costs and their abundance, even if they have a lower biosorption capacity than some synthetic adsorbents [14–16]. These wastes may contain major constituents such as cellulose, hemicellulose and lignin, which provide the pollutant with adequate functional groups (aldehyde, carbonyl, hydroxyl, carboxyl, phenolic, ether) able to retain various metals. As a consequence of the specific composition and structure, their use as biosorbents is advantageous for that they can interact with pollutants from various aqueous effluents. Other advantages of these categories of biosorbents lie in the fact that they have a wide bioavailability and involve low costs [17–23]. For example, the sorption capacity of biosorbents prepared from seeds (e.g., *Litchi chinensis*, *Allium cepa*, *Artocarpus heterophyllus*, *Syzygium cumini*, deoiled karanja or soy seed cake) has been tested for the elimination of metals such as Ni(II), Cr(VI), Cd(II), Zn(II), Cu(II), Pb(II) from the environment with good results [12,24–27].

The selection of high-performance biosorbents based on agro-industrial waste and the choice of the conditions under which an efficient process can be ensured makes it necessary to establish the interactions among the factors on which the biosorption process would depend. One of the reliable tools that can be successfully applied for this purpose is represented by Artificial Neural Networks (ANNs). Some research groups applied ANNs successfully for modelling, simulation, optimization of biosorption processes applied for the removal of heavy metals, dyes and other pollutants from aqueous effluents [28–33].

ANNs are a series of semi-parametric models inspired from the mammalian brain in the manner in which knowledge is acquired (learning) and stored (synaptic weights) [34,35]. The history of ANN concept began in 1943 [36], whereas the first practical network, the perceptron, was developed almost ten years after. A neural model is represented by a set of inter-connected neurons, with definite organization and specific connections, which define the ANN structure [37]. Feedforward multilayer perceptron (MLP) can be considered the best-known type of ANN. In a MLP, the neurons are fully interconnected and arranged in layers (at least two: input and output). They are very popular due to their ability to model almost any type of problem, the inner layer introducing non-linearity, as an internal procedure, which lead to the ability to capture non-linear and non-stationary behavior [38].

Since their development, the practical applications in the area of ANNs increased, this approach being efficiently applied to solve a multitude of problems. The main advantages of ANNs that make them suitable for difficult to solve problems, where the use of phenomenological or conventional modes is impractical or cumbersome, consisting of: the ability of modelling non-normal class distribution, being able to work with continuous forms of data, no requirement for inner knowledge about the problem being solved, robustness in the presence of noise, and a high degree of accuracy [39,40].

In the chemical and environmental engineering areas, the fact that ANNs do not require the existence of phenomenological models is an advantage that allowed the construction of good models for complex processes and the chemical and physical laws governing the system are not fully known. Some examples of successful application of ANNs in chemical and environmental engineering areas are: modelling of polymerization process [41–43], food and fermentation technology [44], freeze drying modelling control, monitoring and optimization [45,46], modelling the electrodeposition state of a polymer-supported ultrafiltration–ultradeposition process for heavy metal recovery [47], and the depollution processes of gaseous fluxes [48].

ANNs are applied more and more to experimental data processing, being able to perform the extrapolation to untested data sets with a special preparation step and to identify causal relationships after an efficient training process [49,50]. Moreover, the simplicity of ANNs set-up and use is deceivable because the optimal parameters (topology and internal parameters) are dependent on the properties of the problem being solved. A too large network tends to overfit the training data, and thus leading to generalization problems, whereas a small network has problems in learning the training data [51]. The search for optimal topology is usually performed by a trial-and-error approach [38], but this has a series of disadvantages such as high computational time, dependence on the researcher knowledge, and inability to ensure optimal results.

To elude these difficulties, it is possible to apply evolutionary algorithms (EAs) that are able to evolve ANN parameters. This is termed neuro-evolution and provides a successful platform for simultaneous optimization of performance and architecture [52]. The use of EAs combined with ANNs has the advantage that several of their features can be coded and co-evolved, provided that more flexible performance criteria are considered compared with the error function [53]. Three levels of evolution can be used: connection weights (training), architecture (topology determination), and learning rules.

The EAs applied in this paper are represented by the Differential Evolution (DE) algorithm [54], and a simultaneous training and determination of the topology is performed. DE characterizes a population based on stochastic metaheuristics for global optimization on continuous spaces, its main advantages being efficiency, flexibility and fundamentality [55]. Due to its low number of control parameters, simple structure and easiness of use, it was successfully applied for solving a multitude of problems, this being one of the main reasons for choosing it in detriment of other EAs. DE has been applied, for example, in chemical engineering: to ensure an optimal temperature profile in the naphtha reforming process, so as to improve the simultaneous elimination of hydrogen and aromatics [56]; appraisal of GTL technology from maximum gasoline yield viewpoint of Fischer-Tropsch synthesis [57]; optimization of a thermally coupled dual membrane reactor to ensure concurrent dimethyl ether synthesis and decalin dehydrogenation [58]; optimization of feed-batch fermentation process [59], etc.

In this framework, we have considered a neural network-based approach to analyze, model and compare the performance of soybean seed waste biomass as a biosorbent for the elimination of heavy metal ions from water and wastewater, as well as to optimize the process in view to generating the scientific basis for scaling-up. The inputs include: pH, sorbent dose, initial concentration of metal ion in solution, contact time, temperature, whereas biosorption efficiency of the metal ions by biosorbents, biosorption capacity of sorbents, and residual concentration of metal ions in solution at the final of biosorption are considered as outputs.

The Differential Evolution algorithm, in variant h-SADE-NN, was applied to obtain optimal models. This DE-based hybrid approach includes the opposition-based principle for initialization, and local search algorithms, Backpropagation and Random Search, which have the role of refining the search for the optimal solution. The obtained results prove the efficiency and utility of this study conducted by simulation.

2. Materials and Methods

2.1. Experimental Context

The experimental protocol is described in detail in a previous paper [60], both in terms of the preparation of biosorbent represented by SB and SWB, carrying out experiments and examining the influence of process parameters such as: pH; sorbent dose, DS (g/L); initial concentration of metal ion in solution, c_0 (mg/L); contact time, t_c (min); temperature, T (°C), on the biosorption capacity, q (mg/g), and biosorption efficiency E (%) a considering three metal ions: Pb(II), Cd(II), Zn(II), respectively, in batch single ion system. The calculation method of these quantities is presented by Bulgariu et al. [12].

The experimental studies aimed a comparison between the biosorption capacity and the biosorption efficiency of the two biosorbents, besides examining the influence of the mentioned parameters and finding correlations between them.

2.1.1. Biosorbents

The soybeans used to obtain the biosorbent needed for experimental studies were washed with distilled water (5–6 times) to remove macroscopic impurities, dried in air at room temperature (22 ± 0.5 °C) for 5 days and then ground. The ground soybeans were then used to obtain the two biosorbents by (i) drying at 65 °C for 6 h and then grinding; (ii) by extraction with n-hexane, for 30 h in a Soxhlet extractor, followed by washing with distilled water and drying at 65 °C for 6 h, and then grinding. The materials obtained from these preparation steps (soybean and soybean waste biomasses) were then mechanically sieved so that the particle size was less than 1.5 mm, and stored in the desiccator to maintain a constant humidity.

The characterization of the two materials obtained (soybean biomass and soybean waste) was performed using the following methods of analysis: X-ray dispersion spectrometry (EDX) (with EDAX-TSL 32 spectrometer); IR spectrometry (with Fourier transform Bio-Rad spectrometer); scanning electron microscopy (SEM) (with SIT 3000 N HITACHI microscope with 15 UV). With the help of these methods of analysis, it was possible to identify the functional groups on the surface of the two biosorbents (soybean and soybean waste biomasses) and the existing non-uniformities that represent the binding centers of the metal ions existing in the aqueous solution. Details on the results of these analyzes are provided by Fertu et al. [60].

2.1.2. Metal Solutions

Aqueous solutions of the studied metal ions, Pb(II), Cd(II) and Zn(II) considered as polluting species in this study, of exact known concentration, were used for the experimental studies. These solutions were freshly prepared for each experiment by diluting a given volume of the corresponding stock solution with distilled water [60]. The aqueous solutions of the metal ions were analyzed using an appropriate spectrophotometric analysis method to ensure the selectivity and accuracy of the determinations (Digital Spectrophotometer S 104 D, glass cuvettes thickness = 1 cm). The concentration of metal ions in the analyzed solutions was calculated from the regression equation of the corresponding calibration curve.

2.1.3. Experimental Conditions

Biosorption experiments were performed in batch system to determine the effect of the initial pH of the aqueous solution, biosorbent dose, initial concentration of metal ions, contacting time, working temperature on the process performance (biosorption capacity, biosorption efficiency) of the two biosorbents used in this study. Establishing the experimental conditions and parameter variation ranges was carried out experimentally for each type of metal ion separately, both for biosorption on soybean biomass and soybean waste, using several sets of experiments that followed the influence of the most important experimental parameters. Therefore, there were used various adsorbent dosages ranging between 4–40 g/L, a pH range between 1 and 6.5, different initial concentration of metals (11.66–416.45 mg Pb(II)/L; 9.22–230.54 mg Cd(II)/L; 13.08–209.25 mg Zn(II)/L, respectively), contact time between 0.1–3 h, working temperature of 5; 25; 50 °C.

The experimental data obtained were collected and used to model and optimize the biosorption process of heavy metal ions Pb(II), Cd(II), Zn(II) based on Artificial Neural Networks.

2.2. Modelling

2.2.1. DE Application

A methodology based on ANNs and DE was applied in order to model and optimize the considered biosorption process. The ANN represented the model, whereas DE was the optimizer used to determine the optimal characteristics of the neural network that lead to an acceptable difference between predictions and the experimental data. DE is a population-based metaheuristic inspired by the Darwinian principle of evolution.

DE, similar to all algorithms of its kind, starts with a pool of solutions (that in this work are represented by coded neural networks). The individuals evolve by mutation, crossover, and selection until a stop criterion is reached. The population's initialization is correlated the problem's upper and lower bounds.

The mutation operator is performed after initialization. Depending on the strategy, a base vector and one or more sets of difference vectors are selected. Choosing mutation vectors is usually a random process that ensures that all vectors are statistically threaded equally [61]. The only constraint in this step is the uniqueness of the used vectors, each representing a specific generation vector solution. In the initial DE algorithm proposed by Price et al., the mutation vector is determined by adding to a base vector a single, scaled, differential term [61].

The next step of the method consists of applying the crossover operator, a trial population being obtained by combining the current population with the mutant one. Because the trial population may contain data from mutation vectors outside the accepted value range, a control value method is required. Boundary constraints are enforced by resetting schemes and penalizing methods. If the trial vector does not meet all constraints, a resetting scheme is used to change the parameters. Penalty methods include techniques such as the Brick Wall Penalty, Adaptive Penalty, and Random Re-initialization [61].

After recombination, the best individuals from the current and trial populations are selected for a new generation. The algorithm is repeated until the current generation's index equals the number of generations initialized at the start. This is a stop condition, and, depending on the problem, the evolutionary algorithms can use improvement-based, movement-based, or distributed-based termination criteria [61,62].

To identify the various combinations of steps, a notation DE/base/num/cross is used. 'base' represents the method of selecting the base vector, 'num' is the number of differential terms, and 'cross' is the crossover mechanism [63]. Examples of strategies include DE/rand/1/exp, DE/best/1/exp, DE/rand-to-best/1/exp, DE/best/2/exp, DE/rand/2/exp [64,65].

2.2.2. hSADE-NN

Finding the optimal topology and internal parameters of neural networks is still difficult, especially when there are many parameters, many possible combinations, multiple training algorithms, or no consistent rules [66]. To overcome these issues, we used a neuro-evolutionary approach that involves a hybrid auto-adaptive Differential Evolution with neural networks (hSADE-NN), which is based on simultaneous optimization of topology and training [46,67–69]. In this new context, optimization requires a large search space and a large number of parameters, affecting DE performance. Despite its attractive features and power, the DE algorithm can be inefficient due to slow convergence and low accuracy, especially in noisy environments or when the solution space is difficult to explore [70]. In order to reduce or eliminate these disadvantages, DE was combined with other optimization algorithms [71–73]. This hybridization can be performed at [55]: (i) an individual level; (ii) at the population level; (iii) an external level; (iv) at the meta-level.

This paper used a hybridized DE with two algorithms: Random Search and Backpropagation (BK). This choice increases the optimizer's performance and the likelihood of finding the optimal ANN through individual hybridization. This process, which uses one of the two algorithms selected on the basis of randomly generated numbers, can only improve the best solution obtained for each generation. Therefore, the core DE is also modified: (1) by introducing opposition-based learning (OBL) as a way to improve initialization [74,75]; (2) by a simple self-adaptive principle (the control parameters are included and developed in the algorithm, applying the same mathematical equation as for the individuals they contain); and (3) by reorganizing individuals who participate in the mutation phase based on fitness [45,46,69,73,74]. The simplified schema of the hSADE-NN algorithm is presented in Figure 1.

To encode the ANN models, the present study uses a direct mapping between genotype, i.e., network representation, and phenotype, i.e., current neural network. As this requires a large number of parameters some limitations were considered: (1) the number of hidden layers was limited to two, given that a two-layer ANN can model almost any process with sufficient precision; (2) the maximum number of neurons allowed in the first and second hidden layers of 40 and 20, respectively, was taken into account.

Network performance was determined using the *Fitness* function (Equation (1)) which is based on the mean square error of traning ($MSE_{training}$), where *err_correction* is a constant value (equal to 10^{-10}) used to eliminate the unlikely event when $MSE_{training}$ is equal to 0.

$$Fitness = \frac{1}{MSE_{training} + err\_correction} \tag{1}$$

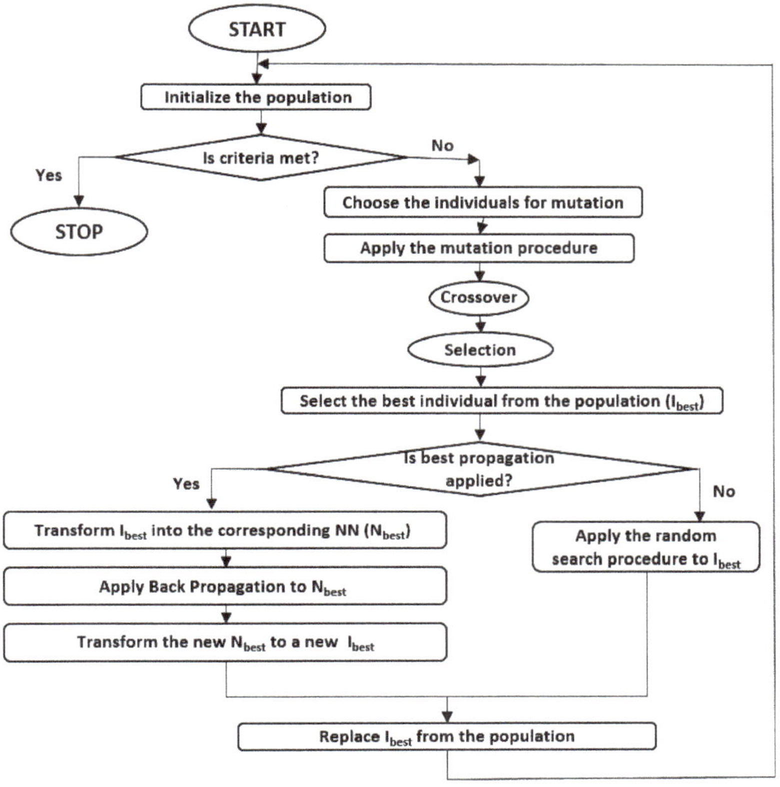

Figure 1. Scheme of the hSADE-NN algorithm: I_{best} represents the best individual in the population and N_{best} is the neural network corresponding to the I_{best} (from [48], with Springer Nature permission, License 5250080341971 of 15 February 2022).

3. Results and Discussions

3.1. Factors Affecting the Biosorption Process

The experimental results acquired from the study addressing the use of soybean biomass and soybean waste biomass to eliminate Pb(II), Cd(II) and Zn(II) ions from aqueous solutions by biosorption were largely discussed and analyzed in a previous paper [60]. It was found that both SB and SWB contain in their structure large amounts of organogenic elements (C, O, P, S), but also a series of ions of alkali and alkaline earth metals (K, Mg, Al) which, due to their high mobility they can easily participate in metal ion exchange processes. Moreover, both biosorbents have in their structure numerous and varied functional groups (such as hydroxyl, carbonyl, carboxyl, phosphate, etc.), which may have an important role in the retention processes of the studied metal ions. Additionally, the morphological non-uniformity of the surfaces of the two materials recommends the use of soybean biomass and soybean waste as biosorbents in the processes of decontamination of environmental components.

The biosorption capacity (q, mg/g) of the two materials depends significantly on the initial pH value of the solution, although the variation is not uniform. The highest values of the biosorption parameters are obtained at a pH of the initial solution of 3.39 where more than 50% of Pb(II), 60% of Cd(II) and 45% of Zn(II), respectively, can be removed from the aqueous solution using these two biosorbents. This behavior is largely determined by the high buffering capacity of the biosorbents, as the pH values measured in the solutions obtained after the completion of the biosorption process are considerably higher than the initial ones. The significant increase in the pH of the final solution (by 2–3 units of pH) causes, in turn, a change in the speciation of metal ions in the aqueous solution, which also influences the efficiency of the biosorption process.

The biosorption efficiency (E, %) of Pb(II), Cd(II) and Zn(II) ions, respectively, in aqueous solutions on SB and SWB also depends on the amount of biosorbent used. The values of the dependent variables q and E calculated for each case showed that a minimum dose of biosorbent of 5 g/L is sufficient for the quantitative retention of the considered metal ions. Additionally, the close values of the biosorption parameters obtained for SB and SWB showed that SWB has about the same efficiency of the biosorption processes as SB, although from an economic point of view their cost is much lower.

The biosorption capacities of SB and SWB increase with the initial concentration of metal ions in the aqueous solution in the studied concentration range. The experimental results showed that the biosorption of Pb(II), Cd(II) and Zn(II) ions by biosorption on SB and SWB, respectively, takes place predominantly through electrostatic interactions. This makes the retention rates moderate for all the cases studied. Therefore, in order to efficiently remove them using the two biosorbents, it is necessary either to perform several successive biosorption steps or to improve the biosorption capacity of the two biosorbents.

As the contact time between phases increases, so does the biosorption capacity for both biosorbents. The close values of the biosorption efficiency obtained for the three metal ions for both SB and SWB are another argument in favor of the hypothesis that the biosorption process is performed by non-selective electrostatic interactions, which makes the speed of these processes not depend on the nature of the metal ion in the aqueous solution. Thus, the time required to reach equilibrium was found of 60 min for Pb(II) and Zn(II) ions, and 30 min for Cd(II) ions, respectively, for both biosorbents.

Temperature has a rather small influence on the biosorption capacity, for both biosorbents and for all metal ions studied. Increasing the temperature at 45 °C causes a rather small increase in the biosorption capacity of the two biosorbents, which was especially visible at high values of the initial concentration of metal ions. Although this variation suggests the endothermic nature of biosorption processes, the retention of Pb(II), Cd(II) and Zn(II) ions on SB and SWB can be successfully achieved at ambient temperature, and these conditions are advantageous both from an economic point of view and from the efficiency of the biosorption process.

From this concise analysis of the biosorption process carried out in an experimental program aiming at removing Pb(II), Cd(II), Zn(II) ions from aqueous solutions using soybean biomass and soybean waste biomass as biosorbents and described in detail in a previous paper [60], it results that both the biosorption capacity (q, mg/g) and the biosorption efficiency (E, %) are in a complex dependence on a series of process parameters such as pH; sorbent dose (DS, g/L); initial concentration of metal ion in solution (c_0, mg/L); contact time (t_c, h); temperature (T, °C). This is why it is very important to examine this process by means of models that can correlate well the dependent and independent variables, and can predict the biosorption capacity and efficiency of Pb(II), Cd(II), Zn(II) ions from aqueous system using SB and SWB, considering the complexity of the process and the interactions between the variables. In this context, a neural network approach was adopted.

3.2. Prediction of Biosorption Capacity and Efficiency Using ANN

As discussed above, the modeling procedure consisted of a hybrid approach that combines an evolutionary algorithm (Differential Evolution—DE) with ANNs. The role of DE is to simultaneously achieve a topological and structural optimization of ANN, whereas ANN acts as a model for the process. Once the optimal ANN model is determined, the same DE-based hybrid approach is applied, which also includes the opposition-based principle for initialization and, further backpropagation and Random Search algorithms for local search, to optimize the process. The aim was to identify the parameters—pH, sorbent dose, *DS* (g/L), initial concentration of metal ion in solution, c_0 (mg/L), contact time, t_c (h) and temperature, *T* (°C)—which would lead to an optimal output signal in terms of biosorption capacity, *q* (mg/g) and biosorption efficiency, *E* (%).

We applied a modified DE version hybridized with two local search algorithms, Random Search and BK, in order to improve the optimization performance and increase the probability of determining an optimal ANN. This hybridization is carried out individually, in a highly selective approach, so that only the best solution found in each generation is improved, using one of the two algorithms chosen based on randomly generated numbers. When applying the BK algorithm for local search, it was noticed that sometimes BK is not able to improve the local network, without clear reasons, although it is applied several times. However, this problem rarely occurred, which is why we did not change the algorithm, but added a second algorithm to improve local search through Random Search. The software used to determine the considered models was implemented in C#, .Net 4.0, Visual Studio. The implementation is freely and openly available at https://elenadragoi.ro/CV/Documents/AITB-%20ANN_DE.7z (accessed on 10 February 2022).

3.3. ANN-Based Modeling

Before starting the modeling procedure, all the experimental data collected went through a set of data processing techniques (normalization, randomization and group selection). The purpose of this procedure was to make sure that the determined models generated the smallest possible errors. Standardization is one of the most common tools used to achieve good results in automatic recognition systems [76]. There are different types of standardization procedures, each with specific applications. In this paper, the goal is to constrain the input feature and reschedule it in that range. This procedure is specific to the min-max normalization type, in which a linear interpolation formula is applied (Equation (2)):

$$v_{norm} = min_t + (max_t - min_t)\frac{v - min_v}{max_v - min_v} \quad (2)$$

where v_{norm} is the normalized value, max_t is the maximum value of the target (in this case 1), min_t is the minimum value of the target (in this case +1), min_v is the minimum value of the entire interval in which v is gathered, and max_t is the maximum value of the entire interval in which v is gathered.

Another data processing technique applied to our experimental data is data splitting. The sampling methodology used can have a significant impact on the quality of the determined neural models [77]. In this case, the available data is split into training and testing sets, the training data being used to determine the ANN parameters, and the testing data are used to evaluate the model performance. For training, 75% of the available experimental data are taken into account, the rest being used for testing. To ensure that training and testing are not performed for a specific region, the experimental data is first arranged randomly. Next, a set of 50 simulations was performed in order to determine the best model (Table 1).

The approach involving the application of the hSADE-NN algorithm had a series of fixed settings related to the maximum allowed topology and the maximum number of iterations for which the models evolve on the DE principle. The purpose of these limitations, which are established on the basis of experience and guidance in the literature is to ensure a compromise between performance and resources consumed. Consequently, the maximum number of generations is 1000, and the maximum allowed topology is 7:20:10:2, where 7 indicates the number of inputs, 20 the number of neurons in the first hidden layer, 10 the number of neurons in the second hidden layer and 2 number of outputs (*q*, mg/g; *E*, %). Depending on the situation, the topology varies, including the number of inputs and outputs. Therefore, in this work, various combinations were performed (Table 1), where SB and SWB indicate the types of biosorbent (soybean biomass, SB or soybean waste biomass, SWB) and Me refers to metal ion.

Table 1. Cases considered for the hSADE-NN approach.

Case	Biosorbent	Input Parameters	Output Parameters
C1	SB	6 (Me, *DS*, pH, c_0, t_c, *T*)	2 (*q*, mg/g and *E*, %)
C2			1 (*q*, mg/g)
C3			1 (*E*, %)
C4	SWB	6 (Me, *DS*, pH, c_0, t_c, *T*)	2 (*q*, mg/g and *E*, %)
C5			1 (*q*, mg/g)
C6			1 (*E*, %)
C7	SB and SWB	7 (tB, Me, *DS*, pH, c_0, t_c, *T*)	2 (*q*, mg/g and *E*, %)
C8			1 (*q*, mg/g)
C9			1 (*E*, %)

Table 2 shows the fitness based on the mean square error in the training phase ($MSE_{training}$), and the topology indicates the general arrangement of the neurons in the layers of the model. For the best model (7:06:02) resulted in the case that combines all the groups and parameters into a single ANN, C7, the average absolute relative error (AARE) computed in the testing case is 46.28% for *q*; and 15.06% for *E*. As the AARE for *q* is unnceptable high, indicating that the identified ANN was not able to efficiently capture the dynamic of the system in all the possible variants, then individual models were considered in various situations in order to identify the best suitable ANN. The correlations obtained for the best models for all the considered cases are presented in Table 3.

Table 2. Modeling results with the hSADE-NN for all the cases considered.

Case	Metric	Fitness	*MSE Training*	*MSE Testing*	Topology
C1	Best	184.5910	0.0054	0.0054	6:05:02
	Worst	66.0597	0.0151	0.0149	6:04:02
	Average	125.8947	0.0088	0.0114	-
C2	Best	1525.2047	0.0007	0.0017	6:11:01
	Worst	279.7430	0.0036	0.0041	6:16:01
	Average	592.4199	0.0024	0.0043	-
C3	Best	244.7126	0.0041	0.0057	6:11:01
	Worst	69.2762	0.0144	0.0138	6:11:01
	Average	111.5186	0.0104	0.0132	-
C4	Best	263.1930	0.0038	0.0067	6:07:02
	Worst	114.3745	0.0087	0.0144	6:11:02
	Average	188.8675	0.0056	0.0132	-
C5	Best	1929.4896	0.0005	0.0009	6:11:01
	Worst	251.5095	0.0040	0.0074	6:09:01
	Average	622.6071	0.0024	0.0026	-
C6	Best	268.7301	0.0037	0.0064	6:06:01
	Worst	111.0732	0.0090	0.0295	6:18:01
	Average	156.7920	0.0070	0.0227	-
C7	Best	190.1452	0.0053	0.0052	7:06:02
	Worst	87.3187	0.0115	0.0129	7:04:02
	Average	123.2298	0.0084	0.0092	-
C8	Best	706.0747	0.0014	0.0014	7:06:01
	Worst	251.4068	0.0040	0.0030	7:08:01
	Average	481.1945	0.0023	0.0023	-
C9	Best	171.8786	0.0058	0.0048	7:05:01
	Worst	85.8678	0.0116	0.0157	7:05:01
	Average	124.9972	0.0083	0.0098	-

Table 3. Correlation for the best models obtained in each case.

Case	*q* Training	*E* Training	*q* Testing	*E* Testing
C1	0.9496	0.8955	0.9317	0.8363
C2	0.9902	-	0.9660	-
C3	-	0.9443	-	0.9217
C4	0.9600	0.9170	0.9534	0.7534
C5	0.9927	-	0.9511	-
C6	-	0.9348	-	0.9117
C7	0.9580	0.8695	0.9169	0.8280
C8	0.9784	-	0.9581	-
C9	-	0.8960	-	0.8948

As it can be observed from Table 3, in all cases considering both *q* and *E* outputs (C1, C4 and C7), the correlations are lower compared with the other cases where individual models were determined for each output. This indicates that, due to the complexity of the interactions between the parameters, individual models are better suited. A similar situation was found by Fertu et al. who modelled experimental data applying response surface methodology [60]. Regarding the concerns for each of the different datasets, the differences between the predictions for the testing phase are presented in Figure 2 for *q* and Figure 3 for *E*.

The data from Figures 2 and 3 indicate that, whereas in the case of the SB dataset (for both *q* and *E*) the combined models (C8 and respectively C9) are capable of closely following the process dynamic, the differences for some particular examples for the SWB dataset are large, indicating that, for the considered process, the best strategy to model the available system is to use individual datasets and outputs (C2, C3, C5 and C6). The mathematical relations that represent each model and their implementation in C# can be downloaded from https://elenadragoi.ro/CV/Documents/Soybean_biosoprtion.cs (accessed on 10 February 2022).

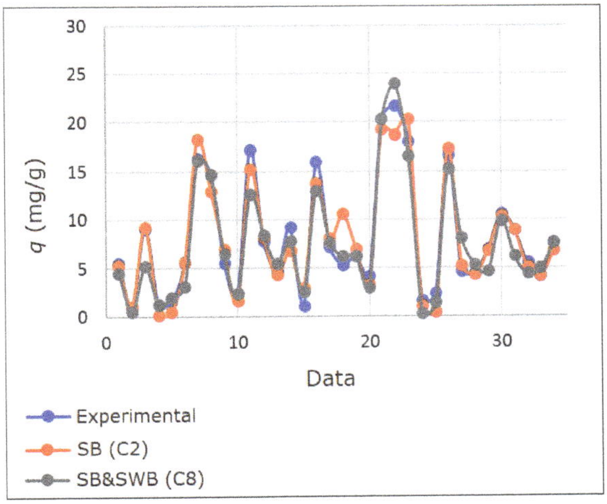

(a)

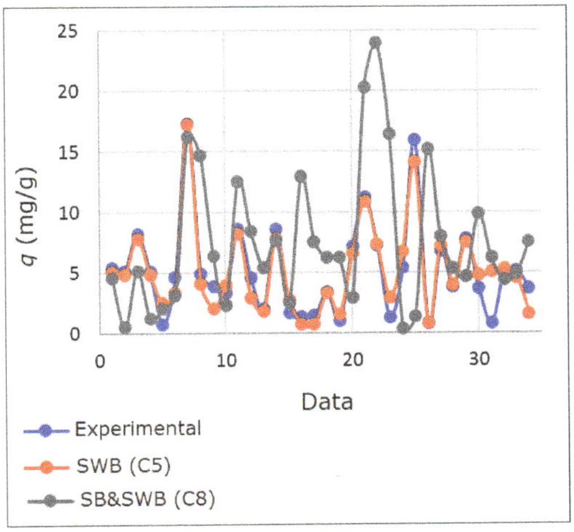

(b)

Figure 2. Comparison between experimental data and predictions for sorption capacity (*q*, mg/g) for datasets: (**a**) soybean biomass; (**b**) soybean waste biomass.

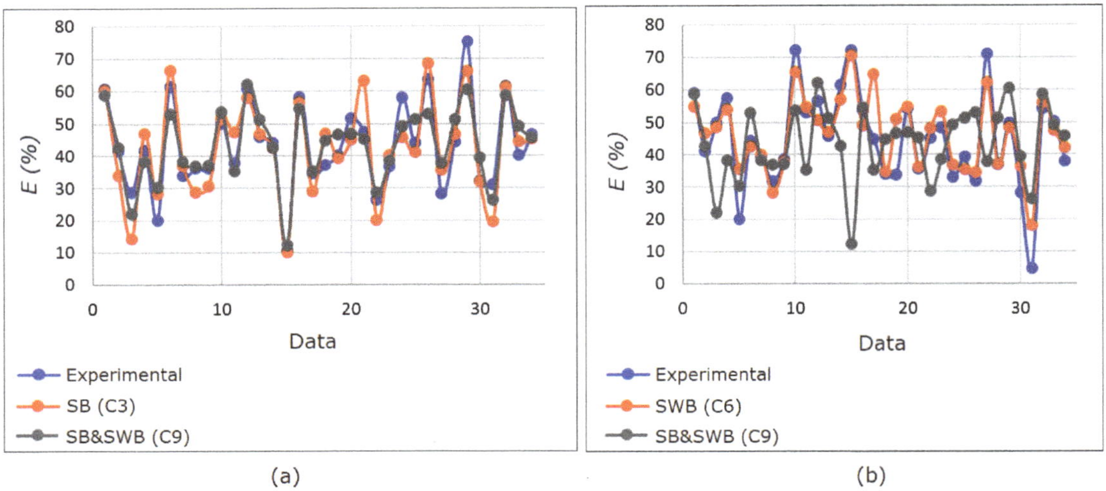

Figure 3. Comparison between experimental and predictions for the *E* (%) for the dataset (**a**) soybean biomass; (**b**) soybean waste.

3.4. Analysis of the Influence of Process Parameters on Biosorption Capacity and Biosorption Efficiency

After the cases that best fit the available experimental data were identified, a series of predictions were performed in order to observe the influence of different inputs on the considered outputs. These predictions were used to generate contour plots for the two datasets considered (biosorption of metal ions on SB and SWB, respectively). These plots were determined by varying two parameters at a time and setting the remaining ones to a fixed value represented by the middle of the experimental interval. Thus, when fixed, the parameters had the following values: DS (g/L) = 22.5; pH = 3.75; c_0 (mg/L) = 117.17; t_c (h) = 12; T (°C) = 27.5.

For example, Figure 4 indicates the simultaneous influence of DS and pH on biosorption dependent variables (q, E) on both SB and SWB, whereas Figure 5 shows the simultaneous influence of t and c_0, respectively, for Pb(II) biosorption. It can be observed that the highest values for q (mg/g) are obtained at low DS and a pH around 6 for SB and at a pH between 3.5–6 for SWB. On the other hand, a similar trend is observed for E (%), for both biosorbents, where the highest efficiency is obtained at higher pH and higher DS. Concerning the t and c_0 influence, Figure 5 shows a distinct behaviour between the SB and SWB.

Figure 6 indicates the simultaneous influence of contact time (t_c, h) and temperature (T, °C) on biosorption dependent variables (q, E) using SB and SWB as biosorbents, whereas Figure 7 shows the simultaneous influence of temperature (T, °C) and pH, respectively, for Cd(II) biosorption process. As it can be observed, a higher t_c does not necessarily ensure a high biosorption capacity, q. When following the influence of T, it can be observed that there are different behaviours in relation with t_c and pH. For example, analyzing E (%) for the soybean waste, the results point out that higher efficiency is obtained at lower T and longer t_c, and at higher T and lower pH. This indicates that, in the multi-dimensional space represented by all the considered parameters, the dependence function is non-linear and has a multi-modal characteristic.

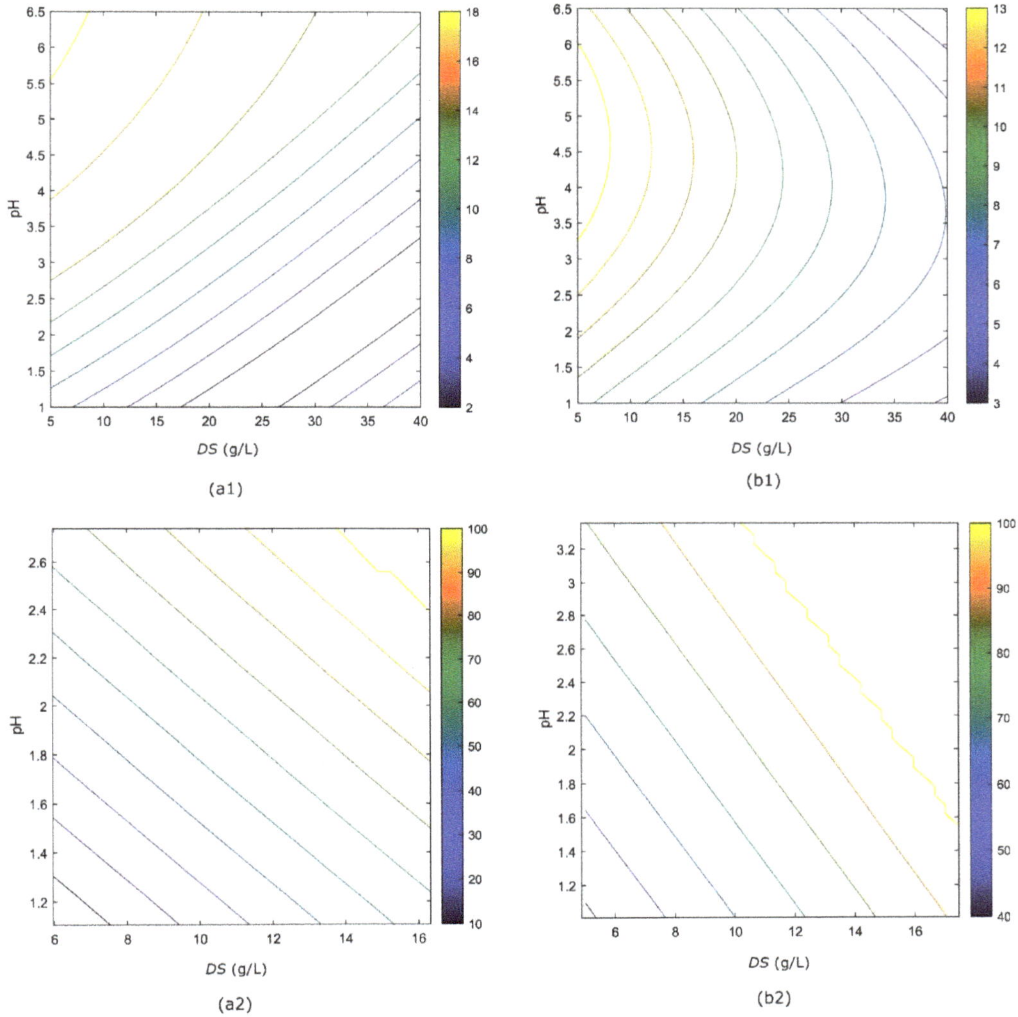

Figure 4. Contour plots for Pb(II) biosorption on SB (**a1**,**a2**) and SWB (**b1**,**b2**), at different values of pH and *DS* (g/L) when the other parameters are kept constant at the middle of the interval of variation: (**a1**) *q* (mg/g) on SB; (**a2**) *E* (%) on SB; (**b1**) *q* (mg/g) on SWB; (**b2**) *E* (%) on SWB.

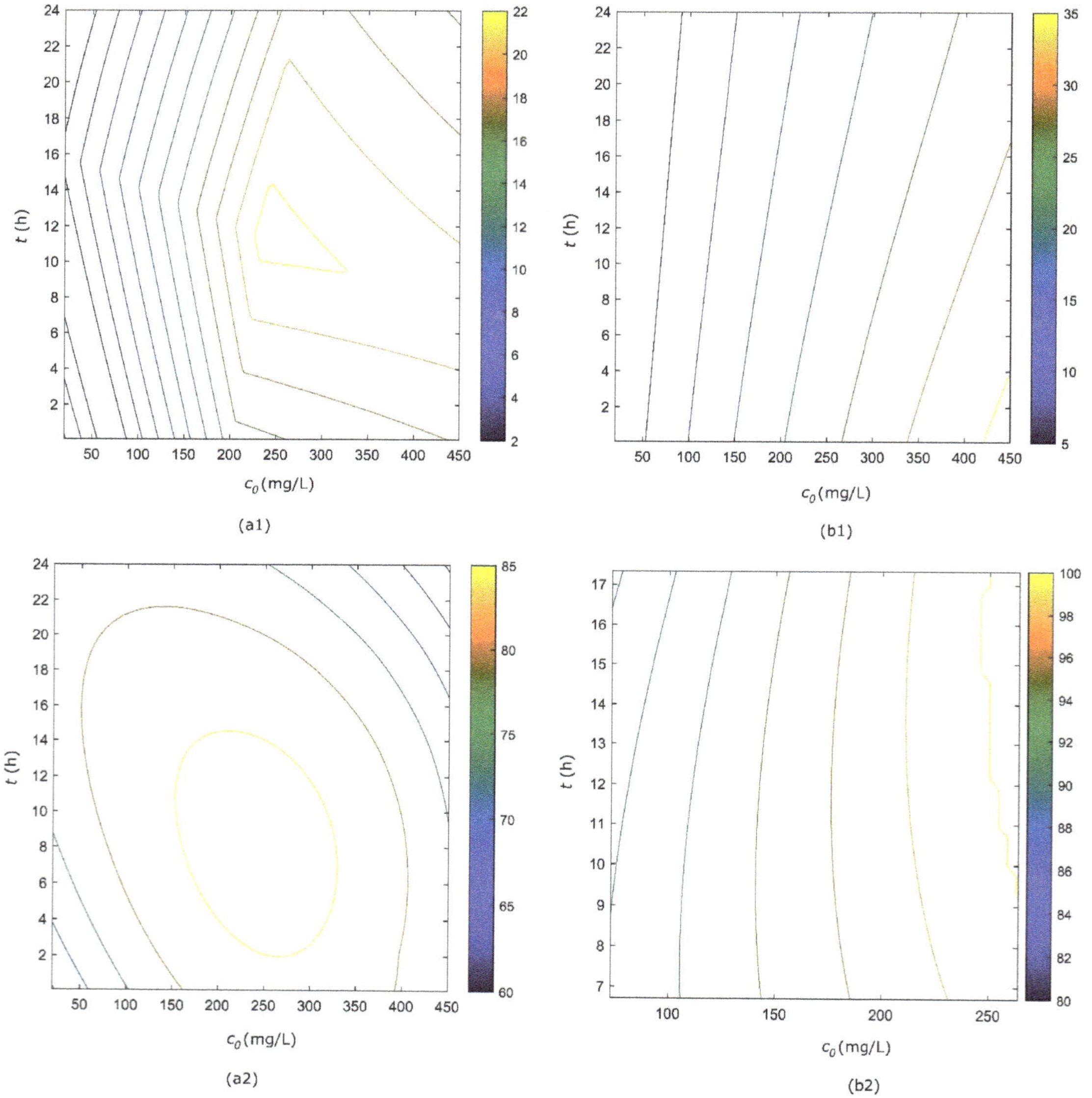

Figure 5. Contour plots for Pb(II) biosorption on SB (**a1**,**a2**) and SWB (**b1**,**b2**), at different values of contact time (t, h) and initial concentration of metal ion in solution (c_0, mg/L) when the other parameters are kept constant at the middle of the interval of variation: (**a1**) q (mg/g) on SB; (**a2**) E (%) on SB; (**b1**) q (mg/g) on SWB; (**b2**) E (%) on SWB.

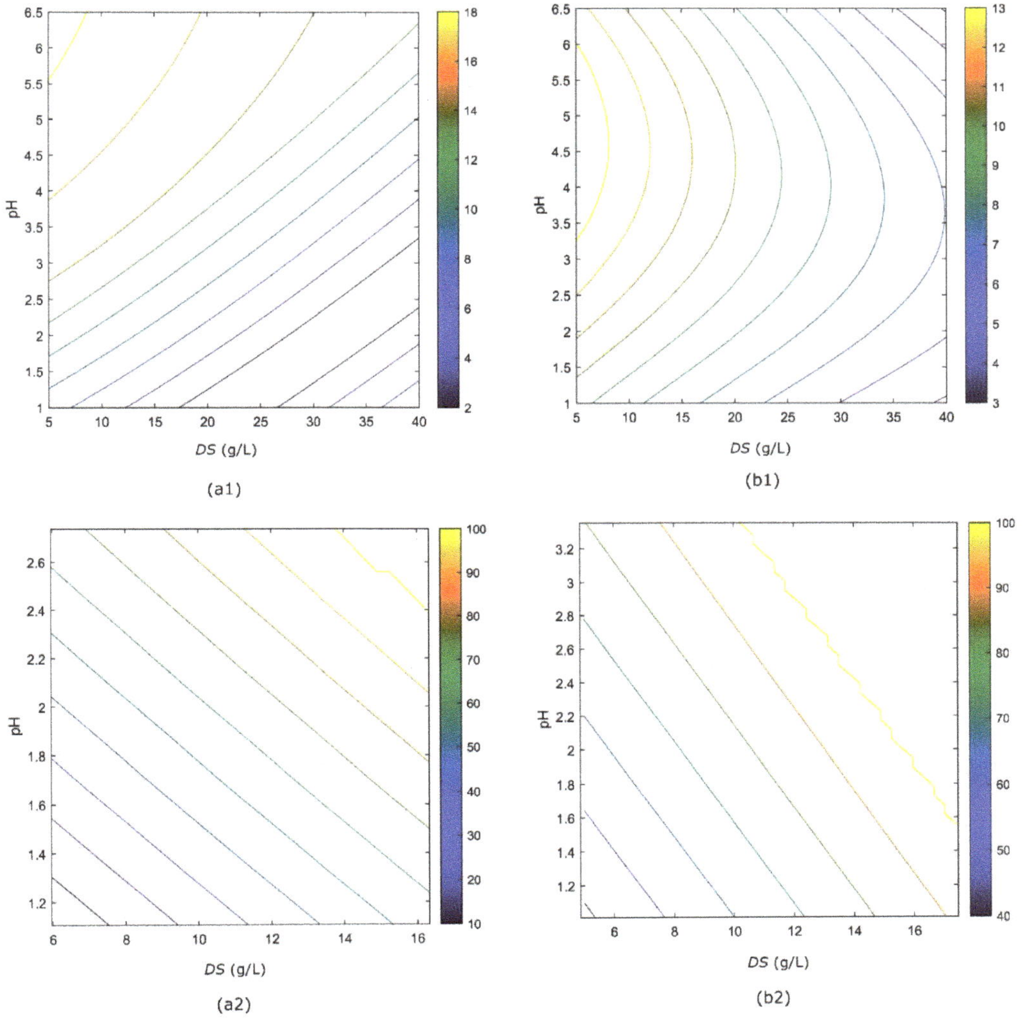

Figure 4. Contour plots for Pb(II) biosorption on SB (**a1**,**a2**) and SWB (**b1**,**b2**), at different values of pH and *DS* (g/L) when the other parameters are kept constant at the middle of the interval of variation: (**a1**) *q* (mg/g) on SB; (**a2**) *E* (%) on SB; (**b1**) *q* (mg/g) on SWB; (**b2**) *E* (%) on SWB.

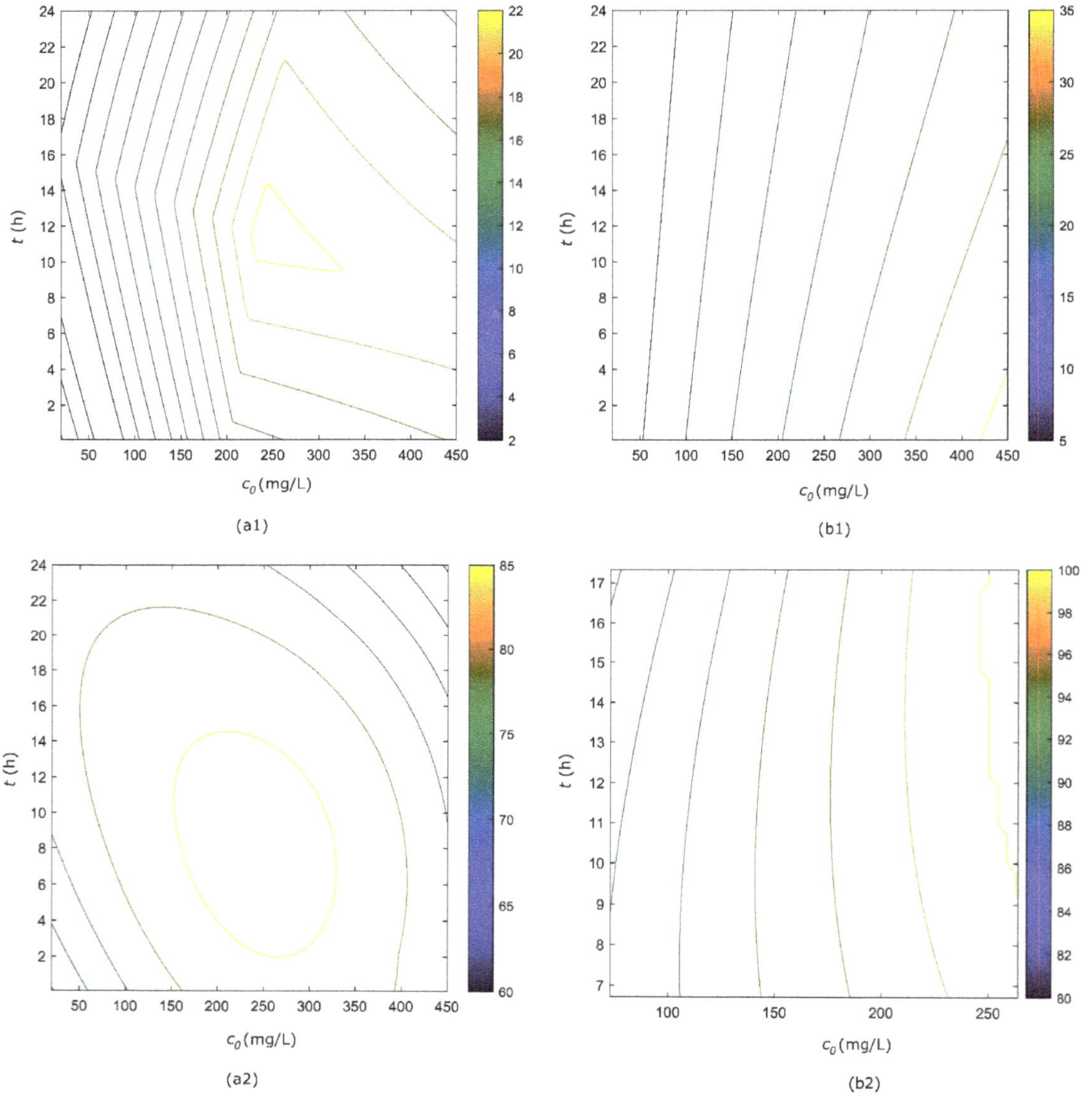

Figure 5. Contour plots for Pb(II) biosorption on SB (**a1**,**a2**) and SWB (**b1**,**b2**), at different values of contact time (t, h) and initial concentration of metal ion in solution (c_0, mg/L) when the other parameters are kept constant at the middle of the interval of variation: (**a1**) q (mg/g) on SB; (**a2**) E (%) on SB; (**b1**) q (mg/g) on SWB; (**b2**) E (%) on SWB.

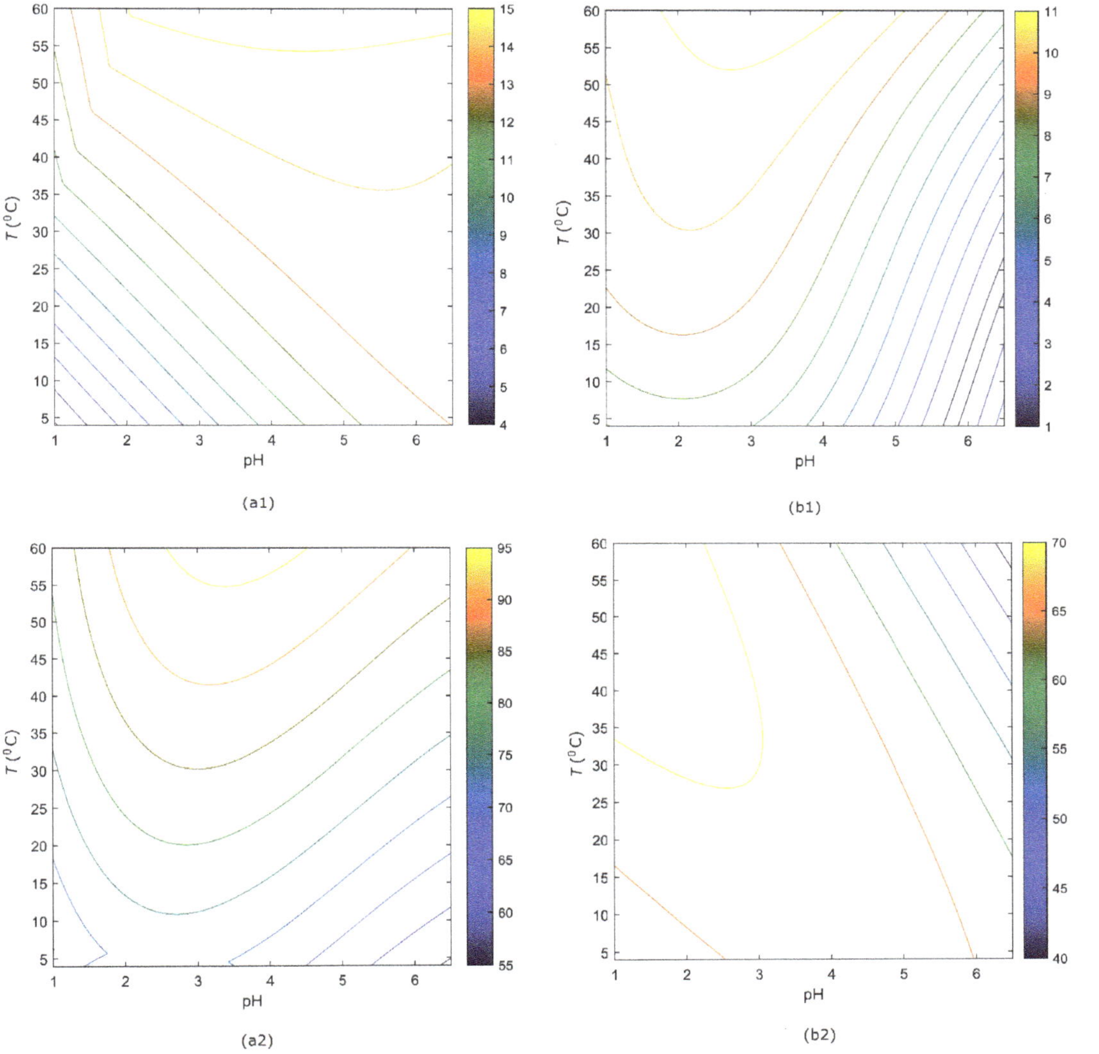

Figure 6. Contour plots for Cd(II) biosorption on SB (**a1**,**a2**) and SWB (**b1**,**b2**), at different values of temperature (T, °C) and pH when the other parameters are kept constant at the middle of the interval of variation: (**a1**) q (mg/g) on SB; (**a2**) E (%) on SB; (**b1**) q (mg/g) on SWB; (**b2**) E (%) on SWB.

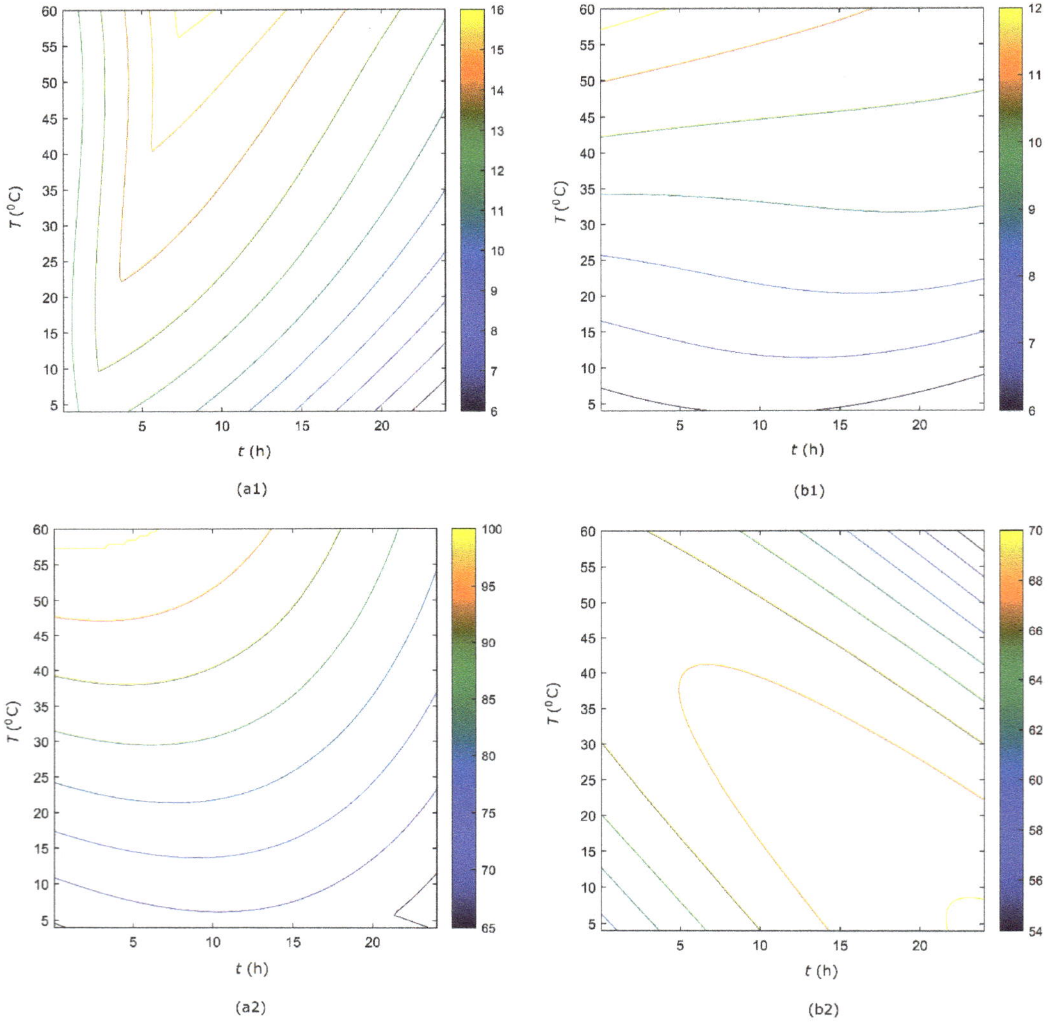

Figure 7. Contour plots for Cd(II) biosorption on SB (**a1**,**a2**) and SWB (**b1**,**b2**), at different values of temperature (T, °C) and contact time (t_c, h) when the other parameters are kept constant at the middle of the interval of variation: (**a1**) q (mg/g) on SB; (**a2**) E (%) on SB; (b1) q (mg/g) on SWB; (**b2**) E (%) on SWB.

Figure 8 shows the simultaneous influence of contact time, t_c and temperature T on biosorption capacity and biosorption efficiency of Zn(II). Figure 9 displays the influence of temperature, T and biosorbent dosage, DS on Zn(II) biosorption, whereas in the case of SB, the highest q values can be obtained for a wide range of temperatures in combination with contact time, only higher T in combination with lower DS indicate a peak. On the other hand, lower T values lead to higher efficiency, E (%).

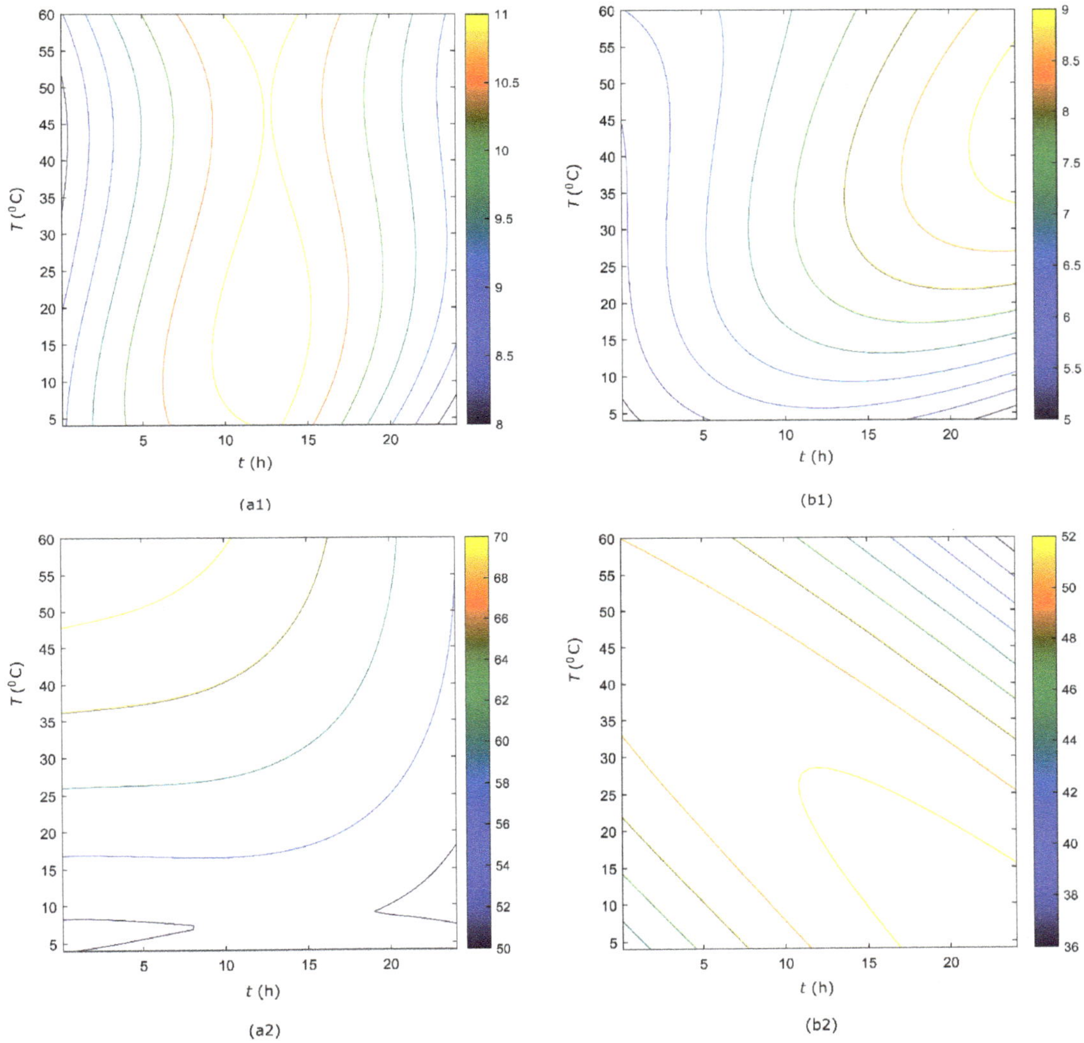

Figure 8. Contour plots for Zn(II) biosorption on SB (**a1**,**a2**) and SWB (**b1**,**b2**), at different values of temperature (T, °C) and contact time (t_c, h) when the other parameters are kept constant at the middle of the interval of variation: (**a1**) q (mg/g) on SB; (**a2**) E (%) on SB; (**b1**) q (mg/g) on SWB; (**b2**) E (%) on SWB.

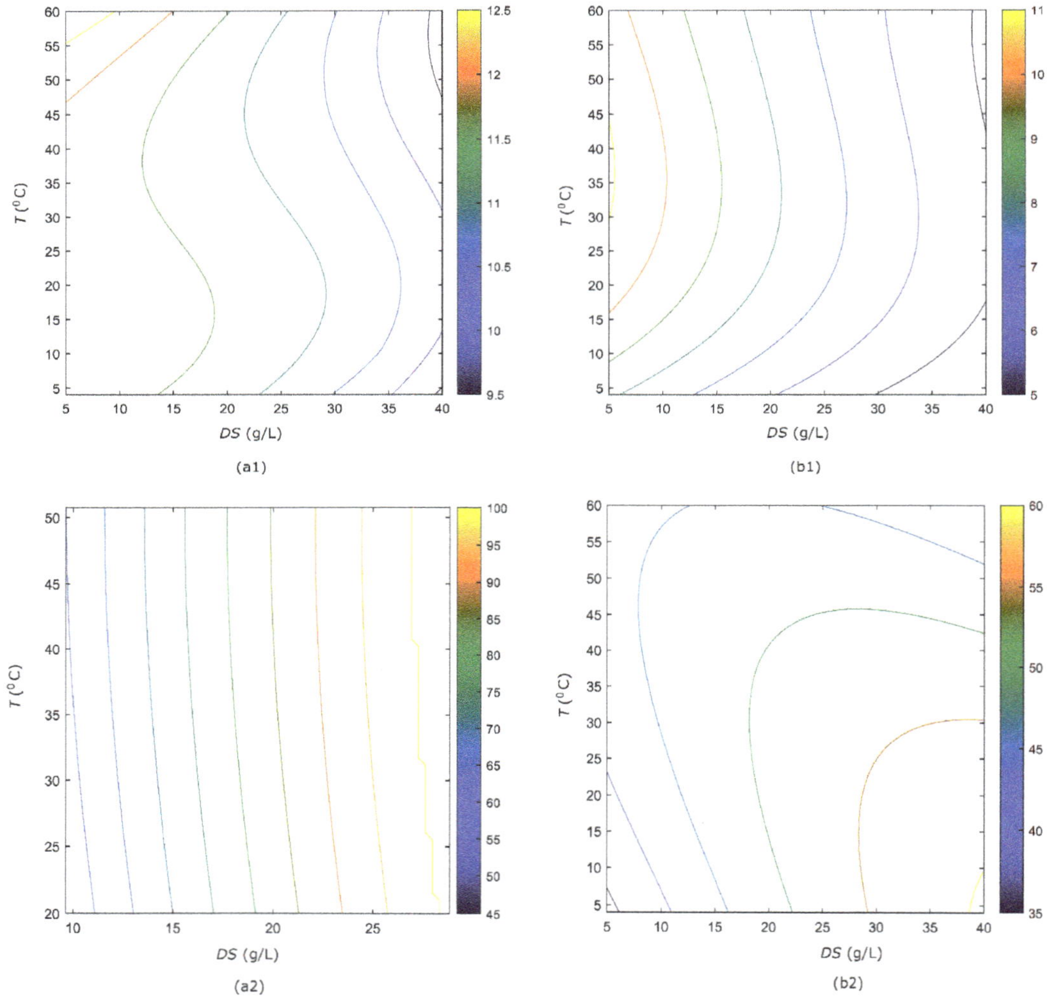

Figure 9. Contour plots for Zn(II) biosorption on SB (**a1**,**a2**) and SWB (**b1**,**b2**), at different values of temperature (*T*, °C) and biosorbent dosage (*DS*, g/L) with the other parameters kept constant at the middle of the interval of variation: (**a1**) *q* (mg/g) on SB; (**a2**) *E* (%) on SB; (**b1**) *q* (mg/g) on SWB; (**b2**) *E* (%) on SWB.

3.5. Optimization

Once the best neural models for the two dependent variables, *q* and *E*, have been determined, the next step is to identify the combinations of input parameters (metal ion, biosorbent, pH, initial metal concentration, dosage, contact time, and temperature) which can lead to an optimal output in terms of process efficiency: maximum of *q* and maximum of *E*. In order to perform this task, a series of 50 simulations were performed for each variable (*q*, *E*) and for each dataset.

From the contour plots it was observed that, overall, the conditions that lead to a higher *E* (%) result in average or lower values for *q*. Thus, optimization becomes the process of identifying parameters for which the combination (*q*, *E*) is highest, without necessarily containing the maximum for either *E* or *q*.

Table 4 shows some series of optimal solutions resulted. As it can be observed, for all metals, various combinations of parameters that lead to a high efficiency were identified, indicating the capability of the selected models to efficiently capture the dynamic of the systems, without the need to individually model each metal.

Table 4. Results of the optimization with the generated optimal neural networks, for which the solutions were verified experimentally.

Biosorbent	Metal	*DS* (g/L)	pH	c_0 (mg/L)	t_c (h)	*T* (°C)	*q* (mg/g)	*E* (%)
SB	Pb(II)	38.55	4.30	263.85	11.49	34.95	19.58	99.60
		34.56	3.41	257.74	13.80	33.67	18.83	98.80
		33.08	4.24	280.39	8.70	34.00	20.18	98.70
		40.90	4.83	304.88	9.02	34.90	19.38	98.00
		32.89	4.55	259.75	10.78	34.91	22.57	97.38
	Cd(II)	28.54	2.99	165.64	7.73	34.99	17.89	89.28
		29.60	2.97	170.27	8.32	34.91	17.75	89.23
		34.96	2.92	208.69	6.78	34.96	15.49	89.15
		27.42	3.14	170.42	7.19	34.97	18.15	89.07
		30.59	2.75	175.85	8.16	33.82	16.96	88.89
	Zn(II)	41.97	6.49	199.26	20.66	34.98	9.56	70.74
		41.05	6.50	202.70	19.31	34.70	9.69	70.42
		41.76	6.49	169.76	15.89	34.80	9.59	70.33
		39.39	6.48	192.77	18.43	33.75	9.88	69.40
		40.31	6.39	152.34	17.68	34.17	9.73	69.26
SWB	Pb(II)	34.22	1.11	34.83	2.82	24.12	8.41	99.91
		31.12	2.81	12.47	8.26	19.96	6.09	99.67
		40.60	1.23	201.41	23.71	28.21	5.10	99.61
		41.63	3.07	140.88	2.38	34.88	10.20	96.36
		40.50	3.24	145.40	3.32	34.36	11.13	95.18
	Cd(II)	41.67	1.04	12.80	13.38	16.86	11.21	89.75
		41.50	1.02	15.02	13.68	15.09	11.23	89.62
		41.94	1.00	10.02	10.05	16.83	12.35	89.54
		41.66	1.02	9.24	9.03	27.42	11.84	89.28
		41.97	1.05	17.33	8.37	24.36	11.53	88.93
	Zn(II)	41.90	1.00	10.86	14.76	15.35	12.51	71.39
		41.72	1.00	9.59	8.44	17.74	13.46	70.40
		41.55	1.11	12.77	10.12	16.73	12.47	70.35
		41.86	1.09	16.22	6.58	19.81	12.98	69.73
		42.00	1.00	9.04	0.11	34.96	16.31	69.69

An analysis of the data in Table 4 shows that there are differences between the biosorption capacity and the biosorption efficiency of the three metals, regardless of the combination of process parameters. The two biosorbents, SB and SWB have different preferences for the three metal ions, since the highest values of *E* are obtained for Pb(II), followed by Cd(II) and, finally, for Zn(II). This situation was explained by Fertu et al. [60], being placed, first of all, on electronegativity differences of the three ions. Pb(II) with the electronegativity of 1.87, can take part easily in ion exchange than

Cd(II) (1.69) and Zn(II) (1.65). In the same context, an important role in preferential biosorption is played by the hydrated radii of these metal ions, which have different sizes (Pb(II) = 4.01 Å < Cd(II) = 4.26 Å < Zn(II) = 4.30 Å), as well as in hydration energies which are different for the there metallic ions ((Pb(II) = −1481 kJ/mol, Cd (II) = −1807 kJ/mol, Zn(II) = −2046 kJ/mol)) [60,78–80].

The two biosorbents offer similar results in terms of biosorption efficiency, the *E* (%) values being over 95% for Pb(II), over 88% for Cd(II) and over 65% for Zn(II). It turns out that both SB and SWB can remove Pb(II) from aqueous solutions in a single biosorption cycle, under the optimal experimental conditions in Table 4, whereas for Cd(II), but especially for Zn(II), depending on the requirements for effluent quality, two cycles of biosorption may be required, with biosorbent regeneration.

These results show that modeling and optimization of Pb(II), Cd(II), Zn(II) biosorption using soybean-based low-cost biosorbents by applying Artificial Neural Networks (ANNs) and Evolutionary Algorithms (EAs) capable of evolving ANN parameters, it is possible to predict biosorption capacity and biosorption efficiency with high accuracy to ensure the quality of effluents resulting from wastewater treatment, according to specific required regulations. Implementation on an industrial scale can improve cost dynamics and facilitate process monitoring and control.

4. Conclusions

In this paper, we have applied modeling and optimization algorithms specific to Artificial Neural Networks to find the interactions among the parameters which affect the biosorption capacity, *q* (mg/g) and biosorption efficiency, *E* (%) of heavy metal ions Pb(II), Cd(II), Zn(II), during their elimination from aqueous solutions by biosorption, using soybean biomass and soybean waste biomass as biosorbents. In this context, evolutionary algorithms (EA) were selected at three levels of evolution: training weights, determining the topology (network architecture) and learning, applying the technique represented by Differential Evolution (DE), when simultaneously training and determining the network topology.

To solve the proposed modeling and optimization problem, the modifications made for DE resulted in a new hybrid self-adaptive Differential Evolution with neural network (hSADE-NN) algorithm, by combining a modified DE version with two algorithms such as Random Search and BK. The resulting hybrid algorithm, hSADE-NN, was applied to model the biosorption process, limiting the number of hidden layers to two, by 40 and 20 neurons, respectively.

The experimental data collected for the three metal ions and randomly arranged went through a set of data processing techniques (normalization, randomization and group selection) to generate a model with the smallest possible error. Various configurations for the process modelling were considered. The results showed that the best configurations focus on different models for biosorption capacity (*q*), biosorption efficiency (*E*) and for the two datasets (soybean biomass and soybean waste biomass), cases C2, C3, C5, C6. Biosorption capacity and biosorption efficiency (*q*, *E*) were determined as function of metal ion type, pH, biosorbent dose (*DS*), initial concentration of metal in solution (c_0), contact time (t_c) and temperature (*T*). For the best models, the correlations in the testing phase are higher than 0.91, indicating that a single model combining all metal ions is suitable to capture the entire process dynamic. This capability was also indicated through the analysis of the surface plots generated using a series of predictions.

In order to optimize and scale up the biosorption process using soy-based biosorbents, the ANNs models were used to determine the conditions that lead to a maximum *q*, *E*, taking into account the specificity of each metal ion. The resulting optimization data showed that the soybean waste biomass is an effective biosorbent for heavy metal ions, enduring a very good removal efficiency, so it can be unreservedly recommended as a biosorbent, in an efficient, low-cost and sustainable way to capitalize on this waste.

Author Contributions: Conceptualization, L.B., S.C. and M.G.; methodology, L.B., E.N.D. and D.I.F.; formal analysis, L.B. and S.C.; investigation, D.I.F., E.N.D. and L.B.; resources, L.B., E.N.D. and D.I.F.; data curation, L.B., E.N.D. and D.I.F.; writing—original draft preparation, L.B., E.N.D. and M.G.; writing—review and editing, L.B., S.C. and M.G.; visualization, D.I.F., E.N.D. and L.B.; supervision, L.B., S.C. and M.G.; project administration M.G.; funding acquisition, M.G. All authors have read and agreed to the published version of the manuscript.

Funding: This work was supported by a grant of the Romanian Ministry of Education and Research, CCCDI—UEFISCDI, project number PN-III-P2-2.1-PED-2019-5239, Contract no. 269PED/2020, within PNCDI III.

Institutional Review Board Statement: Not applicable.

Informed Consent Statement: Not applicable.

Data Availability Statement: Not applicable.

Conflicts of Interest: The authors declare no conflict of interest.

References

1. Sidhu, G.P.S. Heavy metal toxicity in soils: Sources, remediation technologies and challenges. *Adv. Plants Agric. Resour.* **2016**, *5*, 445–446.
2. Toth, G.; Hermann, T.; Da Silva, M.R.; Montanarella, L. Heavy metals in agricultural soils of the European Union with implications for food safety. *Environ. Int.* **2016**, *88*, 299–309. [CrossRef] [PubMed]
3. Farzadkia, M.; Sedeh, M.S.; Ghasemi, A.; Alinejad, N.; Kazemi, M.S.; Jafarzadeh, N.; Torkashvand, J. Estimation of the heavy metals released from cigarette butts to beaches and urban environments. *J. Hazard. Mater.* **2022**, *425*, 1279. [CrossRef]
4. Gonzalez-Acevedo, Z.I.; Garcia-Zarate, M.A.; Nunez-Zarco, E.A.; Anda-Martín, B.I. Heavy metal sources and anthropogenic enrichment in the environment around the Cerro Prieto Geothermal Field, Mexico. *Geothermics* **2018**, *72*, 170–181. [CrossRef]
5. Hussein, M.; Yoneda, K.; Mohd-Zaki, Z.; Amir, A.; Othman, N.A. Heavy metals in leachate, impacted soils and natural soils of different landfills in Malaysia: An alarming threat. *Chemosphere* **2021**, *267*, 128874. [CrossRef] [PubMed]
6. Diganta, M.T.M.; Sharmi, T.T.; Saifullah, A.S.M.; Uddin, M.J.; Sajib, A.M. Appraisal of heavy metal contamination in road dust and human health risk in a municipality of Bangladesh. *Environ. Eng. Manag. J.* **2020**, *19*, 2165–2177.
7. Moreira, J.C. Threats by heavy metals: Human and environmental contamination in Brazil. *Sci. Total Environ.* **1996**, *188*, S61–S71. [CrossRef]
8. Du, J.; Zhang, B.; Li, J.; Lai, B. Decontamination of heavy metal complexes by advanced oxidation processes: A review. *Chin. Chem. Lett.* **2020**, *31*, 2575–2582. [CrossRef]
9. Samouhos, M.; Peppas, A.; Bartzas, G.; Taxiarchou, M.; Tsakiridis, P.E. Arsenic release through refractory gold ore processing. Immobilization and decontamination approaches. *Curr. Opin. Environ. Sci. Health* **2021**, *20*, 100236. [CrossRef]
10. Rajendran, S.; Priya, T.; Khoo, K.S.; Hoang, T.K.; Ng, H.-S.; Munawaroh, H.S.H.; Karaman, C.; Orooji, Y.; Show, P.L. A critical review on various remediation approaches for heavy metal contaminants removal from contaminated soils. *Chemosphere* **2022**, *247*, 132369. [CrossRef]
11. Bilal, M. Waste biomass adsorbents for copper removal from industrial wastewater: A review. *J. Hazard. Mater.* **2013**, *263*, 322–333. [CrossRef] [PubMed]
12. Bulgariu, L.; Fertu, D.I.; Cara, I.G.; Gavrilescu, M. Efficacy of alkaline-treated soywaste biomass for the removal of heavy-metal ions and opportunities for their recovery. *Materials* **2021**, *14*, 7413. [CrossRef] [PubMed]
13. Guerin, T.; Ghinet, A.; Hossarte, M.; Waterlot, C. Wheat and ryegrass biomass ashes as effective sorbents for metallic and organic pollutants from contaminated water in lab-engineered cartridge filtration system. *Bioresour. Technol.* **2020**, *318*, 124044. [CrossRef] [PubMed]
14. Karić, N.; Maia, A.S.; Teodorović, A.; Atanasova, N.; Langergraber, G.; Crini, G.; Ribeiro, A.R.; Đolić, M. Bio-waste valorisation: Agricultural wastes as biosorbents for removal of (in) organic pollutants in wastewater treatment. *Chem. Eng. J. Adv.* **2022**, *9*, 100239. [CrossRef]
15. De Gisi, S.; Lofrano, G.; Grassi, M.; Notarnicola, M. Characteristics and adsorption capacities of low-cost sorbents for wastewater treatment: A review. *Sustain. Mater. Technol.* **2016**, *9*, 10–40. [CrossRef]
16. Kamyab, S.M.; Modabberi, S.; Williams, C.D.; Badiei, A. Pure sodalite synthesis, characterization and application for heavy metal ions removal from aqueous solutions. *Environ. Eng. Manag. J.* **2020**, *20*, 687–700.
17. Chen, M.; Wang, X.; Zhang, H. Comparative research on selective adsorption of Pb(II) by biosorbents prepared by two kinds of modifying waste biomass: Highly-efficient performance, application and mechanism. *J. Environ. Manag.* **2021**, *288*, 112388. [CrossRef] [PubMed]
18. Thi Quyen, V.; Pham, T.H.; Kim, J.; Thanh, D.M.; Thang, P.Q.; Van Le, Q.; Jung, S.H.; Kim, T. Biosorbent derived from coffee husk for efficient removal of toxic heavy metals from wastewater. *Chemosphere* **2021**, *284*, 131312. [CrossRef]

19. Yaashikaa, P.R.; Kumar, P.S.; Saravanan, A.; Vo, D.-V.N. Advances in biosorbents for removal of environmental pollutants: A review on pretreatment, removal mechanism and future outlook. *J. Hazard. Mater.* **2021**, *420*, 126596. [CrossRef]
20. Singh, S.; Kumar, V.; Datta, S.; Dhanjal, D.S.; Sharma, K.; Samuel, J.; Singh, J. Current advancement and future prospect of biosorbents for bioremediation. *Sci. Total Environ.* **2020**, *709*, 135895. [CrossRef]
21. Gavrilescu, M. Biomass—A resource for environmental bioremediation and bioenergy. In *Recent Developments in Bioenergy Research*; Gupta, V.K., Treichel, H., Kuhad, R.C., Rodriguez-Cout, S., Eds.; Elsevier: Amsterdam, The Netherlands, 2020; pp. 19–63.
22. Hlihor, R.M.; Apostol, L.C.; Gavrilescu, M. Environmental bioremediation by biosorption and bioaccumulation: Principles and applications. In *Enhancing Cleanup of Environmental Pollutants*; Anjum, N., Gill, S., Tuteja, N., Eds.; Springer: Cham, Switzerland, 2017; pp. 289–315.
23. Chaouki, Z.; Zaitan, H.; Nawdali, M.; Vasarevicius, S.; Mazeikiene, A. Oil removal from refinery wastewater through adsorption on low cost natural biosorbents. *Environ. Eng. Manag. J.* **2020**, *19*, 105–112.
24. Flores-Garnica, J.G.; Morales-Barrera, L.; Pineda-Camacho, G.; Cristiani-Urbina, E. Biosorption of Ni(II) from aqueous solutions by *Litchi chinensis* seeds. *Bioresour. Technol.* **2013**, *136*, 635–643. [CrossRef] [PubMed]
25. Sheikh, Z.; Amin, M.; Khan, N.; Khan, N.M.; Sami, S.K.; Khan, S.B.; Hafeez, I.; Ali Khan, S.; Bakhs, E.M.; Cheng, C.K. Potential application of Allium Cepa seeds as a novel biosorbent for efficient biosorption of heavy metals ions from aqueous solution. *Chemosphere* **2021**, *279*, 130545. [CrossRef]
26. Giri, D.D.; Alhazmi, A.; Mohammad, A.; Haque, S.; Srivastava, N.; Thakur, V.K.; Gupta, V.K.; Pal, D.B. Lead removal from synthetic wastewater by biosorbents prepared from seeds of *Artocarpus heterophyllus* and *Syzygium cumini*. *Chemosphere* **2022**, *287*, 132016. [CrossRef] [PubMed]
27. Varala, S.; Dharanija, B.; Satyavat, B.; Basava Rao, V.V.; Parthasarathy, R. New biosorbent based on deoiled karanja seed cake in biosorption studies of Zr(IV): Optimization using Box–Behnken method in response surface methodology with desirability approach. *Chem. Eng. J.* **2016**, *302*, 786–800. [CrossRef]
28. Prakash, N.; Manikandan, S.A.; Govindarajan, L.; Vijayagopal, V. Prediction of biosorption efficiency for the removal of copper(II) using artificial neural networks. *J. Hazard. Mater.* **2008**, *153*, 1268–1275. [CrossRef]
29. Liu, Z.-W.; Liang, F.-N.; Liu, Y.-Z. Artificial neural network modeling of biosorption process using agricultural wastes in a rotating packed bed. *Appl. Therm. Eng.* **2018**, *140*, 95–101. [CrossRef]
30. Ghaedi, A.M.; Vafaei, A. Applications of artificial neural networks for adsorption removal of dyes from aqueous solution: A review. *Curr. Opin. Environ. Sci. Health* **2017**, *245*, 20–39. [CrossRef] [PubMed]
31. Mohammadi, F.; Yavari, Z.; Rahimi, S.; Hashemi, M. Artificial Neural Network modeling of Cr(VI) biosorption from aqueous solutions. *J. Water Chem. Technol.* **2019**, *41*, 219–227. [CrossRef]
32. De Leon, B.V.; de Negreiros, B.A.F.; Brusamarello, G.Z.; Petroli, G.; Di Domenico, M.; de Souza, F.B. Artificial neural network for prediction of color adsorption from an industrial textile effluent using modified sugarcane bagasse: Characterization, kinetics and isotherm studies. *Environ. Nanotechnol. Monit. Manag.* **2020**, *14*, 100387. [CrossRef]
33. Hlihor, R.M.; Diaconu, M.; Leon, F.; Curteanu, S.; Tavares, T.; Gavrilescu, M. Experimental analysis and mathematical prediction of Cd(II) removal by biosorption using support vector machines and genetic algorithms. *N. Biotechnol.* **2015**, *32*, 358–368. [CrossRef] [PubMed]
34. Hunter, A.; Kennedy, L.; Henry, J.; Ferguson, I. Application of neural networks and sensitivity analysis to improved prediction of trauma survival. *Comput. Methods Programs Biomed.* **2000**, *62*, 11–19. [CrossRef]
35. Noor, R.A.M.; Ahmad, Z.; Don, M.M.; Uzir, M.H. Modelling and control of different types of polymerization processes using neural networks technique: A review. *Can. J. Chem. Eng.* **2010**, *88*, 1065–1084. [CrossRef]
36. McCulloch, W.; Pitts, W. A logical calculus of ideas immanent in nervous activity. *Bull. Math. Biophys.* **1943**, *5*, 115–133. [CrossRef]
37. Ozbakir, L.; Baykasoglu, A.; Kulluk, S. A soft computing-based approach for integrated training and rule extraction from artificial neural networks: DIFACONN-miner. *Appl. Soft Comput.* **2010**, *10*, 304–317. [CrossRef]
38. Vlahogianni, E.I.; Karlaftis, M.G.; Golias, J.C. Optimized and meta-optimized neural networks for short-term traffic flow prediction: A genetic approach. *Transp. Res. Part C Emerg. Technol.* **2005**, *13*, 211–234. [CrossRef]
39. Lahiri, S.K.; Ghanta, K.C. Artificial neural network model with the parameter tuning assisted by a differential evolution technique: The study of the hold up of the slurry flow in a pipeline. *Chem. Ind. Chem. Eng. Q.* **2009**, *15*, 103–117. [CrossRef]
40. Yardimci, A. Soft computing in medicine. *Appl. Soft Comput.* **2009**, *9*, 1029–1043. [CrossRef]
41. Curteanu, S. Direct and inverse neural network modeling in free radical polymerization. *Cent. Eur. J. Chem.* **2004**, *2*, 113–140. [CrossRef]
42. Fernandes, F.A.N.; Lona, M.F. Neural network applications in polymerization processes. *Braz. J. Chem. Eng.* **2005**, *22*, 401–418. [CrossRef]
43. Gonzaga, J.C.B.; Meleiro, L.A.C.; Kiang, C.; Maciel, R. ANN-based soft-sensor for real-time process monitoring and control of an industrial polymerization process. *Comput. Chem. Eng.* **2009**, *33*, 43–49. [CrossRef]
44. Bhotmange, M.; Shastri, P. Application of Artificial Neural Networks to food and fermentation technology. In *Artificial Neural Networks—Industrial and Control Engineering Applications*; Suzuki, K., Ed.; InTech: Rijeka, Croatia, 2011; pp. 201–222.
45. Dragoi, E.N.; Curteanu, S.; Fissore, D. Freeze-drying modeling and monitoring using a new neuro-evolutive technique. *Chem. Eng. Sci.* **2012**, *72*, 195–204. [CrossRef]

46. Dragoi, E.N.; Curteanu, S.; Fissore, D. On the use of Artificial Neural Networks to monitor a pharmaceutical freeze-drying process. *Dry. Technol.* **2013**, *31*, 72–81. [CrossRef]
47. Llanos, J.; Rodrigo, M.A.; Canizares, P.; Furtuna, R.P.; Curteanu, S. Neuro-evolutionary modelling of the electrodeposition stage of a polymer-supported ultrafiltration-electrodeposition process for the recovery of heavy metals. *Environ. Model. Softw.* **2013**, *42*, 133–142. [CrossRef]
48. Curteanu, S.; Suditu, G.D.; Buburuzan, A.M.; Dragoi, E.N. Neural networks and differential evolution algorithm applied for modelling the depollution process of some gaseous streams. *Environ. Sci. Pollut. Res.* **2014**, *21*, 12856–12867. [CrossRef]
49. Mi, X.; Zou, Y.; Wei, W.; Ma, K. Testing the generalization of artificial neural networks with cross-validation and independent-validation in modelling rice tillering dynamics. *Ecol. Modell.* **2005**, *181*, 493–508. [CrossRef]
50. Puig-Arnavat, M.; Bruno, J.C. Artificial Neural Networks for thermochemical conversion of biomass. In *Recent Advances in Thermo-Chemical Conversion of Biomass*; Sukumaran, A.P.B.S.K., Ed.; Elsevier: Boston, MA, USA, 2015; pp. 133–156.
51. Fe, J.D.; Aliaga, R.J.; Gadea-Girones, R. Evolutionary optimization of neural networks with heterogeneous computation: Study and implementation. *J. Supercomput.* **2015**, *71*, 2944–2962. [CrossRef]
52. Fernandez, J.C.; Hervas, C.; Martinez-Estudillo, F.J.; Gutierrez, P.A. Memetic Pareto Evolutionary Artificial Neural Networks to determine growth/no-growth in predictive microbiology. *Appl. Soft Comput.* **2011**, *11*, 534–550. [CrossRef]
53. Floreano, D.; Durr, P.; Mattiussi, C. Neuroevolution: From architectures to learning. *Evol. Intell.* **2008**, *1*, 47–62. [CrossRef]
54. Storn, R. Differential Evolution research—Trends and open questions. In *Advances in Differential Evolution. Studies in Computational Intelligence*; Chakraborty, U., Ed.; Springer: Berlin/Heidelberg, Germany, 2008; Volume 143, pp. 1–31.
55. Feoktistov, V. *Differential Evolution: In Search of Solutions*; Springer: Berlin, Germany, 2006.
56. Iranshahi, D.; Bahmanpour, A.M.; Paymooni, K.; Rahimpour, M.R.; Shariati, A. Simultaneous hydrogen and aromatics enhancement by obtaining optimum temperature profile and hydrogen removal in naphtha reforming process; a novel theoretical study. *Int. J. Hydrogen Energy* **2011**, *36*, 8316–8326. [CrossRef]
57. Arabpour, M.; Rahimpour, M.R.; Iranshahi, D.; Raeissi, S. Evaluation of maximum gasoline production of Fischer-Tropsch synthesis reactions in GTL technology: A discretized approach. *J. Nat. Gas Sci. Eng.* **2012**, *9*, 209–219. [CrossRef]
58. Samimi, F.; Kabiri, S.; Mirvakili, A.; Rahimpour, M.R. Simultaneous dimethyl ether synthesis and decalin dehydrogenation in an optimized thermally coupled dual membrane reactor. *J. Nat. Gas Sci. Eng.* **2013**, *14*, 77–90. [CrossRef]
59. Rocha, M.; Mendes, R.; Rocha, O.; Rocha, I.; Ferreira, E.N.C. Optimization of fed-batch fermentation processes with bio-inspired algorithms. *Expert Syst. Appl.* **2014**, *41*, 2186–2195. [CrossRef]
60. Fertu, D.I.; Bulgariu, L.; Gavrilescu, M. Modelling and optimization of heavy metals biosorption by low-cost sorbents using Response Surface Methodology. *Processes* **2022**, *10*, 523. [CrossRef]
61. Price, K.; Storn, R.; Lampinen, J. *Differential Evolution. A Practical Approach to Global Optimization*; Springer: Berlin, Germany, 2005.
62. Zielinski, K.; Laur, R. Stopping criteria for Differential Evolution in constrained single-objective optimization. In *Advances in Differential Evolution*; Chakraborty, U., Ed.; Springer: Berlin/Heidelberg, Germany, 2008; Volume 143, pp. 111–138.
63. Takahama, T.; Sakai, S. Solving difficult constrained optimization problems by the ε constrained differential evolution with gradient-based mutation. In *Constraint-Handling in Evolutionary Optimization. Studies in Computational Intelligence*; Mezura-Montes, E.R., Ed.; Springer: Berlin/Heidelberg, Germany, 2009; Volume 198, pp. 51–72.
64. Davendra, D.; Onwubolu, G. Forward backward transformation. In *Differential Evolution: A Handbook for Global Permutation-Based Combinatorial Optimization*; Onwubolu, G., Davendra, D., Eds.; Springer: Berlin/Heidelberg, Germany, 2009; Volume 175, pp. 35–80.
65. Adeyemo, J.; Enitan, A. Optimization of fermentation processes using evolutionary algorithms—A review. *J. Sci. Res. Essay* **2011**, *6*, 1464–1472.
66. Curteanu, S.; Cartwright, H. Neural networks applied in chemistry. I. Determination of the optimal topology of multilayer perceptron neural networks. *J. Chemom.* **2012**, *25*, 527–549. [CrossRef]
67. Chandra, A.; Yao, X. Ensemble learning using Multi-Objective Evolutionary Algorithms. *J. Math. Model. Algorithms* **2006**, *5*, 417–445. [CrossRef]
68. Subudhi, B.; Jena, D. An improved differential evolution trained neural network scheme for nonlinear system identification. *Int. J. Autom. Comput.* **2009**, *6*, 137–144. [CrossRef]
69. Dragoi, E.N.; Curteanu, S.; Lisa, C. A neuro-evolutive technique applied for predicting the liquid crystalline property of some organic compounds. *Eng. Optim.* **2012**, *44*, 1261–1277. [CrossRef]
70. Zarth, A.; Ludermir, T.B. Optimization of neural networks weights and architecture: A multimodal methodology. In Proceedings of the Ninth International Conference on Intelligent Systems Design and Applications (ISDA '09), Pisa, Italy, 30 November–2 December 2009.
71. Brest, J. Constrained real-parameter optimization with e-Self-Adaptive Differential Evolution. In *Constraint-Handling in Evolutionary Optimization*; Mezura-Montes, E., Ed.; Springer: Berlin/Heidelberg, Germany, 2009; Volume 198, pp. 73–93.
72. Das, S.; Suganthan, P.N. Differential evolution a survey of the state-of-the-art. *IEEE Trans. Evol. Comput.* **2011**, *15*, 4–31. [CrossRef]
73. Dragoi, E.N.; Curteanu, S.; Leon, F.; Galaction, A.I.; Cascaval, D. Modeling of oxygen mass transfer in the presence of oxygen-vectors using neural networks developed by differential evolution algorithm. *Eng. Appl. Artif. Intell.* **2011**, *24*, 1214–1226. [CrossRef]

74. Dragoi, E.N.; Curteanu, S.; Galaction, A.I.; Cascaval, D. Optimization methodology based on neural networks and self-adaptive differential evolution algorithm applied to an aerobic fermentation process. *Appl. Soft Comput.* **2013**, *13*, 222–238. [CrossRef]
75. Tizhoosh, H.R. Opposition-Based Learning: A new scheme for machine intelligence. In Proceedings of the International Conference on Computational Intelligence for Modelling, Control and Automation and International Conference on Intelligent Agents, Web Technologies and Internet Commerce (CIMCA-IAWTIC'06), Vienna, Austria, 28–30 November 2005; pp. 695–701.
76. Priddy, K.; Keller, P. *Artificial Neural Networks: An Introduction*; SPIE Press: Washington, DC, USA, 2005.
77. May, R.J.; Maier, H.R.; Dandy, G.C. Data splitting for artificial neural networks using SOM-based stratified sampling. *Neural Netw.* **2010**, *23*, 283–294. [CrossRef] [PubMed]
78. Mobasherpour, I.; Salahi, E.; Pazouki, M. Comparative of the removal of Pb(II), Cd(II) and Ni(II) by nano crystallite hydroxyapatite from aqueous solutions: Adsorption isotherm study. *Arab. J. Chem.* **2012**, *5*, 439–446. [CrossRef]
79. Al Hamouz, O.C.S.; Ali, S.A. Removal of Zinc and Cadmium ions using a cross-linked polyaminophosphonate. *J. Macromol. Sci. Part A* **2013**, *50*, 375–384. [CrossRef]
80. Bashir, A.; Manzoor, T.; Malik, L.A.; Qureashi, A.; Pandith, A.H. Enhanced and Selective Adsorption of Zn(II), Pb(II), Cd(II), and Hg(II) Ions by a Dumbbell- and Flower-Shaped Potato Starch Phosphate Polymer: A Combined Experimental and DFT Calculation Study. *ACS Omega* **2020**, *5*, 4853–4867. [CrossRef] [PubMed]

processes MDPI

Article

Modeling and Optimization of Heavy Metals Biosorption by Low-Cost Sorbents Using Response Surface Methodology

Daniela Ionela Fertu [1,2], Laura Bulgariu [1,*] and Maria Gavrilescu [1,3,*]

1 Department of Environmental Engineering and Management, "Cristofor Simionescu" Faculty of Chemical Engineering and Environmental Protection, "Gheorghe Asachi" Technical University of Iasi, 73 Prof. D. Mangeron Blvd., 700050 Iasi, Romania; danafertu2004@yahoo.com
2 Department of Pharmaceutical Sciences, Faculty of Medicine and Pharmacy, "Dunarea de Jos" University of Galati, 800002 Galati, Romania
3 Academy of Romanian Scientists, 3 Ilfov Street, 050044 Bucharest, Romania
* Correspondence: lbulg@tuiasi.ro (L.B.); mgav@tuiasi.ro (M.G.)

Abstract: This paper exploits, through modeling and optimization, the experimental laboratory data on the biosorption of heavy metal ions Pb(II), Cd(II), and Zn(II) from aqueous media using soybean and soybean waste biomasses. The biosorption modeling was performed using the Response Surface Methodology, followed by optimization based on numerical methods. The aim of the modeling was to establish the most probable mathematical relationship between the dependent variables (the biosorption efficiency of the biosorbents when adsorbing metal ions, $R(\%)$, and the biosorption capacity of sorbents, $q(mg/g)$) and the process parameters (pH; sorbent dose, DS (g/L); initial metal ion concentration in solution, c_0 (mg/L); contact time, t_c (min); temperature, T (°C)), validated by methodologies specific to the multiple regression analysis. Afterward, sets of solutions were obtained through optimization that correlate various values of the process parameters to maximize the objective function. These solutions also confirmed the performance of soybean waste biomass in the removal of heavy metal ions from polluted aqueous effluents. The results were validated experimentally.

Keywords: ANOVA; heavy metals; second-degree function; soybean biomass; waste

Citation: Fertu, D.I.; Bulgariu, L.; Gavrilescu, M. Modeling and Optimization of Heavy Metals Biosorption by Low-Cost Sorbents Using Response Surface Methodology. *Processes* **2022**, *10*, 523. https://doi.org/10.3390/pr10030523

Academic Editors: Andrea Petrella, Marco Race and Danilo Spasiano

Received: 29 January 2022
Accepted: 4 March 2022
Published: 6 March 2022

1. Introduction

Environmental quality has deteriorated, mainly as a consequence of the diversification of anthropogenic activities, population growth, unplanned urbanization, rapid industrialization, and the irrational exploitation of resources. That is why, in recent decades, efforts to improve some innovative and ground-breaking processes have been intensified so as to make it possible to eliminate the pollutants from the environment with increased efficiencies. These processes should be capable of removing pollutants from the environment and, at the same time, protect human health by combining the two major challenges, resource consumption and pollution, by harnessing natural resources and, in particular, waste, and by using waste to improve the quality of the environment by decontaminating its components [1–3].

In parallel with environmental pollution, our current patterns of resource use, production, consumption, and prevention of waste generation and pollution are unsustainable. The Earth has finite resources, and the extensive use of these resources increases the pressure on the natural environment, resulting in global warming, pollution, and the degradation of ecosystems and biodiversity. Natural resources are fundamental to any economy and for ensuring human prosperity. They provide raw materials, energy, food, water, and soil, as well as environmental and social services. Therefore, humanity is faced with two challenges: ensuring the sustainable consumption of natural resources and preventing the pollution of the environment [4–6].

The increased awareness of the link between generating environmental impacts (with short-, medium-, and long-term consequences) and current environmental strategies has led the scientific community to focus on developing sound, economically viable, and environmentally friendly processes. This requires stepping up efforts to integrate pollution prevention and control and the application of the principles of the circular economy. One of the appropriate ways to achieve these requirements for ensuring sustainability involves the use of waste in the depollution of some environmental components [7–9].

The harmful effects of organic and inorganic pollutants on ecosystems and on human health are well known, which is why sustained efforts are being made to develop treating methods to prevent or limit pollution. Heavy metal pollution is one of the most important environmental problems today [10–13]. Various fields of activity, such as mining and steel, the metallurgical industry, the metal surface finishing industry, energy and fuel production, the fertilizer and pesticide industry, galvanizing, electrolysis, electro-osmosis, leather processing, the photographic industry, the electrical equipment industry and electronics, metal surface treatment, aerospace industry, etc., produce and discharge into the environment various industrial wastes and effluents containing heavy metals. Thus, metals, although usually regarded as resources, also become important pollutants of the environment, endangering the health of humans and the ecosystem [14,15]. The intensification of the use of metals and their various chemical compounds in industrial processes has the consequence of generating large amounts of effluents containing high levels of toxic metals. The presence of heavy metals in these effluents can induce major environmental impacts and risks, mainly due to the toxicity of most heavy metals and their persistence in the environment [16,17]. Environmental and management specialists and decision makers are faced with a constant challenge associated with the research and development of advanced but inexpensive technologies for the removal of heavy metals from industrial and municipal effluents.

The range of conventional methods for removing heavy metals from aqueous solutions may include techniques such as [18–23]: chemical precipitation, oxidation or chemical reduction, ion exchange, filtering, membrane techniques, etc. The literature considers that processes based on these techniques are usually expensive and inefficient, especially for solutions containing ions with concentrations between 1 and 100 mg/L [24–26]. Another major disadvantage of conventional techniques for removing metals from industrial effluents is associated with the production of toxic chemical sludge, the treatment and storage of which involves a number of additional costs and major environmental impacts.

Adsorption is considered to be a feasible alternative for drinking and industrial water treatment and wastewater treatment because it is usually a suitable operation that is easy to apply, and the design of adsorption systems is relatively simple, although the costs are relatively high [27–30]. The properties of adsorbates and adsorbents (solids) are relatively specific and depend on the components in their structure. The interaction between the solid surface and the adsorbed molecules can be of physical (physisorption) or chemical (chemisorption) in nature. Activated carbon is currently considered to be the most commonly used and most effective adsorbent for the removal of various pollutants from water, such as dyes and heavy metals, although there are some issues related to the regenerative capacity of the adsorbent or its elimination at the end of its life, based on different disposal strategies other than storage [31–33]. In addition to activated carbon, a wide variety of approved adsorbents to date have been examined for both low-cost (low-cost sorbents) and their ability to remove various types of pollutants from liquid effluents. The general trend is to replace activated carbon—which is the so-called state of the art—with natural materials or waste/by-products from various fields of activity, such as agriculture and industry. Based on literature studies, low-cost sorbents from natural sources have been classified into the following groups [34–37]: (i) agricultural and household waste, (ii) waste and by-products, (iii) sludges, (iv) marine materials, (v) soils and ore tailings, and (vi) new low-cost adsorbents.

Biosorption is a particular alternative in the group of sorption processes that is based on the ability of certain categories of biomass (active–viable, or non-viable) to bind/retain and concentrate heavy metals or organic pollutants, even from dilute solutions. Biomass capitalizes on this property, acting as a chemical or as an ion exchanger of biological (vegetable) origin. This technique does not necessarily have to replace existing treatment processes, but it can be a complementary alternative. Some studies have shown that using biosorption as a technique to remove pollutants from contaminated environments (aqueous effluents) can reduce capital costs by about 20%, operating costs by 36%, and total costs by about 28% compared to conventional systems [38,39]. The first work in the field of biosorption was published in 1951 [40]. Since then, substantial efforts have been made to discover cost-effective materials that can be applied as biosorbents to treat wastewater and other aqueous solutions contaminated with pollutants.

Many researchers have found that by using this new method, in which different categories of biomass can be used as biosorbents, toxic pollutants can be selectively removed from aqueous solutions at desired residual concentration levels [18,39,41–44]. As a broader spectrum of properties suitable for biosorbent requirements has been discovered for biomass, the concept of biosorption has become increasingly attractive in various fields. Bulgariu et al. have reported on the use of soybean waste biomass resulting from oil extraction and modified during alkaline treatment as a biosorbent for the removal of Pb(II), Cd(II), and Zn(II) ions from aqueous media, with very promising results [45].

The biosorption studies in the literature analyze, for the most part, the elimination efficiency of some heavy metals, such as cadmium (Cd), chromium (Cr), mercury (Hg), nickel (Ni), lead (Pb), and zinc (Zn), since they may pose significant risks to public health and the environment. Of these metals, we have selected three, lead, cadmium, and zinc, for biosorption studies, both due to their toxicity and to highlight differences in the biosorption capacity of soybean biomass and soybean waste biomass, and particularly due to their different electronegativity. Thus, Pb(II), which has the highest electronegativity (1.87), can participate more easily in ion exchange interactions than Cd(II) (1.69) and Zn(II) (1.65). Moreover, the hydrated radii of these metal ions have different sizes (Pb(II) = 4.01 Å < Cd(II) = 4.26 Å < Zn(II) = 4.30 Å), while their hydration energies also have different values (Pb(II) = −1481 kJ/mol, Cd(II) = −1807 kJ/mol, Zn(II) = −2046 kJ/mol), and these lead to differences in the sorption abilities of these metal ions [46–48]. This context offers the opportunity for a comparative study of biosorbents' retention capacities and efficiencies, and their preferences for these metals. Additionally, these differences make possible a worthy comparison of the biosorption capacity and the biosorption efficiency of some biosorbents.

Although several publications have studied biosorption for the removal of heavy metals in batch systems, fewer studies have focused on modeling and optimizing this process for scaling-up purposes [49–52]. The biosorption capacity of copper by dried *Chlorella pyrenoidosa* was modelled by Rezende Moreira et al. using Response Surface Methodology (RSM) built based on a Box–Behnken design and artificial neural networks (ANN) [53]. A parallel sigmoidal (PS) model was applied by Blagojev et al. to describe the biosorption process and confirm its applicability to different types of biomass and various types of heavy metal ion [54]. Selva Filho et al. developed a two-parameter model based on a central composite rotatable design to find the optimal oil–water separation efficiency, involving floating macrophytes of the species *Eichhornia crassipes* in constructed wetlands [55]. The Box–Behnken experimental design model was applied by Jaafari and Yaghmaeian to plan biosorption experiments with heavy metals adsorbing onto the freshwater algae *Chlorella coloniales* to determine the effects of independent parameters, such as metal concentration, time of reaction, and algae dose, as well as to optimize these variable [56].

In this paper, the empirical mathematical modeling of the biosorption process of Pb(II), Cd(II), and Zn(II) ions from aqueous media on soybean and soybean waste biomasses was performed using Response Surface Methodology, followed the optimization using numerical methods. The aim of the modeling was to establish the most probable mathematical relationship between the dependent variables, including the biosorption efficiency of the biosorbent when adsorbing the metal ion, $R(\%)$, and the biosorption capacity of the sorbent, $q(mg/g)$, and the process parameters (i.e., the independent variables: pH; sorbent dose, DS (g/L); initial metal ion concentration in solution, c_0 (mg/L); contact time, t_c (min); temperature, T (°C)), validated by using a methodology specific to multiple regression analysis. Subsequently, through optimization, sets of solutions were obtained that correlate various values of the process parameters to maximize the objective function and demonstrate the performance of waste biomass in biosorption process. The results were validated experimentally.

2. Materials and Methods

2.1. Experimental Context

This paper discusses the removal of heavy metals, namely, Pb(II), Cd(II), and Zn(II) ions, from simulated wastewater samples through biosorption, using soybean and soybean waste biomasses as biosorbents [45]. The experimental studies aimed to describe the biosorptive potential of these biosorbents for the removal of Pb(II), Cd(II), and Zn(II) ions from aqueous solutions, assessed based on the most relevant biosorption parameters: the biosorption capacity of the sorbent, $q(mg/g)$, and the biosorption efficiency, $R(\%)$.

2.2. Preparation and Characterization of Biosorbents

The soybeans used to obtain the biosorbent needed for experimental studies were selected from a local farm in Iasi, Romania. After harvesting, the soybeans (1 kg) were washed with distilled water (5–6 times) to remove macroscopic impurities, dried in air at room temperature (22 ± 0.5 °C) for 5 days, and then ground. The material obtained from these operations was prepared further in the following modes: (1) dried at 65 °C for 6 h, then mechanically sieved so that the particle size was less than 1.5 mm, and stored in the desiccator to maintain a constant humidity, thus resulting in a soybean biosorbent; (2) through extraction with n-hexane for 30 h in a Soxhlet extractor, followed by washing with distilled water and drying at 65 °C for 6 h and then grinding, thus resulting in a soybean waste biosorbent.

In order to highlight the structural and morphological features, which are in direct correlation with the biosorptive properties, it was necessary to characterize the materials used as biosorbent in this study. The characterization was performed using the following methods of analysis by applying standard procedures for samples preparation:

- X-ray dispersive spectrometry (EDX)—with an EDAX-TSL 32 spectrometer—which allows the determination of the elemental composition of the analyzed material. The samples were prepared and analyzed according to the protocol described in [57].
- IR spectrometry—with a Bio-Rad Spectrometer with Fourier transform—which revealed the main types of functional groups that were found on the surface of each biosorbent used. For sample preparation, a small quantity of finely powdered solid sample was mixed with 100 times its weight of KBr and compressed into a thin transparent tablet using a hydraulic press. These tablets are transparent to IR radiation, and they were used for analysis.
- Scanning Electronic Microscopy (SEM) (performed using an S-3000 N HITACHI microscope with 15 UV). Microscopic images were recorded in low vacuum conditions, with several orders of magnitude, and their analysis allows the appreciation of the roughness of the material surface adsorbent. The samples were prepared and analyzed following the procedure described in [58,59].

2.3. Preparation and Analysis of the Studied Metal Ions

The aqueous solutions of the selected metal ions considered as the polluting species in this study were freshly prepared with exactly known concentrations for each experiment by diluting a given volume of the corresponding stock solution with distilled water. The solutions used in experiments were analyzed using an appropriate spectrophotometric analysis method to ensure the selectivity and accuracy of the determinations (Digital

Spectrophotometer S 104 D, glass cuvettes thickness = 1 cm) according to the specific analysis method (Table 1).

Table 1. Analytical characteristics of the spectrophotometric methods used to determine the concentration of metal ions.

Parameter	Pb(II)	Cd(II)	Zn(II)
Color reagent	4-(2-piridilazo)-resorcinol	Xilonolorange	Xilonolorange
pH	10.0	6.0	6.0
Buffer solution	Ammoniacal	HMTA + HNO_3	Acetat
λ_{max}, nm	530	575	570
ε_{max}, L/mol cm	1.95×10^4	2.15×10^4	2.64×10^4
Reference sample	witness test	witness test	witness test
Calibration sensitivity, mg/L	0.1694	0.1718	0.2563
Detection limit, ppm	0.1985	0.1325	0.1554
Linearity range used, mg/L	0.75–2.93	0.93–3.73	0.65–2.62
RSD%	0.44%	0.23%	0.28%

HMTA—hexamethylenetetramine.

The concentration of metal ions in the analyzed solutions was calculated from the regression equation of the corresponding calibration curve.

2.4. Experimental Methodology Used for Batch Biosorption Studies

Although the use of batch techniques for biosorption studies is sometimes considered in the literature to be an empirical method that does not accurately present the existing conditions in the treatment of industrial effluents, it is important, especially, in the evaluation of the biosorptive potential of a material for certain polluting species in aqueous solutions.

On a laboratory scale, the use of batch systems for the study of biosorption processes involved contacting a given volume of aqueous solution containing the polluting species (25 mL solution containing Pb(II), Cd(II), or Zn(II) ions of the given concentration) with an exactly weighed amount of biosorbent (soybean biomass or soybean waste). The mixture is stirred intermittently for a certain period of time (which is well established), after which the two phases are separated by filtration (using a quantitative filter paper with large pores), followed by the analysis of the two phases (filtrate—aqueous solution, and solid phase—biosorbent loaded with metals) using the specific methods of analysis discussed above.

2.5. Establishing the Experimental Conditions and Parameter Variation Ranges

Sets of preliminary tests facilitated the selection of a variation range of independent parameters for further working, considering also information from the literature. The best experimental conditions were established for each type of metal ion separately, both for biosorption on soybean biomass and soybean waste, using several sets of experiments that followed the influence of the most important experimental parameters, such as the pH of the initial solution, the dose of biosorbent (*DS*), the initial concentration of metal ion (c_0), the contact time (t_c), and the temperature (T), on the efficiency of the biosorption process. The evaluation of the influence of each experimental parameter was performed keeping constant the values of the other parameters considered in the experimental program.

2.5.1. pH

To determine the suitable pH value of the initial solution for the biosorption of Pb(II), Cd(II), and Zn(II) ions on the two types of soy-based biosorbents, two sets of experiments were performed (for each biosorbent), in which the pH of the aqueous solution (25 mL) containing metal ions varied between 1.0 and 6.5, while the values of the other experimental parameters were kept constant (Table 2). This pH range was chosen so that the studied metal ions could exist in solution predominantly in the form of free ions, and their precipitation in the form of hydroxides could be excluded. The required pH values were adjusted using solutions of different concentrations of HNO_3 (to avoid the introduction into the solution of another anion that could influence the biosorption process).

Table 2. Values of working parameters used in the experimental study on the influence of the initial pH of aqueous solution on the biosorption process.

Me(II)	pH Range	Biosorbent Dose *DS*, g/L	Initial Concentration of Metal Ion $c_{0Me(II)}$, mg/L	Contact Time, t_c, h	Temperature, T °C
Pb(II)	**1.0–6.5**	5.0	83.29	24	23.0
Cd(II)	**1.0–6.5**	5.0	46.11	24	24.5
Zn(II)	**1.0–6.5**	5.0	52.31	24	22.5

pH values were measured exactly with the help of a pH/ion-meter MM + 873, equipped with a combined glass electrode. The pH value for which the biosorption efficiency is at its maximum was considered suitable for the biosorption of metal ions considered for the two types of soy-based biosorbents and was kept constant in all other experiments.

2.5.2. Dose of Biosorbent

The experimental study on the influence of this parameter on biosorption was performed as follows: a volume of 25 mL aqueous solution of a metal ion of an exactly known concentration was brought into contact with the biosorbent (soybean biomass and soybean waste biomass) at a particular pH, contact time, and constant temperature (Table 3). After the required time had elapsed, the two phases were separated and analyzed. The optimal amount of biosorbent was established for each case (metal ion and type of biosorbent) based on the value of the quantitative parameters of the biosorption process, with the best value being considered the one for which the efficiency of the process is maximum.

Table 3. Values of working parameters used in the experimental study of the influence of biosorbent dose on the biosorption process.

Me(II)	pH Range	Biosorbent Dose *DS*, g/L	Initial Concentration of Metal Ion $c_{0Me(II)}$, mg/L	Contact Time, t_c, h	Temperature, T °C
Pb(II)	3.40	**4.0–40.0**	83.29	24	22.0
Cd(II)	3.40	**4.0–40.0**	46.11	24	22.5
Zn(II)	3.40	**4.0–40.0**	52.31	24	21.0

2.5.3. Initial Concentration of Metal Ions

The influence of this parameter was studied by varying the concentration of each metal ion in a range of at least two orders of magnitude (Table 4), and keeping constant the values of the other parameters corresponding to the suitable values. Metal ion concentrations were obtained by accurately measuring volumes of the metal ion stock solution, which were then diluted to a 25 mL volumetric flask.

Table 4. Values of working parameters used in the experimental study of the influence of the initial concentration of the metal ions on the biosorption process.

Me(II)	pH Range	Biosorbent Dose, *DS*, g/L	Initial Concentration of Metal Ion $c_{0Me(II)}$, mg/L	Contact Time, t_c, h	Temperature, T °C
Pb(II)	3.40	5.0	**11.66–416.45**	24	23.0
Cd(II)	3.40	5.0	**9.22–230.54**	24	24.0
Zn(II)	3.40	5.0	**13.08–209.25**	24	22.0

The results of these experiments allowed, on the one hand, the evaluation of the maximum amount of metal ion that can be retained in the most suitable experimental conditions on each type of biosorbent and, on the other hand, the concentration range for which the two biosorbents studied can be considered to be effective in removing Pb(II), Cd(II), and Zn(II) ions from aqueous solutions.

2.5.4. Contact Time

For these experimental studies, exactly weighed amounts of biosorbent (corresponding to the suitable values) were contacted with 25 mL of aqueous solution containing a metal ion (the concentration being chosen from the studied concentration range), at the most suitable pH value of the initial solution and at a constant temperature (Table 5). After well-defined periods of time (between 5 and 180 min), with intermittent stirring, the two phases were separated by filtration and the concentration of the metal ions in the solution was determined. This way, it was possible to establish the minimum contact time of the two phases necessary to achieve the balance of the biosorption process for each type of biosorbent and metal ion studied.

Table 5. The values of the working parameters used in the experimental study of the influence of the contact time on the biosorption process.

Me(II)	pH Range	Biosorbent Dose, DS, g/L	Initial Concentration of Metal Ion $c_{0Me(II)}$, mg/L	Contact Time, t_c, h	Temperature, T °C
Pb(II)	3.40	5.0	83.29	**5–180**	24.0
Cd(II)	3.40	5.0	46.11	**5–180**	26.0
Zn(II)	3.40	5.0	52.31	**5–180**	24.5

2.5.5. Temperature

The study of temperature influence on the biosorption efficiency for Pb(II), Cd(II), and Zn(II) ion retention of the selected biosorbents encompassed the analysis of their performance, which was assessed in three sets of experiments at three different temperatures for each biosorbent and metal ion for every possible combination, when the pH of the initial solutions and the dose of biosorbent were kept constant at the most suitable values, determined during the previous experiments, while the concentration of metal ions varied for the entire concentration range studied (Table 6).

Table 6. The values of the working parameters used in the experimental study of the influence of temperature on the biosorption process.

Me(II)	pH Range	Biosorbent Dose, DS, g/L	Initial Concentration of Metal Ion $c_{0Me(II)}$, mg/L	Contact Time, t_c, h	Temperature, T °C
Pb(II)	3.40	5.0	**11.66–416.45**	3	**5; 25; 50**
Cd(II)	3.40	5.0	**9.22–230.54**	3	**5; 25; 50**
Zn(II)	3.40	5.0	**13.08–209.25**	3	**5; 25; 50**

In this case, the phase separation was performed after 3 h, a period of time considered to be sufficient to reach equilibrium, and the solutions obtained after filtration were analyzed in order to determine the concentration of metal ions.

The experimental results from each set of experiments were used to quantitatively evaluate the biosorptive performance of the studied biosorbents, using the following quantitative indicators [45]:

- Biosorption capacity, *q(mg/g)*, which represents the amount of metal ion retained per unit mass of biosorbent under given experimental conditions, and which is calculated using Equation (1).

$$q = \frac{(c_0 - c)\ V}{m} \tag{1}$$

- Biosorption efficiency, $R(\%)$, which represents the percentage of metal ion retained in the biosorption process, and which is given by Equation (2).

$$R = \frac{(c_0 - c)}{c_0} \times 100 \tag{2}$$

where: c_0 and c—the initial and equilibrium concentration of the studied metal ion (mg/L), respectively; m—the mass of the biosorbent (g); V—the volume of aqueous solution used in the biosorption studies (mL).

3. Application of Response Surface Methodology for Modeling and Optimization of the Biosorption Process Using Natural Soy-Based Biosorbents

3.1. Preliminary Assessment of Variation Intervals for Independent Variables

The set of preliminary experiments facilitated the evaluation of the suitability of soybean biomass and soybean waste as biosorbents for the selected heavy metal ions (Pb(II), Cd(II), and Zn(II)) and the process efficiency, and also helped to determine the variation intervals of the process parameters. In this context, the following process parameters were chosen as independent variables for modeling: pH, sorbent dose (DS, g/L), metal ion concentration (c_0, mg/L), contact time (t_c, min), and temperature (T, °C). The variation intervals are shown in Table 7 [60].

Table 7. Independent variables of the biosorption process and variation intervals.

x_1		x_2		x_3						x_4		x_5	
pH (A)		***DS* (g/L) (B)**		**$c_{0\ Pb(II)}$ (mg/L) (C)**		**$c_{0\ Cd(II)}$ (mg/L) (C)**		**$c_{0\ Zn(II)}$ (mg/L) (C)**		**t_c (min) (D)**		**T (°C) (E)**	
min	max	min	max	min	max	min	max	min	max	min	max	min	max
1	6.5	5	40	11.66	416.45	9.22	230.54	13.08	209.25	5	180	5	50

3.2. Experiments Design

The experimental program aimed to identify the conditions in which the best response of the analyzed system is obtained. The design of the experiments in an appropriate manner was used to optimize the processes based on the analysis of the individual and interactive effects of the independent variables, simultaneously, on the entire designed experimental space. From among the most relevant multivariate statistical analysis techniques, Response Surface Methodology (RSM) has been frequently applied as a technique for programming experiments. As mentioned above, RSM is a combination of mathematical and statistical techniques that describe the relationship between a series of independent variables and one or more responses of the investigated experimental system. In this sense, the answer is examined on the whole space of the independent variables, where the answer has the best value [61]. The behavior of the system is usually described by a second-degree polynomial function (Equation (3)).

$$y_i = A_0 + \sum_{i=1}^{n} A_i x_i + \sum_{i=1}^{n} A_{ii} x_i^2 + \sum_{\substack{i \neq 1 \\ j = 1}}^{n} A_{ij} x_i x_j + \varepsilon \tag{3}$$

where y_i is the predicted system response, x_i and x_j are independent variables, A_0 is the constant coefficient (free term), A_i, A_{ii}, and A_{ij} are linear, square (second order) interaction coefficients, n is the number of independent variables, and ε is a random error.

The concordance of the model was evaluated using the correlation coefficient (R^2). The Fisher (F) test and the probability value (Prob > F) were applied to evaluate the significance of the model terms. In this paper, RSM was applied for modeling the biosorption efficiency (R, %) of heavy metal ions Pb(II), Cd(II), and Zn(II) from aqueous solutions using soybean biomass and soybean waste biomass from the industrial process of extracting soybean oil (for biodiesel). Additionally, we modelled the biosorption capacity of the studied biosorbents (q, mg/g). In the second part, the optimization of the biosorption efficiency (R, %) as well as the biosorption capacity (q, mg/g), was performed. The five independent variables (Table 7) were studied at five levels, designated as $-\alpha, -1, 0, 1, \alpha$. The matrix of the experimental program was the one corresponding to a full rotatable central composite design 2^5, with 45 experiments, of which 3 were in the central point (Table 8). Experimental data processing was performed in the Design-Expert software environment.

Table 8. Experimental matrix of the full rotatable central composite design 2^5 (coded values) (independent variables are presented in Table 7).

Experiment	Coded Independent Variables				
	A	B	C	D	E
1	−1	−1	−1	−1	−1
2	1	−1	−1	−1	−1
3	−1	1	−1	−1	−1
4	1	1	−1	−1	−1
5	−1	−1	1	−1	−1
6	1	−1	1	−1	−1
7	−1	1	1	−1	−1
8	1	1	1	−1	−1
9	−1	−1	−1	1	−1
10	1	−1	−1	1	−1
11	−1	1	−1	1	−1
12	1	1	−1	1	−1
13	−1	−1	1	1	−1
14	1	−1	1	1	−1
15	−1	1	1	1	−1
16	1	1	1	1	−1
17	−1	−1	−1	−1	1
18	1	−1	−1	−1	1
19	−1	1	−1	−1	1
20	1	1	−1	−1	1
21	−1	−1	1	−1	1
22	1	−1	1	−1	1
23	−1	1	1	−1	1
24	1	1	1	−1	1
25	−1	−1	−1	1	1
26	1	−1	−1	1	1
27	−1	1	−1	1	1
28	1	1	−1	1	1
29	−1	−1	1	1	1
30	1	−1	1	1	1
31	−1	1	1	1	1
32	1	1	1	1	1
33	−2.38	0	0	0	0
34	2.38	0	0	0	0
35	0	−2.38	0	0	0
36	0	2.38	0	0	0
37	0	0	−2.38	0	0
38	0	0	2.38	0	0
39	0	0	0	−2.38	0
40	0	0	0	2.38	0
41	0	0	0	0	−2.38
42	0	0	0	0	2.38
43	0	0	0	0	0
44	0	0	0	0	0
45	0	0	0	0	0

4. Results and Discussions

The removal of metal ions from aqueous solutions via biosorption takes place with maximum efficiency only under certain well-specified experimental conditions, regardless of the nature of the metal ion or biosorbent used. Consequently, in the evaluation of the biosorption performance of a material for certain metal ions, the first step is to establish the optimal experimental conditions with which to achieve the biosorption process.

The fulfillment of this desideratum involves the study of the influence of the main experimental parameters (such as the initial pH of the solution containing metal ions, the biosorbent dose, the initial concentration of metal ions in aqueous solution, the contact time, and the temperature) on the biosorption capacity of the material used as a biosorbent for each metal ion separately. The paramount conditions for the development of the studied process are obtained experimentally and correspond to those values of the parameters for which the retention of metal ions from aqueous solutions takes place with the highest efficiency.

4.1. Characterization of Biosorbents

The results of the EDX analysis (Figure 1) showed that both soybean and soybean waste biomasses contain large amounts of organogenic elements (C, O, P, and S) in their structure, and also a series of ions of alkaline and alkaline earth metals (K, Mg, and Al) which, due to their high mobility, can easily participate in ion exchange processes. On the other hand, the comparison of the values obtained from the EDX analysis shows that, unlike soybean biomass, in the case of soybean waste, the C, P and S content is significantly lower, while the values of oxygen percentages and mobile ions (K and Al) are higher. These differences allow us to say that, following the n-hexane extraction step, there are still many functional groups in the composition of soybean waste, even if some of the organic compounds (most likely lipids and fatty acid salts) have been removed.

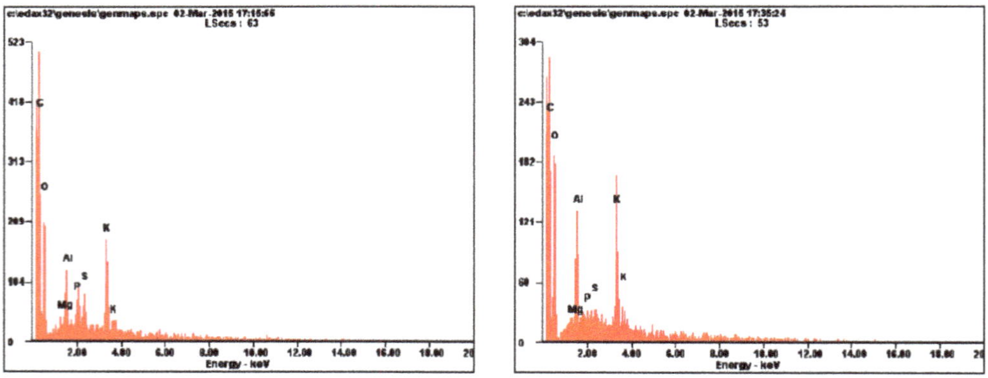

(a) (b)

Figure 1. EDX spectra recorded for soybean biomass (**a**) and soybean waste biomass (**b**).

The nature of the functional groups on the surface of these biosorbents and the differences that appear in the structure of the soybean biomass after the extraction stage were highlighted with the help of IR spectrometry. A detailed analysis of the IR spectra recorded on the dry matter (Figure 2) showed that, in the soybean biomass spectrum (spectrum 1), the most important absorption bands correspond to the functional groups of proteins and polysaccharides in the cell walls. Thus, the absorption band at 3407 cm^{-1} (which corresponds to the stretching vibration of the O–H bond from aliphatic alcohols and water molecules), the bands at 1652 and 1543 cm^{-1} (attributed to the stretching vibration of the C=O bond of carbonyl compounds), the bands of 1244 and 1053 cm^{-1} (corresponding to the symmetrical and asymmetric tensile vibration of the P=O bond in the phosphate groups), and the bands between 1200 and 690 cm^{-1} (which can be attributed to the vibrations of the C–OH bonds and C–O–C in polysaccharides) [45] indicate the presence of functional groups, such as hydroxyl, carbonyl, carboxyl, phosphate groups, etc., which may play an important role in the retention processes of the studied metal ions.

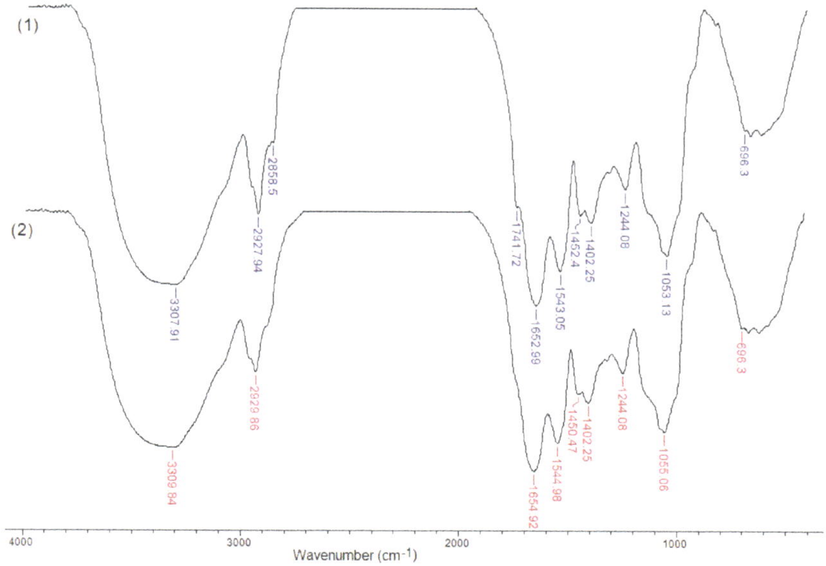

Figure 2. IR spectra recorded for soybean biomass (1) and soybean waste (2).

After the extraction stage, in the IR spectrum of soybean waste (Figure 2, spectrum 2), most of the absorption bands are slightly shifted towards higher wave numbers (2–3 cm^{-1}), while the absorption bands at 2858 cm^{-1}, 1741 cm^{-1}, 1402 cm^{-1} and 1244 cm^{-1}, which are characteristic of lipids, have disappeared or decreased in intensity (without changing too much the position of the maximum absorption). These differences show that, at the extraction stage, only this type of component (lipids) was removed from the soybean biomass composition, while most proteins and polysaccharides remained in the soybean biomass composition, and their functional groups will be binding centers for metal ions in the aqueous solution.

The SEM images (Figure 3) show that, from a morphological point of view, both soybean biomass and soybean waste have a rough and irregular surface, which presents numerous non-uniformities. However, it can be seen that, after the extraction stage, the surface of the soybean biomass (Figure 3b) becomes more porous and has more "breaks", and this is probably determined by the rupture of the cell walls that takes place during extraction time. This increase in the roughness of the biosorbent surface after the solvent extraction step can greatly influence the efficiency of the biosorption process in the case of soybean waste, as it causes an increase in its specific surface area.

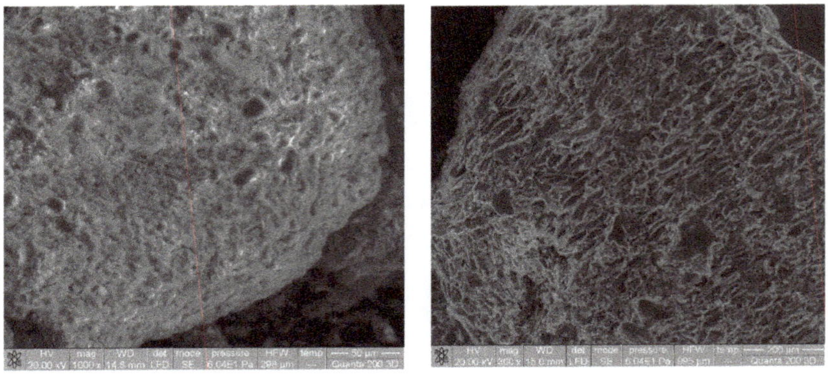

(a) (b)

Figure 3. SEM images recorded for soybean biomass (**a**) and soybean waste (**b**).

All this experimental evidence, namely, (i) the presence of mobile ions in the composition of the two soy-based biosorbents, (ii) the presence of a large number of various functional groups on the surface of soybean biomass and soybean waste, but also (iii) the morphological non-uniformity of the surface, recommends the use of soybean biomass and soybean waste as biosorbents in the processes of the decontamination of environmental components.

4.2. Influence of Process Parameters on Biosorption Efficiency

4.2.1. Initial pH of the Solution Containing Heavy Metal Ions

One of the most important experimental parameters that can significantly influence the biosorptive characteristics of a given material is the initial pH of the aqueous solution, since it affects both the form of speciation and the solubility of the metal ions present in the solution, as well as, especially, the degree of dissociation of the functional groups on the surface of the biosorbent.

The considered pH range (Table 2) was selected because: (i) it covers the acidic and slightly acidic to neutral range, which results in a significant variation in the degree of dissociation of the functional groups on the surface of the two biosorbents; (ii) the speciation change of the studied metal ions (Pb(II), Cd(II), and Zn(II)) is avoided, because in this interval they are present predominantly in the form of free divalent ions. The variation of the biosorption capacity of soybean biomass and soybean waste as a function of the initial pH of the aqueous solution, for each metal ion separately, is shown in Figure 4.

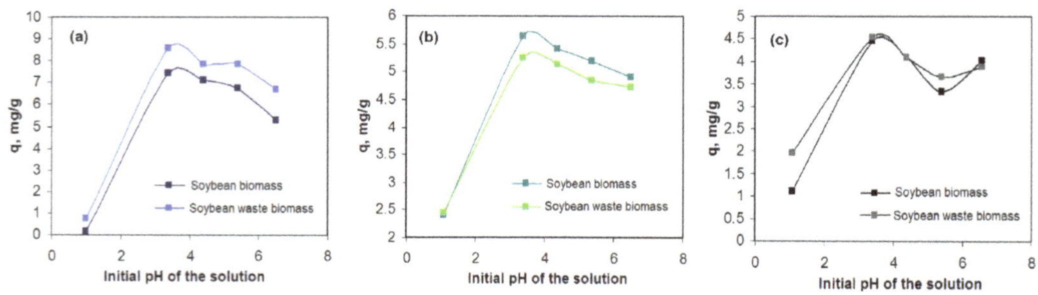

Figure 4. Variation of the biosorption capacity of the two biosorbents (soybean biomass and soybean waste biomass) with the initial pH of the solution in the case of the retention of (**a**) Pb(II), (**b**) Cd(II), and (**c**) Zn(II) ions.

As can be seen from Figure 4, the biosorption capacity of the two materials depends significantly on the value of the initial pH of the solution, although the variation is not uniform. Thus, the biosorption capacity of soybean biomass and soybean waste biomass intensifies considerably with the increasing initial pH of the solution in the pH range between 1.09 and 3.39 and then decreases for all studied metal ions. The highest values of the biosorption parameters are obtained at a pH of the initial solution of 3.39, where more than 50% of Pb(II), 60% of Cd(II), and 45% of Zn(II) can be removed from the aqueous solution using these two biosorbents. It should also be noted that when using soybean waste, the efficiency of the biosorption process increases only in the case of Pb(II) ions (from 50% to 60%), while Cd(II) (from 61% to 60%) and Zn(II) ions (from 45% to 46%) remain approximately constant compared to the use of soybean biomass as a biosorbent.

Other research found that the initial pH of a metal ion solution influences significantly the binding sites on biosorbent surfaces and metal ion behavior [62–64]. These remarks are also related to those reported by other authors for the biosorption of some heavy metals on algae [65–67]. Studies on the important role pH plays in the biosorption process highlighted the mechanisms responsible for biosorption in relation with pH value, namely, complexation, oxidation reduction, and separation by hydrolysis, considering that the pH determines the charge of the biosorbent surface, the degree of ionization, etc. [45,68–72]. Choloco-Gonzales et al. found that the biosorption capacity of Pb(II), Cd(II), and Zn(II) on agave bagasse reaches a maximum at around pH 5.5, beyond the point of zero charge (pH_{PZC}) [73]. Salem et al. studied the biosorption of Pb(II), Cd(II), and Zn(II) on loquat bark (Eriobotrya japonica) and observed the maximum adsorption around pH 4.0 [74]. The researchers found that the surface area of the biosorbent they used was negatively charged as the pH increased from 2 to 4, whereby the functional groups became deprotonated and became available for Pb(II), Zn(II), and Cd(II) ions. Lezcano et al. used different categories of biomass as biosorbents identified from eutrophic ecosystems and tree leaves [75]. For initial pH values between 3 and 5, the biomass behaved like a proton acceptor and the biosorption capacity reached maximum values, but at initial pH values higher than 6, the biomass behaved like a proton donor.

Unlike most of the biosorbents used to remove metal ions from aqueous solutions, where the maximum biosorption capacity is obtained in the weak acid to neutral range (pH = 4.0–6.0), in the case of soybean biomass and soybean waste, this maximum is obtained at much lower pH values (pH = 3.39). This is probably due to the high buffering capacity of the two biosorbents and offers the possibility to use these biosorbents in acidic wastewater treatment processes (such as those from the metal coating industry). Based on these observations, it can be said that the biosorption of the selected ions on soybean biomass and soybean waste biomass takes place with the highest efficiency at a pH of 3.39, and this value was considered to be the most suitable and was used in all other experimental studies.

4.2.2. Dose of Biosorbent

The experimental results on the influence of biosorbent dose on the biosorption capacity of soybean biomass and soybean waste biomass for Pb(II), Cd(II), and Zn(II) ions, tested using different amounts of biosorbent (Table 3), are shown in Figure 5. The increase in the amount of biosorbent in the studied range (4.0–40.0 g/L) caused a slightly significant increase in the percentages of removal of metal ions from 44% to 90% in the case of Pb(II), from 57% to 73% in the case of Cd(II), and from 40% to 56% in the case of Zn(II), which does not differ significantly according to the nature of the biosorbent. At the same time, the biosorption capacity decreases significantly with the increase in the amount of biomass added, from 10.62 to 1.81 for Pb(II), from 6.56 to 0.83 for Cd(II), and from 5.76 to 0.67 for Zn(II), values that are insignificantly influenced by the type of biosorbent (soybean biomass or soybean waste).

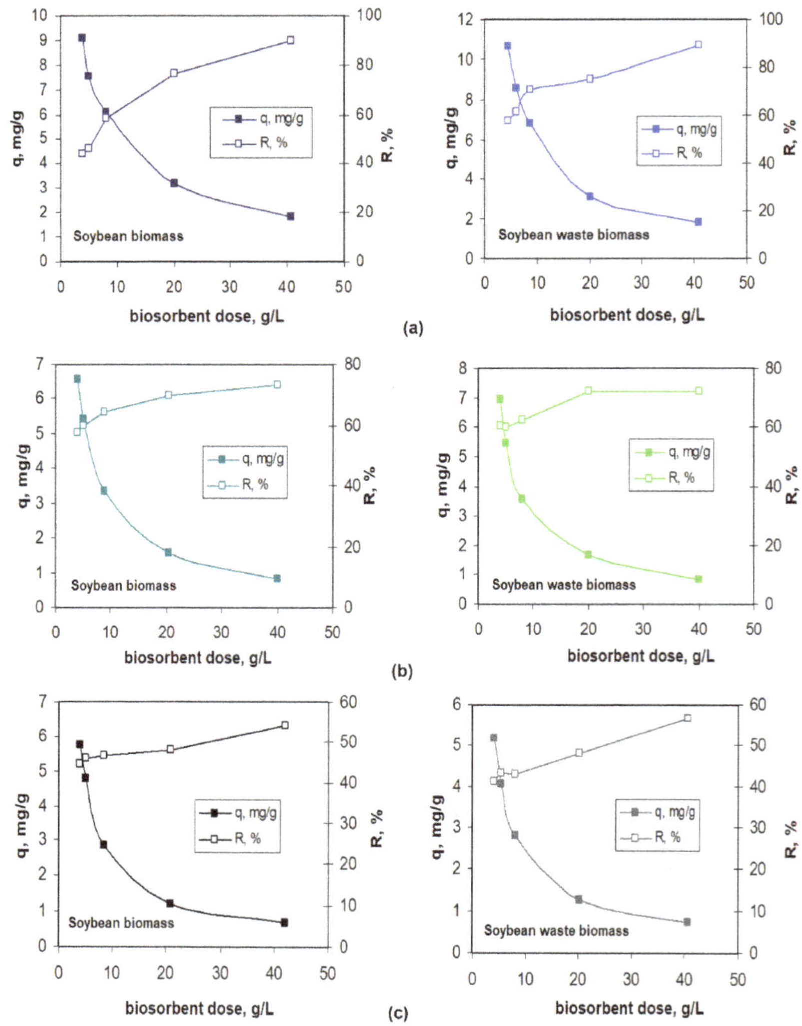

Figure 5. Influence of biosorbent dose on the biosorption efficiency of (**a**) Pb(II), (**b**) Cd(II), and (**c**) Zn(II) ions on soybean biomass and soybean waste (pH = 3.39; c_0 = 46.11 mg/L; contact time = 24 h; temperature = 22 °C).

Comparing the biosorption capacity (*q, mg/g*) and the biosorption efficiency (*R*, %), it was established that a dose of biosorbent of 5 g/L can be considered to be the best for the retention of Pb(II), Cd(II), and Zn(II) ions in aqueous solutions for both soybean biomass and soybean waste, and this value was used in subsequent experimental studies.

4.2.3. Initial Concentration of Metal Ions

The dependence of the biosorption capacity of the two biosorbents on the initial concentration of Pb(II), Cd(II), and Zn(II) ions was studied in concentration ranges described in Table 4, at the selected values of solution pH (3.39) and biosorbent dose (5 g/L). The experimental results obtained in this case for each metal ion are shown in Figure 6. From these graphical representations, it can be seen that, as the concentration of metal ions increases in the studied concentration range, the biosorption capacity q values increase, while the values of the biosorption efficiency R decrease. This variation in opposite direction of q and R is valid for all studied metal ions and for both biosorbents, although the limits of variation of the two parameters depend on the nature of the metal ion and the biosorbent used in the biosorption process.

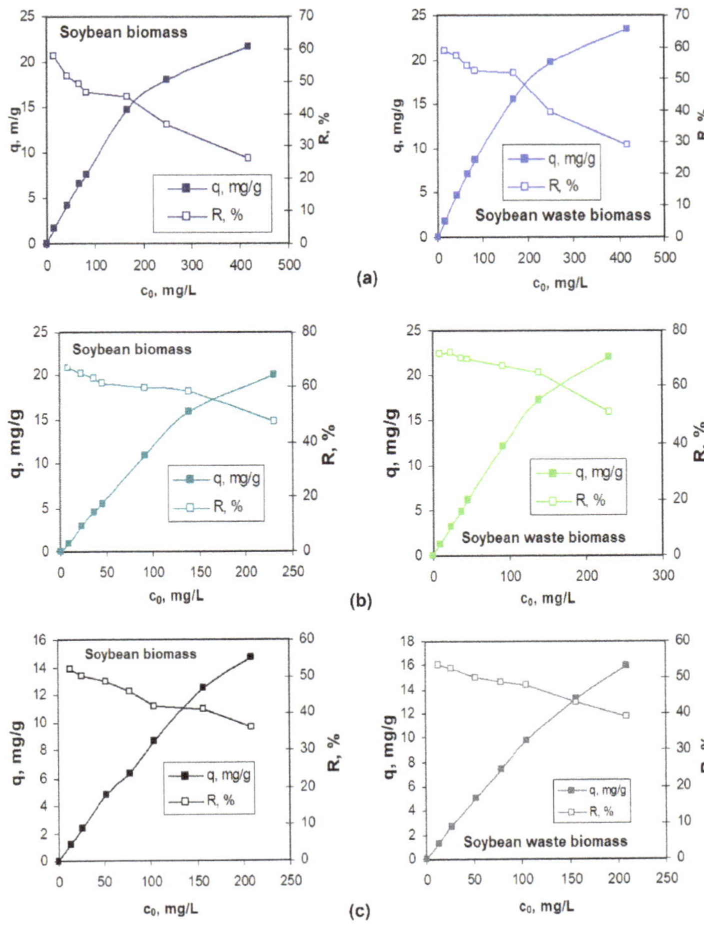

Figure 6. Influence of the initial concentration of metal ions on the biosorption capacity and biosorption efficiency of (**a**) Pb(II), (**b**) Cd(II), and (**c**) Zn(II) ions on soybean biomass and soybean waste (pH = 3.39; biosorbent dose = 5.0 g/L; contact time = 24 h; temperature = 22 °C).

When the concentration of metal ions in the aqueous solution is high, most of the functional groups on the biosorbent surface are already occupied and, therefore, their transition to free (unreacted) functional groups, which are found in the inner planes or even inside the particles of the biosorbent, is hampered by diffusion resistances.

Amer et al. attributed the increase in biosorption efficiency as the metal ion concentration increases to the more intense interaction between metal ions and sequestration sites [76]. They used different initial metal concentrations (Cu (II), Ni(II), Zn(II), and Pb(II) ions) of 5, 20, 40, 60, 80, and 100 mg/L, with *Sophora japonica* pods powder as the biosorbent. Kamar et al. analyzed the effect of initial metal ion concentration on the biosorption of Pb(II), Cu(II), and Cd(II) onto cabbage leaves powder, and found that absorption efficiency decreased slightly with increasing initial concentrations from 10 to 50 mg/L, and a rapid decrease in the initial concentrations ranging from 50 to 100 mg/L [77]. They attributed this behavior to the decrease in the number of available sorption sites on the surface area of the sorbent as the initial concentration of metal ions in the solution increased. Similar behaviors were reported in other works, as well [78,79].

The retention of Pb(II), Cd(II), and Zn(II) ions by biosorption on soybean biomass and soybean waste biomass, respectively, can be predominantly caused by electrostatic interactions. Similar assumptions were made by [80,81]. The efficiency of the biosorption process is moderate, because the electrostatic interactions responsible for the retention of metal ions are non-selective, and the metal ions will interact with the functional groups that have the highest availability. After occupying these available groups on the surface, spatial obstructions are generated, which limit the penetration of other ions from the solution into the free functional groups. Consequently, the biosorption efficiency has moderate values following the series Cd(II) > Pb(II) > Zn(II) for both biosorbents. However, the biosorption efficiency values for all studied metal ions are higher in the case of soybean waste than in the case of soybean biomass up to 6%, which shows that the breaking of cell walls in soybean biomass during the oil extraction process is important for biosorption intensification. Following the solvent extraction step, an increase in the availability of surface functional groups to interact with metal ions in the aqueous solution can be assumed.

4.2.4. Contact Time

The contact time between the two phases necessary to reach the equilibrium state is another important parameter of the biosorption process, especially when the implementation of such a process, e.g., for the treatment of industrial wastewater, is pursued. Too long a contact time increases the costs of the biosorption process, while too low a value of this parameter leads to a drastic decrease in the efficiency of the process as such. The influence that the contact time between the two phases (solid phase—biosorbent, and liquid phase—aqueous solution containing the studied metal ions) has on the efficiency of the biosorption under the considered experimental conditions is shown in Figure 7. As the contact time between phases increases, so does the biosorption capacity (q) for both biosorbents. This increase is more evident in the initial stage, in the first 30 min, when more than 66% of the total amount of Pb(II), 60% of the total amount of Cd(II), and 57% of the total amount of Zn(II) are retained on soybean biomass, and 67% of the total amount of Pb(II), 61% of the total amount of Cd(II), and 64% of the total amount of Zn(II) are retained on soybean waste biomass.

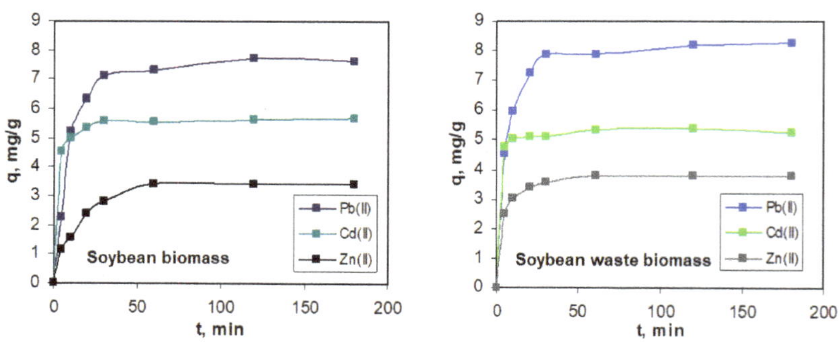

Figure 7. Influence of contact time on sorption capacity of Pb(II), Cd(II), and Zn(II) ions in aqueous solutions by soybean biomass and soybean waste (pH = 3.39; DS = 5 g/L; c_0 = 83.29 mg Pb(II)/L; 46.11 mg Cd(II)/L; 52.31 mg Zn(II)/L; temperature T = 24 °C).

Such a variation is most likely determined by the number and availability of active centers able to interact with metal ions in the solution. Thus, when the number of active centers is high (the first moments of the biosorption process) and more geometrically and spatially available (the case of soybean waste compared to soybean biomass), the interaction between functional groups at the surface and metal ions in aqueous solution is easier to achieve and the biosorption rate is high. On the other hand, the close values of the biosorption efficiency obtained for the three metal ions studied, both in the case of soybean biomass and soybean waste (with differences of up to 7%) is another argument in favor of the hypothesis that the biosorption process is achieved via non-selective electrostatic interactions, and that the process rates do not depend on the nature of the metal ion in the aqueous solution. After the initial stage, the biosorption rate decreases significantly near the steady state, which is obtained practically after 60 min in the case of Pb(II) and Zn(II) ions and after 30 min in the case of Cd(II) ions for both types of biosorbents. Based on these experimental observations, it can be concluded that a contact time of 3 h is sufficient to reach equilibrium in the case of biosorption of the three metal ions, both on soybean biomass as well as on soybean waste biomass.

4.2.5. Temperature

The influence of temperature on the biosorption capacity of soybean biomass and soybean waste for Pb(II), Cd(II), and Zn(II) ions was studied, as mentioned before, at three different temperature values, ranging from 5 to 50 °C for different concentrations of each metal ion separately, considering a contact time of 3 h (Table 6). The experimental results of these studies are presented in Figure 8 and show that temperature has a relatively small influence on the biosorption capacity, both in the case of soybean biomass and in the case of soybean waste, for all metal ions studied. Increasing the temperature at 45 °C causes an increase in the biosorption capacity of the two biosorbents, but only with a few units, and this is visible especially at high values of the initial concentration of metal ions in the aqueous solution. Depending on the nature of the metal ion, it can be seen that the major variation in biosorption capacity resulted in the case of Pb(II), followed by Cd(II), and then Zn(II). This hierarchy is observed for both soybean biomass and soybean waste, although it should be noted that in the case of soybean waste, the variation in biosorption capacity is higher than in the case of soybean biomass for all studied metal ions.

The increase in the biosorption capacity of soybean biomass and soybean waste with increasing temperatures is suggestive of the endothermic nature of the biosorption processes of Pb(II), Cd(II), and Zn(II) ions on the two biosorbents. However, the relatively small variation of q values at relatively large temperatures (45 °C) shows that performing the biosorption process at high temperatures is not necessarily an advantage. The biosorption efficiency of the studied ions on soybean biomass and soybean waste can be successfully achieved at ambient temperatures, and these conditions are advantageous considering both the costs and the efficiency of the biosorption process.

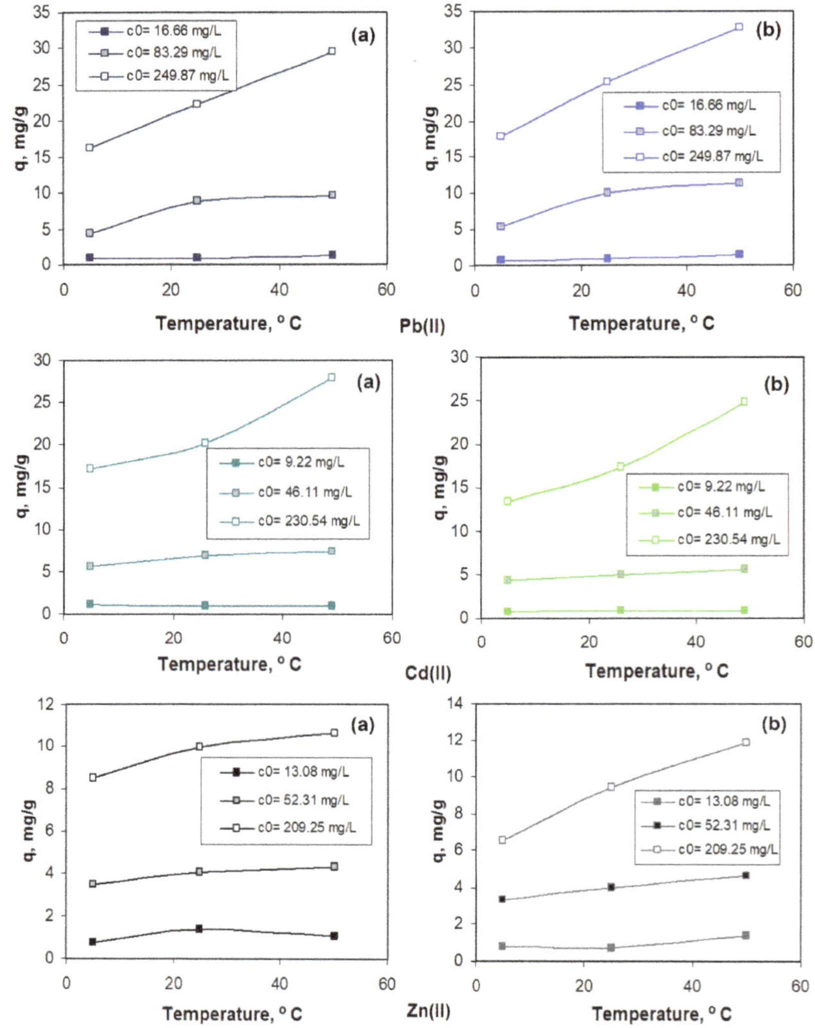

Figure 8. Influence of temperature on the biosorption capacity of soybean biomass (**a**) and soybean waste (**b**) for the retention of the metal ions (pH = 3.39; 5 g biosorbent/L; 3 h).

4.3. Modeling and Optimization of Biosorption Process by Response Surface Methodology

4.3.1. Model Development and Validation

Equations of the model generated in Design-Expert software environment after the interpretation of the experimental data for the biosorption of Pb(II), Cd(II), and Zn(II) ions by soybean biomass and soybean biomass waste following the variables and experimental matrix showed in Tables 7 and 8 are presented in Table S1 (Equations (1)–(12) (a, b)). It is found that these Equations are second-degree polynomial functions, which contain significant terms, the rest being eliminated in the process of verifying the concordance of the model and the significance of the coefficients because, following the statistical analysis for a 95% confidence interval, some of the coefficients were considered to be insignificant.

Equations (4)–(15) (a, b) (Table S1) show the extent to which the individual independent variables or the interactions between them affect the biosorption efficiency (R, %) and the biosorption capacity (q, mg/g). Negative coefficient values indicate that individual factors and/or interactions between them negatively affect dependent variables (R, q), while positive coefficients suggest that dependent variables are positively influenced (increase) under the action of those independent variables. The compatibility of the developed models, of a polynomial type (second order—quadratic), is justified in Table S2 (after adequacy testing, which consists of comparing the set of performances obtained with the help of the model with the performances collected experimentally).

Based on this information, the suitability of the selection of the recommended polynomial (quadratic) model for all six experimental cases (three metal ions and two tested biosorbents) was confirmed. Additionally, the independent variables were selected to find the best set of predictors in each case and, therefore, the simplest form of the mathematical model that was in good agreement with the experimental data. This selection was made in order to reduce random errors in estimating the dependent variables and to ensure the accuracy of the degrees of freedom. In selecting the variables of the model, the compliance of their hierarchy was taken into account (lower order terms were possibly excluded from the model) after the higher order ones (for the same variable). The elimination of insignificant predictors was performed using the "backward" procedure because this is recommended in models with collinear variables as the most robust compared to the "forward" or "stepwise" techniques, with only those factors that have a significant influence of the response variable being retained in the model. By applying the "backward" procedure to a critical threshold $\alpha_{\text{crit}} = 0.05$, modeling was started with all the predictors (independent variables) in the model; the predictors for which the highest values of Prob > α_{crit} have been eliminated; the model was then reconsidered and the iteration continued until the Prob < α_{crit} values.

4.3.2. ANOVA Analysis

ANOVA analysis was performed for the biosorption efficiency R and the biosorption capacity q of the metal ions Pb(II), Cd(II), and Zn(II) models (Tables S3 and S4). Significance tests for second-order polynomial models—with RSM reduced according to the results of the ANOVA analysis for R and q—are presented in Table 9. In the case of Pb(II) biosorption on soybean biomass, the model is characterized by the value $F = 53.11$, which indicates that the model is significant (Equations (1) a, b, Table S1), and there is only a 0.01% probability that the system answer may be affected by accidental errors. "Prob > F" values less than 0.05 indicate that the terms of the model are significant. In this case, A, B, C, D, A2, and B2 are significant terms. The value 76.25 for "lack of correlation" for F implies that it is significant, meaning that there is a chance of only 1.3% that the answer will be affected by random errors. In the case of Pb(II) biosorption on soybean waste biomass, the F value of 68.37 implies that the model is significant and that there is a probability of only 0.01% that the system response may be affected by random errors. As with the soybean biomass biosorbent, "Prob > F" values less than 0.05 indicate that the model terms (A, B, C, D, A2, and B2) are significant. Similar situations were found in the case of the models for Cd(II) and Zn(II) biosorption on soybean biomass and soybean waste biomass, respectively.

The ANOVA confirms the adequacy of the developed models because Prob > F is less than 0.05. The developed models have small standard deviations and values of the coefficient of determination R^2 generally over 0.9. The high values of R^2_{adj} show that the total variations of $R(\%)$ and $q(mg/g)$ can be described by the selected model type. The value of the signal-to-chance ratio (adequate accuracy) is greater than 4, which indicates that the measurement results are not random, but are consistent with the actual values. The values of R^2_{pred} are in good agreement with R^2_{adj} (which appreciates the correctness of the model by the fact that it is well selected and not due to the large number of variables). R^2_{pred} indicates the extent to which the regression model predicts the response for new observations (measurements). This statistical measure supports the assessment of the model in the sense that it is consistent with experimental data.

Table 9. Significance tests for RSM models developed for $R(\%)$ and $q(mg/g)$ during biosorption of metal ions Pb(II), Cd(II), and Zn(II) on soybean biomass and soybean waste biomass.

Measure	Value	Measure	Value
	Soybean biomass, Pb(II), $R(\%)$		
Standard deviation	4.36	R^2	0.8935
Mean	56.67	R^2_{adj}	0.8766
C.V.	7.70	R^2_{pred}	0.8181
PRESS	1235.51	Accuracy	35.286
	Soybean waste biomass, Pb(II), $R(\%)$		
Standard deviation	3.79	R^2	0.9152
Mean	56.37	R^2_{adj}	0.9018
C.V.	6.72	R^2_{pred}	0.8611
PRESS	893.27	Accuracy	36.682
	Soybean biomass, Cd(II), $R(\%)$		
Standard deviation	3.82	R^2	0.9236
Mean	65.68	R^2_{adj}	0.9092
C.V.	5.81	R^2_{pred}	0.8873
PRESS	794.87	Accuracy	38.598
	Soybean waste biomass, Cd(II), $R(\%)$		
Standard deviation	3.76	R^2	0.9147
Mean	65.63	R^2_{adj}	0.8897
C.V.	5.73	R^2_{pred}	0.7760
PRESS	1265.08	Accuracy	32.177
	Soybean biomass, Zn(II), $R(\%)$		
Standard deviation	3.68	R^2	0.8934
Mean	50.04	R^2_{adj}	0.8698
C.V.	7.35	R^2_{pred}	0.7975
PRESS	925.45	Accuracy	28.876
	Soybean waste biomass, Zn(II), $R(\%)$		
Standard deviation	3.55	R^2	0.9179
Mean	50.84	R^2_{adj}	0.9024
C.V.	6.97	R^2_{pred}	0.8201
PRESS	1019.70	Accuracy	39.604
	Soybean biomass, Pb(II), $q(mg/g)$		
Standard deviation	1.58	R^2	0.9268
Mean	13.80	R^2_{adj}	0.9105
C.V.	11.45	R^2_{pred}	0.8705
PRESS	159.00	Accuracy	28.458
	Soybean waste biomass, Pb(II), $q(mg/g)$		
Standard deviation	1.53	R^2	0.9503
Mean	16.77	R^2_{adj}	0.9393
C.V.	9.12	R^2_{pred}	0.9115
PRESS	150.02	Accuracy	34.585
	Soybean biomass, Cd(II), $q(mg/g)$		
Standard deviation	0.87	R^2	0.9838
Mean	11.62	R^2_{adj}	0.9784
C.V.	7.45	R^2_{pred}	0.9658
PRESS	52.29	Accuracy	51.854

Table 9. *Cont.*

Measure	Value	Measure	Value
	Soybean waste biomass, Cd(II), $q(mg/g)$		
Standard deviation	1.06	R^2	0.9857
Mean	12.68	R^2_{adj}	0.9803
C.V.	8.35	R^2_{pred}	0.9677
PRESS	80.87	Accuracy	46.917
	Soybean biomass, Zn(II), $q(mg/g)$		
Standard deviation	0.71	R^2	0.9413
Mean	6.42	R^2_{adj}	0.9193
C.V.	11.09	R^2_{pred}	0.8056
PRESS	53.76	Accuracy	24.146
	Soybean waste biomass, Zn(II), $q(mg/g)$		
Standard deviation	0.88	R^2	0.9576
Mean	8.74	R^2_{adj}	0.9398
C.V.	10.07	R^2_{pred}	0.8555
PRESS	81.77	Accuracy	27.449

C.V., Cross-validation; PRESS, Predicted Residual Error Sum of Squares.

Details on the diagnosis of the statistical properties of the developed models were obtained from the analysis of some graphical representations (not presented here). The so-called adjustment errors and residuals are estimates of errors. Small residues expressed the goodness of fit of the experimental data. The graphical representation of the probability as a function of the error scattering estimate (σ^2) has shown that it is approximately linear, demonstrating the normal distribution hypothesis for the chosen 95% confidence interval. The residue limit is a maximum of $\pm 3\sigma$. From the diagrams residual values vs. calculated values for all the analyzed experimental situations, there is no deviation from normality or from the hypothesis that the errors have a relatively constant dispersion. Since the residues are in the range $(-3, 3)$ it means that there are no aberrant observations associated with the developed models. The diagrams' residual values—independent variables—have revealed that there are no aberrant observations regarding each independent variable that can be induced in the experimental program and the elaborated models, respectively. Extreme observations, which do not fall within the general trend of the outliers are absent, as all observations fall within the range of a maximum of three standard deviations from the average.

To identify the points that significantly influence the regression, the Cook distance was analyzed (as defined in the specialized works) and which is a measure of the influence of the ith observation on the predicted values. This distance is another variable for which the aberrant values are determined according to the rule of the three standard deviations. It was found that for all the elaborated models, this condition is fulfilled. The identification of possible aberrant values was also performed by analyzing the standardized residual values, to identify possible extreme values in the y-space without changing the position of the regression curve to the extreme value—the so-called leverage points. The extreme points y associated with the mean values x do not change the regression curve, with the values being located in the range of 0–1, usually below the value of the standard deviation. In this context, it is found that there is a good concordance between the experimental values of the variables R and q and those estimated using the RSM model (Figures 9 and 10).

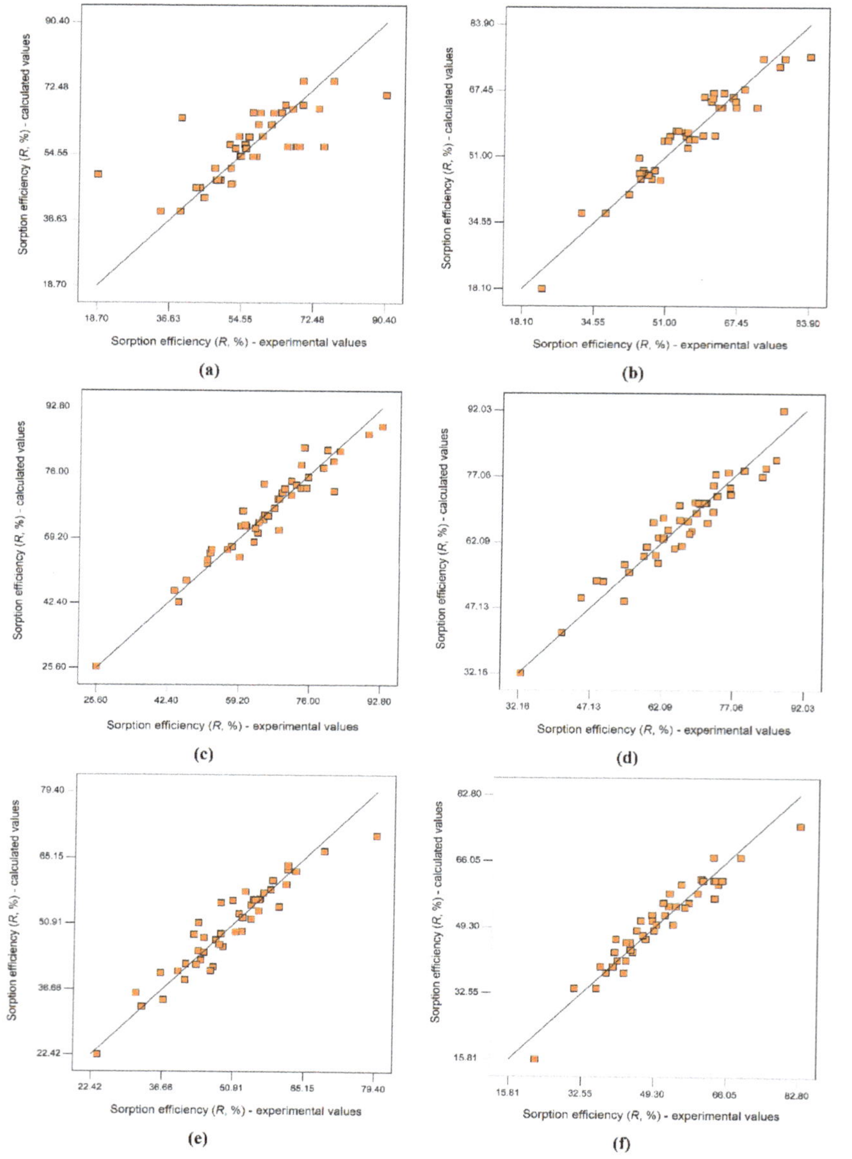

Figure 9. Graphical representation of the concordance between experimental and calculated (predicted) values using the mathematical models developed for biosorption efficiency (*R*, %) of Pb(II), Cd(II), and Zn(II) on soybean biomass and soybean waste biomass: (**a**) Pb(II) on soybean biomass; (**b**) Pb(II) on soybean waste biomass; (**c**) Cd(II) on soybean biomass; (**d**) Cd(II) on soybean waste biomass; (**e**) Zn(II) on soybean biomass; (**f**) Zn(II) on soybean waste biomass.

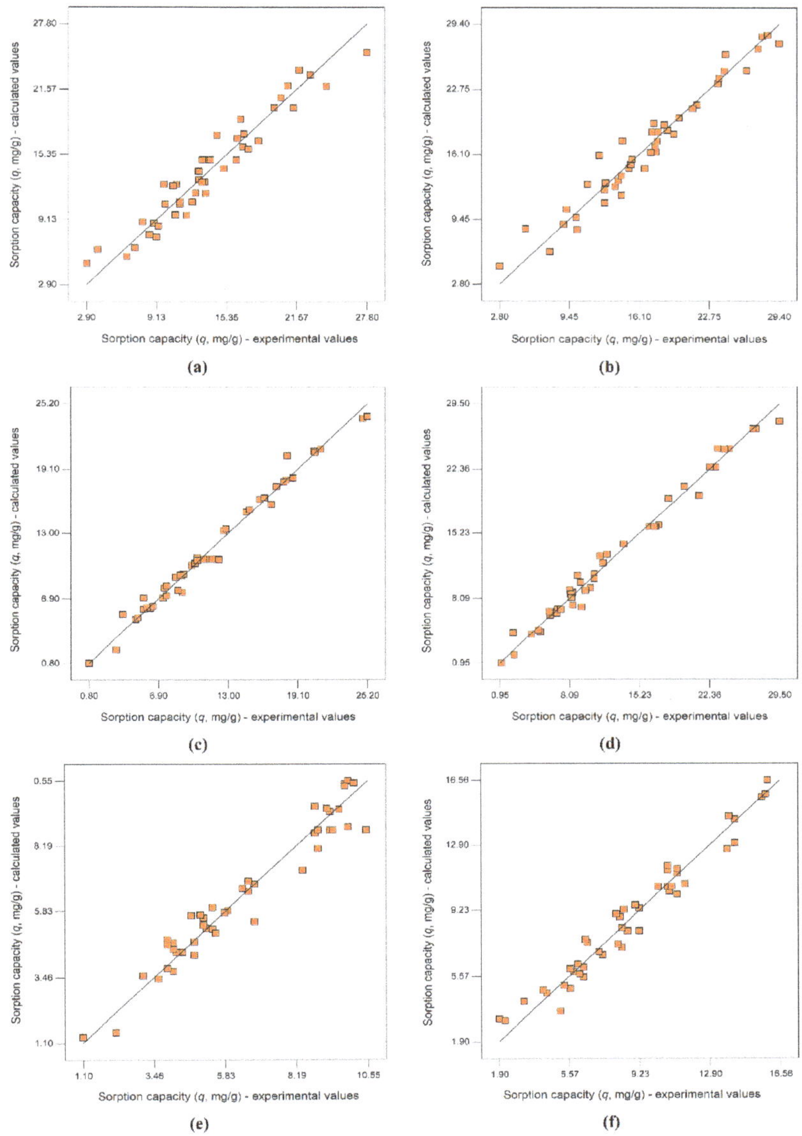

Figure 10. Graphical representation of the concordance between the experimental values and those calculated (predicted) with the mathematical models developed for the biosorption capacity (*q*, *mg/g*) of soybean biomass and soybean waste biomass when adsorbing metal ions (**a**) Pb(II) on soybean biomass; (**b**) Pb(II) on soybean waste biomass; (**c**) Cd(II) on soybean biomass; (**d**) Cd(II) on soybean waste biomass; (**e**) Zn(II) on soybean biomass; (**f**) Zn(II) on soybean waste biomass.

To verify the collinearity/multicollinearity of the independent variables (possibly the overestimation of the correlation or determination coefficient as well as the dispersion of the estimated coefficients), the tolerance of the variables xi was tested by the relation (4) and the variance inflation factor, denoted by *VIF* (relation 5), which shows the extent to which the variance of the model coefficients may increase due to the lack of orthogonality (multicollinearity) of the independent variables.

$$\sigma_i = 1 - R_i^2 \quad (4)$$

$$VIF_i = \frac{1}{\sigma_i} \quad (5)$$

The standard error of the coefficient of a model increases proportionally to the square root of *VIF*. If a coefficient of a parameter is orthogonal to the other terms of the model, the *VIF* value is unitary. We found that, in most of the cases studied, *VIF* = 1. In no case is *VIF* greater than or equal to 10 (which would indicate the existence of multicollinearity, which occurs when a group of independent variables are strongly correlated with each other, and the rest no longer provide significant information, with an overestimation of the coefficient of determination R^2) [81,82].

4.3.3. Sensitivity Analysis

A comparison of the effects of all the factors influencing the independent variables can be seen in Figures 11 and 12. The answer is represented graphically by changing the value of one factor on the chosen range of variation (generating a disturbance) and keeping the other factors constant. The slope or curvature associated with a factor suggests that the response is sensitive to that factor. A relatively flat line suggests that this factor does not significantly influence the response. In this way, the sensitivity analysis of the elaborated mathematical models was performed.

It can be seen that in the case of Pb(II) biosorption on soybean biomass, the change in pH and initial metal concentration significantly influences the system response (*R*,%). The same finding is true for Pb(II) biosorption on soybean waste biomass, but the sensitivity of the response is higher with the variation of these two parameters. In the case of Cd(II) biosorption on soybean biomass, the pH and concentration of the metal in the initial solution, and also the sorbent dose and the contact time, significantly influence the biosorption efficiency. Temperature has little influence. In the case of Cd(II) biosorption on soybean waste biomass, the pH, sorbent dose, and contact time significantly influence the system response, especially to values of these factors to the left of the reference point. Temperature also plays an important role in the biosorption efficiency of Cd(II) soybean waste biomass.

When Zn(II) is retained on soybean biomass, it is significantly influenced by pH (upwards, especially in the area to the left of the reference point) and metal ion concentration in the initial solution (downwards). The sensitivity of the *R*(%) response to the variation in sorbent dose, contact time, and temperature around the reference point is relatively small. However, in the case of Zn(II) biosorption on soybean biomass waste, the sensitivity of the system to changes in pH and the initial concentration of the metal ions are significant. The influence of temperature is not significant. The biosorption capacity *q* is sensitive to variations in pH, sorbent dose, and initial concentration of Pb(II) during biosorption on soybean biomass, but the sensitivity is higher in the case of biosorption capacity values of Pb(II) on soybean waste biomass. The biosorption capacity of soybean biomass for Cd(II) is sensitive to changes in ion concentration in the initial solution and sorbent dose. The pH, contact time, and temperature have a less significant influence on the system response (*q*, *mg/g*). In the case of the analysis of the sensitivity of biosorption capacity of soybean waste biomass for Cd(II), it can be seen that the sorbent dose and the initial ion concentration in the solution significantly influence *q*, especially in the right field of the reference point. Additionally, *q* is relatively sensitive to changes in pH, contact time, and temperature.

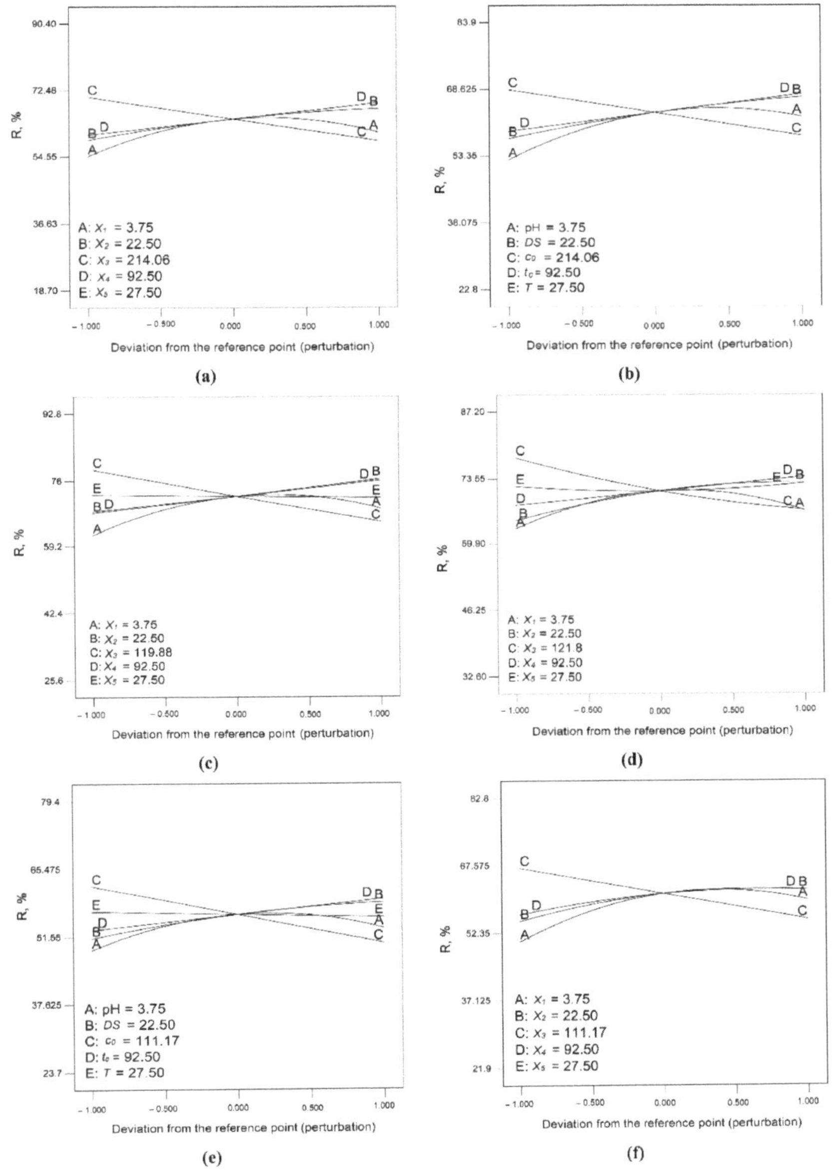

Figure 11. Perturbation diagrams resulting from the sensitivity analysis for the mathematical models developed for biosorption efficiency of heavy metal ion on soybean biomass and soybean waste biomass (the meaning of the notations (**a–f**) is the same as those from Figure 9).

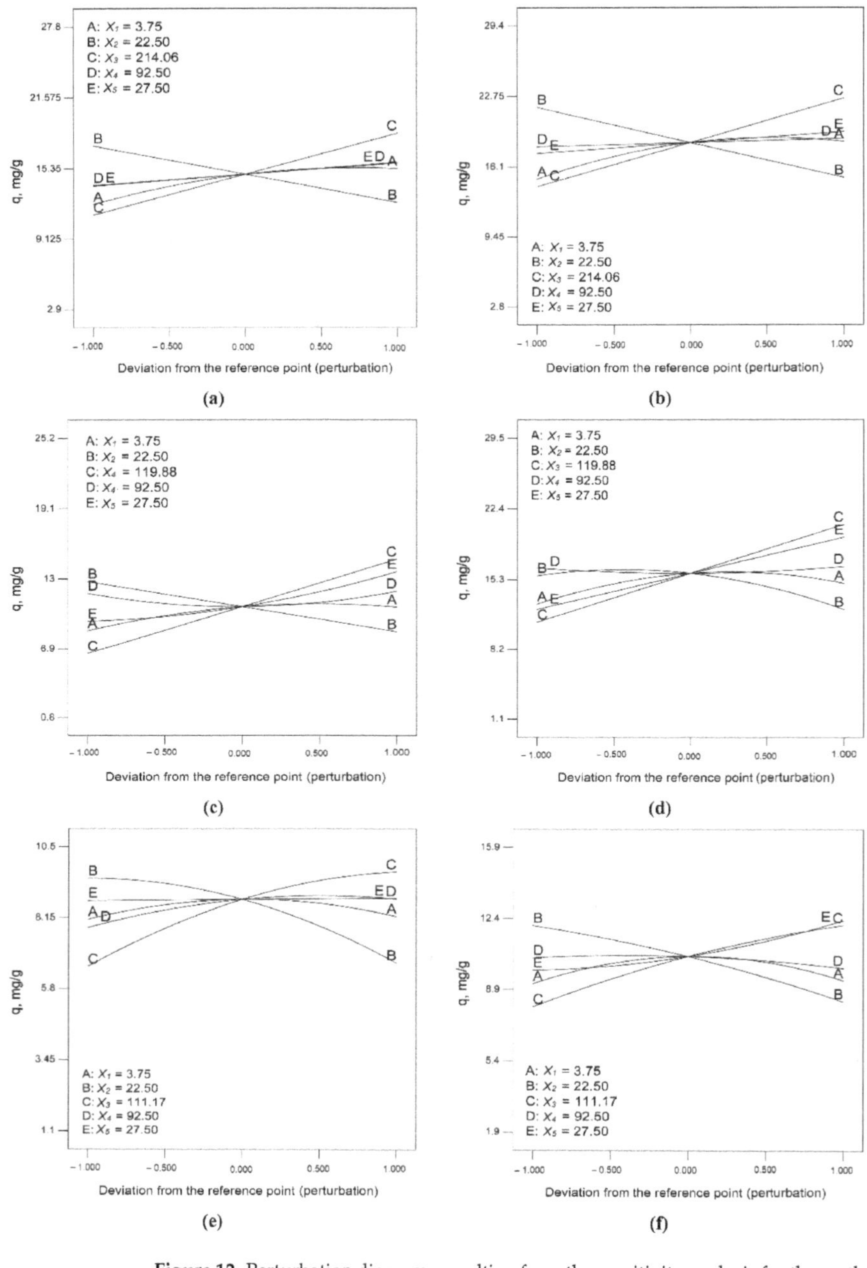

Figure 12. Perturbation diagrams resulting from the sensitivity analysis for the mathematical models developed for biosorption capacity of soybean biomass and soybean waste biomass (the meaning of the notations (**a–f**) is the same as those from Figure 9).

In the case of Zn(II) biosorption on soybean biomass, the biosorption capacity q is strongly sensitive to variations in sorbent dose and ion concentration in the solution, relatively sensitive to changes in pH and contact time, and a little sensitive to temperature variation. In the case of Zn(II) biosorption on soybean waste biomass, the biosorption capacity is significantly influenced by the sorbent dose and the ion concentration in the solution followed by pH and contact time, as in the case of soybean biomass. Temperature has a significantly higher effect on q than in the case of soybean biomass.

4.3.4. Analysis of Response Surfaces and Optimization of the Biosorption Process

Both the contour curves (not shown) and the response surfaces presented in Figures 13 and 14 demonstrate that there is a close correlation between process parameters (pH, DS, c_0, t_c, and T) and system responses (biosorption efficiency of Pb(II), Cd(II), and Zn(II) (R, %) and the biosorption capacity of the biosorbents (q, mg/g)). In order to analyze the best applicable variants of the biosorption process in a sustainable way, it is necessary to optimize the process by finding the most favorable combination of process parameters that maximizes the system response, followed by the experimental validation of the solution or group of solutions.

In this context, the following optimization alternatives can be considered: maximizing the response; minimizing the response; achieving a certain imposed target in the experimental field; establishing sets of values of independent variables to obtain a certain value of the system response (which may minimize costs and environmental impacts) or a certain amount of desirability. To analyze the objective or performance functions in relation to the optimal operating variants of the biosorption system, we considered the variation of the process parameters (independent variables) and the influence thereof on the previously established experimental field, provided that the biosorption efficiency (R, %) and biosorption capacity (q, mg/g) are maximized. The optimization procedure was applied to obtain the maximum for each of the purpose functions associated with Equations (1–12, a, b) (Table S1).

From the graphical representation in three-dimensional coordinates (response surfaces, Figures 13 and 14), it can be seen that the optimal values of the parameters to be optimized are either in the experimental field or on its borders, or they may differ from the global optimum of the target function [82,83]. In the context of this paper, optimization means establishing the best decision (solution), called the optimal decision (optimal solution), by maximizing the objective function (desirability). To determine the optimal (most favorable) solution for system operation from the individual point of view of the two dependent variables (R, q), the minimum and maximum levels of each independent variable were established and a weight was associated to adjust the shape of the objective (desirability) function. This function is zero outside the system limits and one when the optimization goal is reached.

For each answer $y_i(x)$, the desirability function $d_i(y_i)$ assigns numbers between 0 and 1 to the possible values of y_i: when $d_i(y_i) = 0$, this represents a completely undesirable value of y_i; when $d_i(y_i) = 1$, this represents an ideal (desirable) response value. The individual desirable values are then combined using the geometric mean, which gives the general desirability D (Equation (6)):

$$D = (d_1(y_1)\, d_2(y_2)\, \ldots\, d_k(y_k))^{1/k} \quad (6)$$

where k is the number of possible answers.

Our purpose was to maximize these objective functions, beginning with a starting point to search for the maximum, going step by step through the experimental field on a suitable slope until the maximum value was reached. There may be several variants of favorable solutions, since the response surface does not have a constant curvature, but this varies depending on the starting point and the slope traveled step by step. This variant of identifying a set of favorable solutions was chosen to maximize the chances of finding the "best local maximum" [84,85]. Therefore, we applied a multi-response method to optimize the combinations of the five independent variables (pH, DS, c_0, t_c, and T). In each case, the optimization was started from 10 starting points chosen in the experimental field, resulting in 10 solutions for each experimental variant (three heavy metals and two biosorbents), presented in Tables 10 and 11.

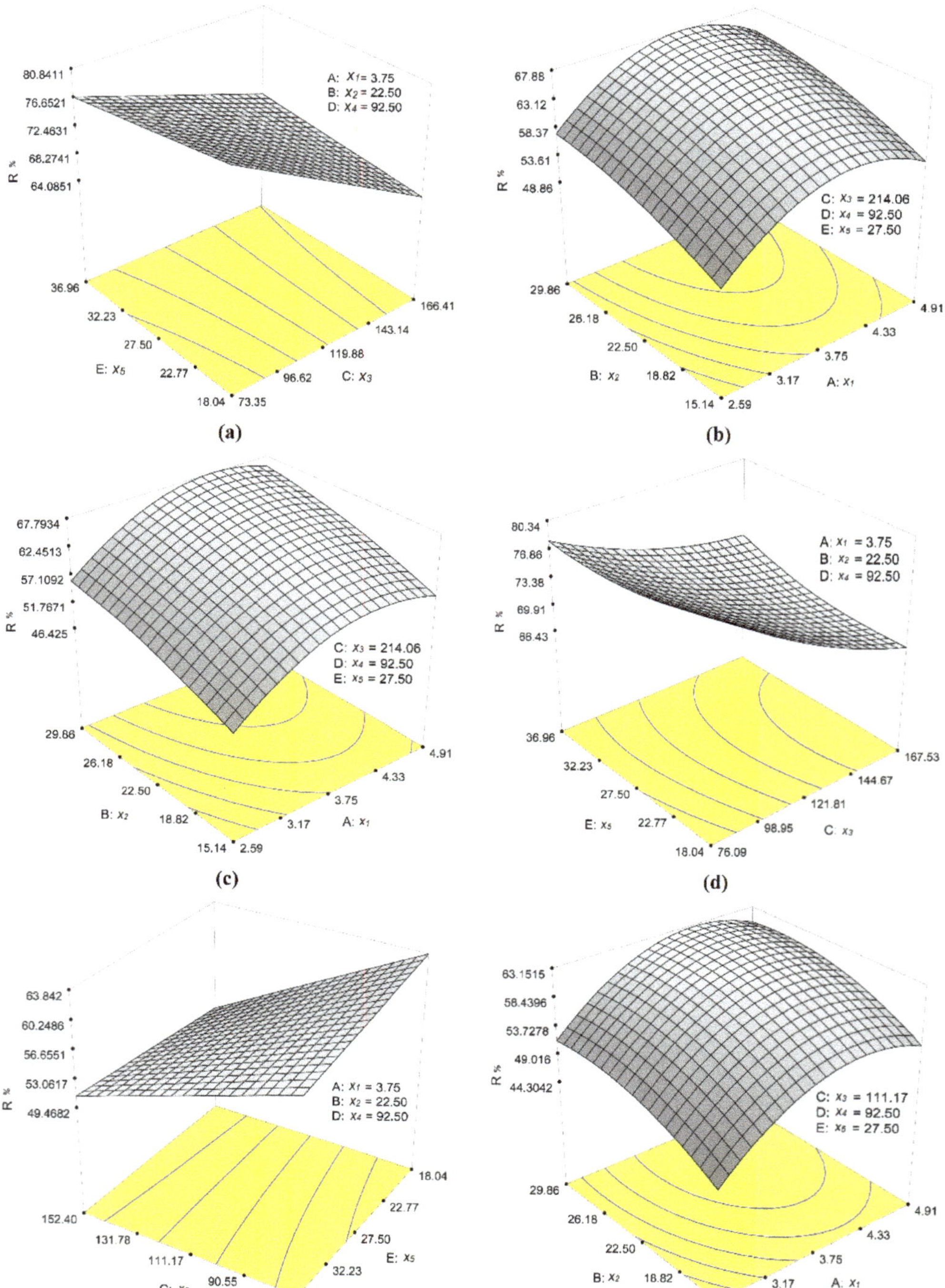

Figure 13. Response surfaces for mathematical models developed for biosorption efficiency of metal ions (R, %) (the meaning of the notations (**a–f**) is the same as those from Figure 9).

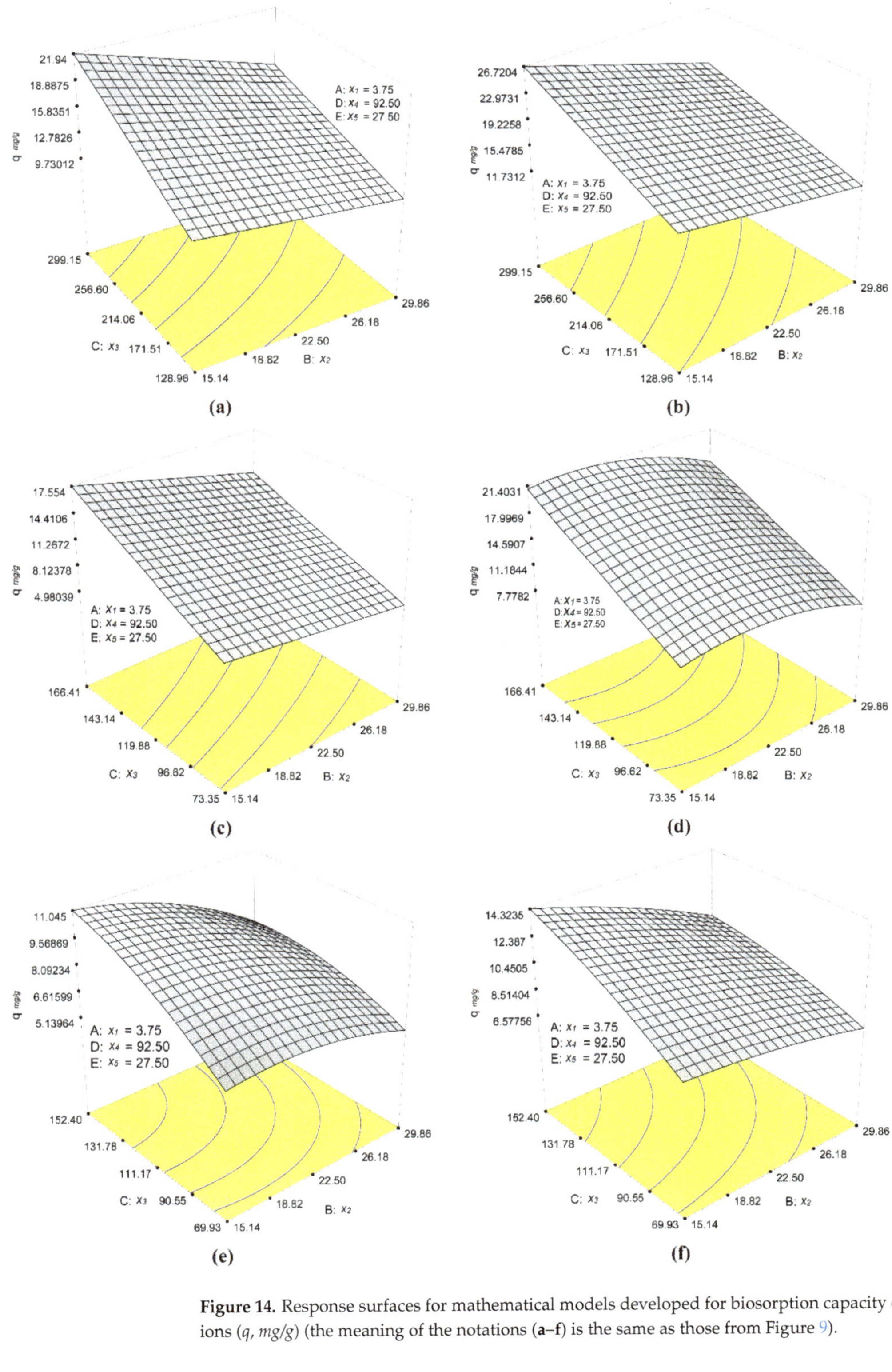

Figure 14. Response surfaces for mathematical models developed for biosorption capacity of metal ions (q, mg/g) (the meaning of the notations (**a–f**) is the same as those from Figure 9).

Table 10. Solutions for the operation in favorable conditions for reaching the maximum value of the biosorption efficiency (R, %) on soybean biomass and soybean waste biomass.

Soybean Biomass, Pb(II), *R*(%)							
Constraints							
Variabile	**Variation range**	**Lower limit**	**Upper limit**	**Minimum weight**	**Maximum weight**	**Importance**	
x_1	experimental	2.59	4.91	1	1	3	
x_2	experimental	15.14	29.86	1	1	3	
x_3	experimental	128.96	299.15	1	1	3	
x_4	experimental	55.71	129.29	1	1	3	
x_5	experimental	18.04	36.96	1	1	3	
R	maximum	2.59	4.91	1	1	3	
Solutions							
No.	x_1	x_2	x_3	x_4	x_5 *	R(%)	**Probability**
1	**4.02**	**29.86**	**129.03**	**129.29**	**32.79**	**78.13**	**0.829**
2	4.03	29.84	129.00	129.29	36.48	78.13	0.829
3	4.05	29.84	128.96	129.29	25.55	78.13	0.829
4	4.06	29.86	128.98	129.29	18.04	78.13	0.829
5	4.01	29.86	128.96	129.06	29.65	78.11	0.828
6	4.01	29.86	128.96	128.98	34.73	78.10	0.828
7	4.07	29.33	128.96	129.27	18.26	78.02	0.827
8	4.10	28.43	128.96	129.29	18.04	77.79	0.824
9	4.06	29.86	128.96	124.90	18.41	77.61	0.821
10	4.21	25.42	128.96	129.29	28.96	76.63	0.808
* has no effect on optimization results							
Soybean waste biomass, Pb(II), *R*(%)							
Constraints							
Variabile	**Variation range**	**Lower limit**	**Upper limit**	**Minimum weight**	**Maximum weight**	**Importance**	
x_1	experimental	2.59	4.91	1	1	3	
x_2	experimental	15.14	29.86	1	1	3	
x_3	experimental	128.97	299.15	1	1	3	
x_4	experimental	55.71	129.29	1	1	3	
x_5	experimental	18.04	36.96	1	1	3	
R	maximum	22.8	83.9	1	1	3	
Solutions							
No.	x_1	x_2	x_3	x_4	x_5 *	R(%)	**Probability**
1	**4.23**	**29.86**	**128.96**	**129.29**	**34.37**	**77.40**	**0.894**
2	4.24	29.86	128.96	129.27	32.98	77.40	0.894
3	4.23	29.86	129.02	129.29	27.99	77.39	0.894
4	4.23	29.82	128.96	129.29	32.86	77.39	0.894
5	4.44	29.32	128.96	129.29	21.86	77.03	0.888
6	4.17	29.84	128.96	124.02	36.96	76.77	0.883
7	3.94	28.58	128.96	128.46	36.96	76.49	0.879
8	4.51	29.86	128.96	124.07	18.04	76.45	0.878
9	4.56	28.68	128.96	124.52	18.04	75.96	0.87
10	4.44	28.59	128.97	122.14	36.74	75.93	0.87
* has no effect on optimization results							

Table 10. *Cont.*

Soybean biomass, Cd(II), $R(\%)$							
Constraints							
Variabile	**Variation range**	**Lower limit**	**Upper limit**	**Minimum weight**	**Maximum weight**	**Importance**	
x_1	experimental	2.59	4.91	1	1	3	
x_2	experimental	15.14	29.86	1	1	3	
x_3	experimental	73.35	166.41	1	1	3	
x_4	experimental	55.71	129.29	1	1	3	
x_5	experimental	18.04	36.96	1	1	3	
R	maximum	25.6	92.8	1	1	3	
Solutions							
No.	x_1	x_2	x_3	x_4	x_5	$R(\%)$	**Probability**
1	**4.06**	**29.86**	**73.35**	**129.28**	**18.25**	**89.62**	**0.953**
2	4.02	29.86	73.35	128.89	18.11	89.60	0.952
3	3.83	29.25	73.35	129.29	18.04	89.04	0.944
4	4.22	28.08	73.35	129.29	18.04	88.47	0.936
5	4.62	28.81	73.35	129.29	18.13	87.4	0.919
6	4.25	29.85	73.39	129.29	30.11	86.85	0.911
7	3.20	29.86	73.35	129.29	18.15	85.99	0.899
8	4.27	24.14	73.35	129.29	18.32	85.98	0.898
9	4.76	29.86	73.35	115.78	18.12	85.68	0.894
10	3.84	29.80	73.35	66.33	18.04	82.57	0.848
Soybean biomass waste, Cd(II), $R(\%)$							
Constraints							
Variabile	**Variation range**	**Lower limit**	**Upper limit**	**Minimum weight**	**Maximum weight**	**Importance**	
x_2	experimental	15.14	29.86	1	1	3	
x_3	experimental	76.095	167.53	1	1	3	
x_4	experimental	55.71	129.29	1	1	3	
x_5	experimental	18.04	36.96	1	1	3	
R	maximum	32.6	87.2	1	1	3	
Solutions							
No.	x_1	x_2	x_3	x_4	x_5	$R(\%)$	**Probability**
1	**3.93**	**29.86**	**76.09**	**128.04**	**18.06**	**85.03**	**0.960**
2	4.03	28.73	76.09	126.19	18.04	84.89	0.958
-3	3.94	28.50	78.89	125.74	18.04	84.29	0.947
4	4.24	26.63	78.30	121.66	18.04	83.45	0.931
5	4.03	27.94	76.09	129.29	36.96	82.64	0.916
6	3.96	28.94	77.41	129.29	36.96	82.53	0.914
7	4.14	29.27	76.09	129.29	36.95	82.53	0.914
8	3.08	28.14	76.09	129.28	18.24	81.73	0.899
9	3.76	27.06	76.09	62.49	18.04	79.55	0.859
10	3.87	23.43	76.09	74.48	18.04	79.51	0.859

Table 10. *Cont.*

Soybean biomass, Zn(II), R(%)							
Constraints							
Variabile	**Variation range**	**Lower limit**	**Upper limit**	**Minimum weight**	**Maximum weight**	**Importance**	
x_1	experimental	2.59	4.91	1	1	3	
x_2	experimental	15.14	29.86	1	1	3	
x_3	experimental	69.93	152.41	1	1	3	
x_5	experimental	18.04	36.96	1	1	3	
R	maximum	23.7	79.4	1	1	3	
Solutions							
No.	x_1	x_2	x_3	x_4	x_5	R(%)	**Probability**
1	**4.18**	**29.86**	**70.60**	**129.22**	**18.04**	**69.39**	**0.82**
2	3.71	29.75	69.93	129.29	18.04	69.3	0.819
3	3.83	29.86	69.98	122.99	18.04	68.98	0.813
4	3.92	29.79	69.93	121.40	18.04	68.93	0.812
5	3.92	29.86	69.93	129.29	21.43	68.9	0.812
6	3.52	29.51	69.93	129.29	18.04	68.71	0.808
7	4.26	26.23	69.93	129.28	18.04	68.5	0.804
8	3.81	29.86	70.63	104.65	18.04	67.27	0.782
9	4.02	29.86	69.93	129.29	35.61	65.97	0.759
10	3.97	29.86	69.93	102.36	36.20	63.54	0.715
Soybean waste biomass, Zn(II), R(%)							
Constraints							
Variabile	**Variation range**	**Lower limit**	**Upper limit**	**Minimum weight**	**Maximum weight**	**Importance**	
x_1	experimental	2.59	4.91	1	1	3	
x_2	experimental	15.14	29.86	1	1	3	
x_3	experimental	69.93	152.4	1	1	3	
x_4	experimental	55.71	129.29	1	1	3	
x_5	experimental	18.04	36.96	1	1	3	
R	maximum	21.9	82.8	1	1	3	
Solutions							
No.	x_1	x_2	x_3	x_4	x_5	R(%)	**Probability**
1	**4.18**	**27.30**	**69.93**	**118.59**	**33.02**	**69.97**	**0.789**
2	4.19	27.31	69.93	118.51	34.70	69.97	0.789
3	4.18	27.28	69.93	118.68	36.72	69.97	0.789
4	4.19	27.25	69.93	118.28	22.52	69.97	0.789
5	4.19	27.28	69.93	118.97	28.27	69.97	0.789
6	4.19	27.24	69.93	118.93	21.21	69.97	0.789
7	4.18	27.24	69.93	119.12	21.51	69.97	0.789
8	4.19	27.31	69.93	117.73	22.54	69.96	0.789
9	4.17	27.51	69.93	118.70	21.78	69.96	0.789
10	4.00	29.85	72.76	101.50	18.04	68.65	0.768

Table 11. Solutions for the operation in favorable conditions for reaching the maximum value of the biosorption capacity (q, mg/g) on soybean biomass and soybean waste biomass.

Soybean Biomass, Pb(II), $q(mg/g)$

Constraints

Variabile	Variation range	Lower limit	Upper limit	Minimum weight	Maximum weight	Importance
x_1	experimental	2.59	4.91	1	1	3
x_2	experimental	15.14	29.86	1	1	3
x_3	experimental	128.96	299.15	1	1	3
x_4	experimental	55.71	129.29	1	1	3
x_5	experimental	18.04	36.96	1	1	3
q	maximum	2.9	27.8	1	1	3

Solutions

No.	x_1	x_2	x_3	x_4	x_5	$q(mg/g)$	Probability
1	**4.50**	**15.14**	**297.55**	**129.23**	**36.96**	**25.04**	**0.889**
2	4.90	15.33	299.15	129.29	36.96	24.96	0.886
3	4.85	15.14	293.04	129.29	36.96	24.76	0.878
4	4.16	15.14	299.15	129.29	30.92	24.31	0.86
5	4.91	15.14	299.15	111.25	35.56	24.13	0.853
6	4.91	15.14	299.12	115.73	27.83	23.47	0.826
7	4.79	15.14	299.15	129.26	19.03	23.13	0.812
8	4.78	15.14	299.15	119.95	18.20	22.64	0.793
9	3.00	16.62	299.14	129.29	36.68	22.22	0.776
10	4.61	22.13	299.15	86.15	18.06	17.99	0.606

Soybean waste biomass, Pb(II), $q(mg/g)$

Constraints

Variabile	Variation range	Lower limit	Upper limit	Minimum weight	Maximum weight	Importance
x_1	experimental	2.59	4.91	1	1	3
x_2	experimental	15.14	29.86	1	1	3
x_3	experimental	128.96	299.15	1	1	3
x_4	experimental	55.71	129.29	1	1	3
x_5	experimental	18.04	36.96	1	1	3
q	maximum	2.8	29.4	1	1	3

Solutions

No.	x_1	x_2	x_3	x_4	x_5	$q(mg/g)$	Probability
1	**4.29**	**15.16**	**299.15**	**73.55**	**36.96**	**28.36**	**0.961**
2	3.91	15.14	299.15	58.68	36.96	28.27	0.958
3	4.26	15.16	298.54	86.18	36.96	28.2	0.955
4	3.81	15.14	299.15	57.03	36.96	28.16	0.954
5	4.55	15.14	299.15	129.29	27.23	27.5	0.929
6	3.83	17.37	299.15	56.66	36.94	26.94	0.907
7	4.67	15.14	287.63	129.29	32.83	26.91	0.906
8	4.91	18.64	299.15	89.55	36.96	25.86	0.867
9	4.91	19.75	299.15	61.52	36.96	25.5	0.854
10	3.38	23.18	299.15	55.71	36.96	22.82	0.753

Table 11. *Cont.*

Soybean biomass, Cd(II), $q(mg/g)$							
Constraints							
Variabile	**Variation range**	**Lower limit**	**Upper limit**	**Minimum weight**	**Maximum weight**	**Importance**	
x_1	experimental	2.59	4.91	1	1	3	
x_2	experimental	15.14	29.86	1	1	3	
x_3	experimental	73.35	166.41	1	1	3	
x_4	experimental	55.71	129.29	1	1	3	
x_5	experimental	18.02	36.96	1	1	3	
q	maximum	0.8	25.2	1	1	3	
Solutions							
No.	x_1	x_2	x_3	x_4	x_5	$q(mg/g)$	**Probability**
1	**4.44**	**15.14**	**166.41**	**129.29**	**36.96**	**24.08**	**0.954**
2	4.04	15.14	166.41	129.29	36.95	23.86	0.945
3	4.83	15.90	166.40	129.29	36.96	23.75	0.94
4	4.02	16.55	166.41	129.29	36.96	23.32	0.923
5	4.22	15.14	166.41	87.18	36.96	22.67	0.896
6	4.73	19.73	166.41	55.71	36.96	22.12	0.874
7	4.91	20.87	166.39	55.79	36.96	21.59	0.852
8	2.94	15.89	166.41	55.71	36.96	21.49	0.848
9	4.82	26.92	166.41	55.71	36.96	19.29	0.758
10	4.85	27.03	166.41	129.29	36.74	19.29	0.758
Soybean waste biomass, Cd(II), $q(mg/g)$							
Constraints							
Variabile	**Variation range**	**Lower limit**	**Upper limit**	**Minimum weight**	**Maximum weight**	**Importance**	
x_1	experimental	2.59	4.91	1	1	3	
x_2	experimental	15.14	29.86	1	1	3	
x_3	experimental	73.35	166.41	1	1	3	
x_4	experimental	55.71	129.29	1	1	3	
x_5	experimental	18.04	36.96	1	1	3	
q	maximum	1.1	29.5	1	1	3	
Solutions							
No.	x_1	x_2	x_3	x_4	x_5	$q(mg/g)$	**Probability**
1	**4.19**	**19.18**	**166.03**	**55.71**	**36.94**	**28.18**	**0.959**
2	3.83	19.74	166.41	129.26	36.96	28.17	0.953
3	3.80	16.76	166.41	55.71	36.96	28.09	0.95
4	3.92	17.97	166.41	121.58	36.96	28.03	0.948
5	3.65	16.45	166.21	55.71	36.96	27.86	0.942
6	4.08	18.51	166.41	104.52	36.96	27.76	0.939
7	3.46	16.98	166.29	129.29	36.96	27.65	0.935
8	3.03	15.30	166.41	55.71	36.96	26.33	0.889
9	2.91	21.76	166.38	129.29	36.96	25.93	0.874
10	2.70	15.52	166.41	55.71	36.96	25.17	0.847

Table 11. *Cont.*

Soybean biomass, Zn(II), $q(mg/g)$							
Constraints							
Variabile	**Variation range**	**Lower limit**	**Upper limit**	**Minimum weight**	**Maximum weight**	**Importance**	
x_1	experimental	2.59	4.91	1	1	3	
x_2	experimental	15.14	29.86	1	1	3	
x_3	experimental	69.92	152.41	1	1	3	
x_4	experimental	55.71	129.29	1	1	3	
x_5	experimental	18.04	36.96	1	1	3	
q	maximum	1.1	10.5	1	1	3	
Solutions							
No.	x_1	x_2	x_3	x_4	x_5	$q(mg/g)$	**Probability**
1	**3.87**	**16.76**	**150.20**	**115.39**	**23.68**	**10.97**	**1**
2	3.14	15.75	147.19	124.67	34.68	10.72	1
3	4.41	16.43	150.60	102.61	26.94	10.8	1
4	4.17	15.44	141.70	90.41	22.14	10.62	1
5	3.94	15.79	146.89	91.32	34.57	10.81	1
6	4.60	16.16	151.60	117.36	21.50	10.8	1
7	3.53	19.65	147.86	121.46	18.98	10.5	1
8	4.20	17.77	152.09	117.92	18.32	10.87	1
9	3.19	15.29	140.39	94.78	36.42	10.62	1
10	3.71	16.71	148.00	123.95	19.30	10.9	1
Soybean waste biomass, Zn(II), $q(mg/g)$							
Constraints							
Variabile	**Variation range**	**Lower limit**	**Upper limit**	**Minimum weight**	**Maximum weight**	**Importance**	
x_1	experimental	2.59	4.91	1	1	3	
x_2	experimental	15.14	29.86	1	1	3	
x_3	experimental	69.93	152.41	1	1	3	
x_4	experimental	55.71	129.29	1	1	3	
x_5	experimental	18.04	36.96	1	1	3	
q	maximum	1.9	15.9	1	1	3	
Solutions							
No.	x_1	x_2	x_3	x_4	x_5	$q(mg/g)$	**Probability**
1	**3.83**	**15.46**	**143.90**	**98.83**	**36.78**	**15.98**	**1**
2	3.96	15.16	152.07	93.99	34.73	15.93	1
3	4.41	15.38	146.40	62.19	36.24	15.98	1
4	3.89	15.68	147.73	61.19	35.16	15.96	1
5	4.42	15.21	147.40	64.41	36.71	16.21	1
6	3.88	15.80	139.80	70.79	36.80	16.04	1
7	3.63	16.06	145.80	84.59	36.82	16.11	1
8	3.56	15.23	147.60	59.05	35.35	16.11	1
9	3.67	15.14	152.40	55.73	31.56	15.25	0.954
10	3.49	15.14	152.40	81.88	24.65	13.77	0.848

The best values for biosorption efficiency (*R*, %) are obtained for Cd(II), both for the biosorption on soybean biomass and waste biomass (85–89%), while the initial concentration of the metal ions is 73–76 mg/L. The Pb(II) ion is absorbed with a maximum biosorption efficiency of 77–78%, both on soybean biomass and soybean waste biomass, starting from an initial concentration of Pb(II) in a solution of 128–129 mg/L. Zn(II) ions are absorbed with the lowest efficiency (*R*~69%), the initial concentration of the metal ions in the solution being around 70 mg/L. Therefore, it is found that soybean waste biomass can ensure a maximum biosorption efficiency equivalent to that demonstrated in the case of soybean biomass. In all cases the working pH values are in the range of 3–4.5 and the contact time is 120–130 min, while temperature influences to a relatively small extent the optimum for R(%).

Regarding the best values for the biosorption capacity of metal ions (q, mg/g), it is found that the maximum value of q for soybean waste biomass is higher than that obtained for soybean biomass (with 12% for Pb(II); 16% for Cd(II); about 45% for Zn(II)). The initial concentration of the metal ions in the solution for which a maximum q is obtained is 72% of the maximum limit of the experimental range in the case of Pb(II) (c_0 = 416.45 mg Pb(II)/L); 72% for Cd(II) (c_0 = 230.54 mg Cd(II)/L); 70% for Zn(II) (c_0 = 209.25 mg Zn(II)/L). The dose of sorbent is about 15–20 g/L (lower than the maximum value of R of ~30 mg/L).

The optimization goal is reached with a probability higher than 0.89.

The optimal solutions that consider some economic constraints (for example: the minimum amount of sorbent and the maximum initial concentration of the metal ions in the solution, which ensures maximum biosorption efficiency (R) and maximum biosorption capacity (q)) while maintaining the other parameters in the experimental field are presented in Tables 12 and 13.

Table 12. Solutions for the operation in favorable conditions for reaching the maximum value of the biosorption efficiency (R, %) of soybean biomass and soybean waste biomass considering a series of constraints (minimum dose of sorbent and maximum concentration of metal ion in solution).

Soybean Biomass, Pb(II), R(%)							
Constraints							
Variabile	Variation range	Lower limit	Upper limit	Minimum weight	Maximum weight	Importance	
x_1	experimental	2.59	4.91	1	1	3	
x_2	**minimum**	**15.14**	**29.86**	**1**	**1**	**3**	
x_3	**maximum**	**128.96**	**299.15**	**1**	**1**	**3**	
x_4	experimental	55.71	129.29	1	1	3	
x_5	experimental	18.04	36.96	1	1	3	
R	**maximum**	**18.7**	**90.4**	**1**	**1**	**3**	
Solutions							
No.	x_1	x_2	x_3	x_4	x_5 *	R(%)	**Probability**
1	**4.03**	**15.14**	**299.15**	**129.29**	**19.77**	**57.78**	**0.817**
2	4.01	15.14	299.12	129.26	19.25	57.78	0.817
3	4.09	15.14	299.15	129.29	21.81	57.76	0.817
4	4.03	15.16	299.11	129.29	33.56	57.80	0.817
5	4.13	15.14	299.15	129.29	28.56	57.72	0.816
6	3.67	15.14	295.31	129.29	21.68	57.39	0.808
7	3.99	15.14	299.15	110.01	21.73	55.51	0.8
8	4.83	15.14	299.06	129.29	36.42	54.46	0.793
9	4.11	15.27	298.58	81.83	18.04	52.31	0.774
10	3.56	15.14	299.15	55.72	27.69	48.00	0.742

* has no effect on optimization results

Table 12. *Cont.*

Soybean waste biomass, Pb(II), $R(\%)$

Constraints

Variabile	Variation range	Lower limit	Upper limit	Minimum weight	Maximum weight	Importance
x_1	experimental	2.59	4.91	1	1	3
x_2	**minimum**	**15.14**	**29.86**	**1**	**1**	**3**
x_3	**maximum**	**128.96**	**299.15**	**1**	**1**	**3**
x_4	experimental	55.71	129.29	1	1	3
x_5	experimental	18.04	36.96	1	1	3
R	**maximum**	**22.8**	**83.9**	**1**	**1**	**3**

Solutions

No.	x_1	x_2	x_3	x_4	x_5 *	$R(\%)$	Probability
1	**4.18**	**15.14**	**299.15**	**129.29**	**29.25**	**57.21**	**0.826**
2	4.22	15.14	299.07	129.29	29.55	57.21	0.826
3	4.22	15.14	299.03	129.29	19.57	57.21	0.826
4	4.28	15.14	297.40	128.42	36.96	57.21	0.823
5	3.94	15.14	299.15	127.44	18.04	56.62	0.821
6	4.59	15.14	299.15	128.16	24.74	56.49	0.82
7	3.81	15.14	299.13	129.29	30.40	56.43	0.819
8	3.74	15.14	299.15	129.29	27.29	56.15	0.817
9	4.12	15.14	299.08	101.16	18.04	53.89	0.798
10	4.42	17.45	299.15	129.29	22.97	59.18	0.795

* *has no effect on optimization results*

Soybean biomass, Cd(II), R(%)

Constraints

Variabile	Variation range	Lower limit	Upper limit	Minimum weight	Maximum weight	Importance
x_1	experimental	2.59	4.91	1	1	3
x_2	**minimum**	**15.14**	**29.86**	**1**	**1**	**3**
x_3	**maximum**	**73.35**	**166.41**	**1**	**1**	**3**
x_4	experimental	55.71	129.29	1	1	3
x_5	experimental	18.04	36.96	1	1	3
R	**maximum**	**25.6**	**92.8**	**1**	**1**	**3**

Solutions

No.	x_1	x_2	x_3	x_4	x_5	$R(\%)$	Probability
1	**4.03**	**15.14**	**166.31**	**129.29**	**34.12**	**66.49**	**0.847**
2	4.42	15.14	166.41	129.29	33.83	65.78	0.842
3	4.03	15.14	166.41	129.05	27.52	65.5	0.841
4	3.95	15.14	166.40	129.29	20.91	64.53	0.834
5	4.17	15.14	166.41	127.01	21.74	64.39	0.833
6	4.09	15.14	166.24	129.29	18.04	64.19	0.831
7	4.10	16.04	166.41	129.24	18.04	64.69	0.817
8	4.33	15.14	166.41	99.57	23.57	61.36	0.81
9	3.86	15.14	164.99	106.79	18.04	61.8	0.81
10	2.59	15.15	165.27	79.44	21.71	48.82	0.699

Table 12. *Cont.*

Soybean waste biomass, Cd(II), $R(\%)$							
Constraints							
Variabile	**Variation range**	**Lower limit**	**Upper limit**	**Minimum weight**	**Maximum weight**	**Importance**	
x_1	experimental	2.59	4.91	1	1	3	
x_2	**minimum**	**15.14**	**29.86**	**1**	**1**	**3**	
x_3	**maximum**	**76.09**	**167.52**	**1**	**1**	**3**	
x_4	experimental	55.71	129.29	1	1	3	
x_5	experimental	18.04	36.96	1	1	3	
R	**maximum**	**32.6**	**87.2**	**1**	**1**	**3**	
Solutions							
No.	x_1	x_2	x_3	x_4	x_5	$R(\%)$	**Probability**
1	**4.00**	**15.18**	**167.52**	**129.29**	**36.96**	**67.08**	**0.857**
2	3.78	15.14	167.51	129.17	36.96	66.93	0.857
3	4.08	15.61	167.53	128.57	36.96	67.44	0.852
4	4.03	15.14	167.52	96.14	36.96	64.36	0.835
5	3.70	15.14	167.53	90.66	36.96	63.71	0.829
6	4.01	15.14	167.53	122.99	26.14	63.16	0.824
7	3.80	15.29	167.53	129.29	18.04	63.45	0.824
8	4.12	15.14	167.52	72.07	36.77	62.25	0.819
9	4.00	15.14	167.53	63.65	36.96	61.79	0.812
10	3.83	15.14	167.53	73.06	19.15	58.75	0.782
Soybean biomass, Zn(II), $R(\%)$							
Constraints							
Variabile	**Variation range**	**Lower limit**	**Upper limit**	**Minimum weight**	**Maximum weight**	**Importance**	
x_1	experimental	2.59	4.91	1	1	3	
x_2	**minimum**	**15.14**	**29.86**	**1**	**1**	**3**	
x_3	**maximum**	**69.93**	**152.4**	**1**	**1**	**3**	
x_4	experimental	55.71	129.29	1	1	3	
x_5	experimental	18.04	36.96	1	1	3	
R	**maximum**	**23.7**	**79.4**	**1**	**1**	**3**	
Solutions							
No.	x_1	x_2	x_3	x_4	x_5	$R(\%)$	**Probability**
1	**4.05**	**15.72**	**152.40**	**129.29**	**35.07**	**49.76**	**0.766**
2	4.05	15.14	152.40	129.29	26.53	48.69	0.765
3	4.00	15.16	152.40	129.27	25.06	48.59	0.764
4	4.01	15.60	152.40	129.26	29.28	49.27	0.763
5	4.38	15.14	152.40	117.63	36.96	47.96	0.756
6	3.72	15.18	152.40	98.05	18.04	45.11	0.726
7	3.81	15.14	152.40	75.24	35.81	44.65	0.722
8	4.33	15.17	152.40	68.19	36.70	43.87	0.712
9	3.80	15.14	152.39	62.25	36.96	43.62	0.71
10	4.00	15.28	152.40	71.49	18.17	43.22	0.703

Table 12. *Cont.*

Soybean biomass, Zn(II), $R(\%)$							
Constraints							
Variabile	Variation range	Lower limit	Upper limit	Minimum weight	Maximum weight	Importance	
x_1	experimental	2.59	4.91	1	1	3	
x_2	**minimum**	**15.14**	**29.86**	**1**	**1**	**3**	
x_3	**maximum**	**69.93**	**152.4**	**1**	**1**	**3**	
x_4	experimental	55.71	129.29	1	1	3	
x_5	experimental	18.04	36.96	1	1	3	
R	**maximum**	**21.9**	**82.8**	**1**	**1**	**3**	
Solutions							
No.	x_1	x_2	x_3	x_4	x_5 *	$R(\%)$	**Probability**
1	**4.19**	**15.14**	**152.40**	**118.38**	**18.67**	**51**	**0.782**
2	4.19	15.14	152.40	118.35	34.03	51	0.782
3	4.18	15.14	152.40	118.86	36.56	51	0.782
4	4.19	15.15	152.40	118.71	25.27	51	0.782
5	4.19	15.15	152.40	117.83	36.89	51	0.782
6	4.19	15.14	152.28	117.96	24.99	51.02	0.782
7	4.22	15.14	152.40	129.14	29.26	50.84	0.78
8	4.25	15.14	152.40	107.70	19.35	50.82	0.78
9	4.21	15.53	152.40	115.01	32.83	51.45	0.779
10	3.52	15.17	152.40	122.77	36.96	48.98	0.763
* has no effect on optimization results							

Table 13. Solutions for the operation in favorable conditions for reaching the maximum value of the biosorption capacity (q, mg/g) of soybean biomass and soybean waste biomass considering a series of constraints (minimum dose of sorbent and maximum concentration of the metal ions in solution).

Soybean Biomass, Pb(II), $q(mg/g)$							
Constraints							
Variabile	Variation range	Lower limit	Upper limit	Minimum weight	Maximum weight	Importance	
x_1	experimental	2.59	4.91	1	1	3	
x_2	**minimum**	**15.14**	**29.86**	**1**	**1**	**3**	
x_3	**maximum**	**128.96**	**299.15**	**1**	**1**	**3**	
x_4	experimental	55.71	129.29	1	1	3	
x_5	experimental	18.04	36.96	1	1	3	
q	**maximum**	**2.9**	**27.8**	**1**	**1**	**3**	
Solutions							
No.	x_1	x_2	x_3	x_4	x_5	$q(mg/g)$	**Probability**
1	**4.03**	**15.14**	**299.15**	**110.38**	**36.96**	**24.07**	**0.947**
2	4.90	15.14	299.13	106.96	36.56	24.06	0.947
3	4.33	15.14	299.10	95.85	36.96	23.65	0.941
4	3.32	15.14	299.15	118.21	36.96	23.35	0.937
5	3.39	15.14	299.15	128.56	30.42	23.21	0.934
6	4.91	15.14	299.15	95.92	33.00	23.19	0.934
7	4.71	15.19	299.15	129.29	18.09	23.02	0.93
8	3.75	15.15	299.10	76.92	36.96	22.31	0.92
9	3.19	15.14	299.15	121.70	18.04	21.12	0.901
10	4.67	15.14	299.10	75.91	18.54	20.81	0.896

Table 13. *Cont.*

Soybean waste biomass, Pb(II), $q(mg/g)$							
Constraints							
Variabile	**Variation range**	**Lower limit**	**Upper limit**	**Minimum weight**	**Maximum weight**	**Importance**	
x_1	experimental	2.59	4.91	1	1	3	
x_2	**minimum**	**15.14**	**29.86**	**1**	**1**	**3**	
x_3	**maximum**	**128.96**	**299.15**	**1**	**1**	**3**	
x_4	experimental	55.71	129.29	1	1	3	
x_5	experimental	18.04	36.96	1	1	3	
q	**maximum**	**2.8**	**29.4**	**1**	**1**	**3**	
Solutions							
No.	x_1	x_2	x_3	x_4	x_5	$q(mg/g)$	**Probability**
1	**4.38**	**15.14**	**299.15**	**67.04**	**36.96**	**28.44**	**0.988**
2	4.24	15.14	299.15	94.63	36.94	28.15	0.984
3	4.16	15.14	299.15	121.50	35.20	27.78	0.979
4	4.57	15.14	299.15	64.80	32.50	27.67	0.978
5	4.33	15.14	299.15	118.99	31.69	27.65	0.978
6	3.64	15.14	299.15	108.99	36.96	27.39	0.974
7	3.81	15.14	299.15	116.03	18.04	26.52	0.962
8	4.03	15.14	299.15	82.37	22.77	26.34	0.96
9	3.85	15.14	299.13	62.19	24.89	26.1	0.957
10	2.87	15.14	299.15	55.71	36.04	25.61	0.95
Soybean biomass, Cd(II), $q(mg/g)$							
Constraints							
Variabile	**Variation range**	**Lower limit**	**Upper limit**	**Minimum weight**	**Maximum weight**	**Importance**	
x_1	experimental	2.59	4.91	1	1	3	
x_2	**minimum**	**15.14**	**29.86**	**1**	**1**	**3**	
x_3	**maximum**	**73.35**	**166.41**	**1**	**1**	**3**	
x_4	experimental	55.71	129.29	1	1	3	
x_5	experimental	18.04	36.96	1	1	3	
q	**maximum**	**0.8**	**25.2**	**1**	**1**	**3**	
Solutions							
No.	x_1	x_2	x_3	x_4	x_5	$q(mg/g)$	**Probability**
1	**4.37**	**15.14**	**166.39**	**128.73**	**36.96**	**24.024**	**0.984**
2	4.91	15.14	166.41	128.82	35.56	23.17	0.971
3	3.54	15.14	166.41	128.11	36.96	23.16	0.972
4	4.49	15.14	166.41	78.10	36.96	22.9	0.968
5	3.83	15.14	166.41	55.71	26.16	18.29	0.895
6	3.84	15.14	166.41	129.29	21.53	16.97	0.872
7	4.19	15.14	166.24	55.71	18.99	16.44	0.862
8	2.59	17.87	165.80	56.24	36.96	19.63	0.855
9	3.36	15.14	166.13	129.22	18.04	15.35	0.841
10	2.67	21.87	166.41	55.71	18.14	10.88	0.608

Table 13. *Cont.*

Soybean waste biomass, Cd(II), $q(mg/g)$							
Constraints							
Variabile	**Variation range**	**Lower limit**	**Upper limit**	**Minimum weight**	**Maximum weight**	**Importance**	
x_1	experimental	2.59	4.91	1	1	3	
x_2	**minimum**	**15.14**	**29.86**	**1**	**1**	**3**	
x_3	**maximum**	**73.35**	**166.41**	**1**	**1**	**3**	
x_4	experimental	55.71	129.29	1	1	3	
x_5	experimental	18.04	36.96	1	1	3	
q	**maximum**	**1.1**	**29.5**	**1**	**1**	**3**	
Solutions							
No.	x_1	x_2	x_3	x_4	x_5	$q(mg/g)$	**Probability**
1	**4.29**	**15.14**	**166.41**	**129.29**	**36.95**	**27.78**	**0.979**
2	4.04	15.14	166.41	74.01	36.96	27.43	0.975
3	4.24	15.14	166.41	89.03	36.96	27.23	0.973
4	3.82	15.14	166.41	87.43	36.96	27.21	0.972
5	4.73	15.14	166.41	129.29	36.95	27.15	0.972
6	3.53	15.14	166.41	67.79	36.96	27.13	0.972
7	3.89	15.49	166.41	120.96	36.96	27.68	0.97
8	4.70	15.14	165.22	55.73	36.96	26.99	0.966
9	4.13	15.14	165.36	129.29	35.45	26.6	0.961
10	4.91	15.14	166.41	70.72	32.11	23.46	0.923
Soybean biomass, Zn(II), $q(mg/g)$							
Constraints							
Variabile	**Variation range**	**Lower limit**	**Upper limit**	**Minimum weight**	**Maximum weight**	**Importance**	
x_1	experimental	2.59	4.91	1	1	3	
x_2	**minimum**	**15.14**	**29.86**	**1**	**1**	**3**	
x_3	**maximum**	**69.93**	**152.4**	**1**	**1**	**3**	
x_4	experimental	55.71	129.29	1	1	3	
x_5	experimental	18.04	36.96	1	1	3	
q	**maximum**	**1.1**	**10.5**	**1**	**1**	**3**	
Solutions							
No.	x_1	x_2	x_3	x_4	x_5	$q(mg/g)$	**Probability**
1	**3.66**	**15.14**	**152.40**	**88.17**	**29.02**	**10.97**	**0.999**
2	3.04	15.14	152.40	96.51	25.09	10.84	0.999
3	3.78	15.14	152.40	86.51	32.88	10.93	0.999
4	4.35	15.14	152.40	74.83	33.35	10.55	0.999
5	4.13	15.14	152.40	69.14	30.73	10.5	0.999
6	3.43	15.14	152.40	71.43	31.49	10.55	0.999
7	3.90	15.14	152.40	67.72	31.00	10.5	0.999
8	4.73	15.14	152.40	125.03	18.82	10.76	0.999
9	4.00	15.14	152.40	114.03	20.21	11.18	0.999
10	2.72	15.21	152.38	55.71	18.87	9.63	0.966

Table 13. *Cont.*

Soybean waste biomass, Zn(II), $q(mg/g)$							
Constraints							
Variabile	**Variation range**	**Lower limit**	**Upper limit**	**Minimum weight**	**Maximum weight**	**Importance**	
x_1	experimental	2.59	4.91	1	1	3	
x_2	**minimum**	**15.14**	**29.85**	**1**	**1**	**3**	
x_3	**maximum**	**69.93**	**152.41**	**1**	**1**	**3**	
x_4	experimental	55.71	129.29	1	1	3	
x_5	experimental	18.04	36.96	1	1	3	
q	**maximum**	**1.9**	**15.9**	**1**	**1**	**3**	
Solutions							
No.	x_1	x_2	x_3	x_4	x_5	$q(mg/g)$	**Probability**
1	**3.73**	**15.14**	**152.40**	**87.21**	**36.40**	**16.55**	**0.999**
2	4.16	15.14	152.40	87.47	34.66	15.91	0.999
3	3.97	15.14	152.40	101.60	35.39	15.99	0.999
4	3.40	15.14	152.40	85.39	36.19	16.37	0.999
5	3.60	15.14	152.40	128.91	36.96	15.66	0.994
6	3.53	15.14	152.06	55.71	30.48	14.89	0.974
7	4.15	15.81	152.27	129.07	36.96	15.34	0.971
8	3.84	15.14	152.40	55.71	26.20	13.96	0.951
9	4.30	15.14	152.40	105.00	24.33	13.47	0.938
10	4.36	15.14	152.36	55.72	23.63	13.14	0.929

Comparing the data in Tables 10 and 12, the following aspects can be seen when constraints are imposed:

- The optimization of the Pb(II) biosorption process using soybean biomass as the biosorbent when considering two constraints, the minimum dose of biosorbent and the maximum Pb(II) concentration in the initial solution (the other parameters taking appropriate values in the experimental field), will result in a 49% decrease in the sorbent dose, a 120% increase in the initial concentration of the metal ions in the solution, and a 40% decrease in the operating temperature. However, the maximum separation efficiency will decrease by about 26%.
- The optimization of the Pb(II) biosorption process using soybean waste biomass as a biomass, considering the two constraints presented above, will result in a decrease in the required amount of sorbent and an increase in the initial concentration of the ion in the solution, as in the case of Pb(II) biosorption on soybean biomass. The maximum separation efficiency will also decrease by 26%.

Therefore, soybean biomass and soybean waste biomass can ensure the same biosorption efficiency of Pb(II) under the same operating conditions.

- The optimization of the Cd(II) biosorption process on soybean biomass when considering two constraints, the minimum dose of biosorbent and the maximum Cd(II) concentration in the initial solution (the other parameters taking appropriate values in the experimental field), results in a decrease in the required amount of sorbent by 49%, increase the initial ion concentration by almost 119%, and also the operating temperature by almost 89%. Under these conditions, the maximum biosorption efficiency decreases by almost 26%.
- A similar situation is obtained in the case of Cd(II) biosorption on soybean waste biomass, in which case, for the two constraints imposed on those in Table 10, the efficiency of Cd(II) biosorption decreases by almost 22%.
- The optimization of the Zn(II) biosorption process on soybean biomass, when imposing the constraints of minimum sorbent dose and maximum Zn(II) concentration in the initial solution (similar to the previous situations), results in reducing the sorbent dose

by almost 47%, an increase in the initial concentration of the metal ions in solution by about 116%, and also an increase in the operating temperature by almost 40%, as well as the reduction of the maximum value of the biosorption efficiency by almost 28%.

- The optimization of the Zn(II) biosorption process on soybean waste biomass under the abovementioned constraints results in a reduction of the biosorbent dose by almost 44%, an increase in the initial concentration of Zn(II) in the initial solution by 118%, and a decrease in the operating temperature by nearly 43%.

If the biosorption process is to be scaled-up, a cost–benefit analysis of the two optimization alternatives is required. Additionally, under these operating conditions with the constraints imposed, the values imposed by the regulations in force regarding the concentrations of heavy metal ions in the treated effluents can be reached, because the saved sorbent can be used in an additional biosorption cycle, ensuring an increased biosorption efficiency to comply with the regulations on the quality of liquid effluents in economic operating conditions.

Imposing these constraints in the process of optimizing the biosorption capacity q (minimization of the sorbent dose and maximization of the initial concentration of metal ions in solution) results in the same optimal values of parameters and the maximum biosorption capacity as in the case of optimization without the two constraints (see Table 13 vs. Table 11). The maximum values of R and q under the operating conditions presented in Tables 10–13 were verified experimentally (for the highest probability of achieving the purpose of the objective function—solution 1) (Tables 14 and 15).

The experimental data presented in Tables 14 and 15, which confirmed the solutions obtained from the optimization process, demonstrate a very important matter, i.e., that soy waste biomass can ensure a biosorption efficiency of the three metal ions similar to that obtained when using soybean biomass as a biosorbent, with great potential for scaling-up. All this information shows that both soybean biomass and soybean waste biomass can be used effectively as biosorbents to remove heavy metal ions from aqueous effluents under certain operating conditions, making it possible to exploit industrial soybean waste biomass, which cannot be used in animal feed due to the content of the extraction solvent.

Table 14. (**a**,**b**) Experimental values of $R(\%)$ and $q(mg/g)$ for validation of optimal solutions (Tables 10 and 11).

(a)

No.	x_1 (pH)		x_2 (DS)		x_3 (c_0)		x_4 (t_c)		x_5 (T)		R_{exp} (%)	R_{calc} (%)
	calc	*exp*	*calc*	*exp*	*calc*	*exp*	*calc*	*exp*	*calc*	*exp*		
Soybean biomass, Pb(II), $R(\%)$												
1	4.02	4.00	29.86	30.00	129.03	124.93	129.29	129.00	32.79	30.00	73.17	78.13
Soybean waste biomass, Pb(II) $R(\%)$												
2	4.23	4.20	29.86	30.00	128.96	124.93	129.29	129.00	34.37	30.00	78.12	77.40
Soybean biomass, Cd(II), $R(\%)$												
3	4.06	4.00	29.86	30.00	73.35	74.96	129.28	129.00	18.25	20.00	72.31	89.62
Soybean waste biomass, Cd(II), $R(\%)$												
4	3.93	4.00	29.86	30.00	76.89	74.96	128.04	128.00	18.06	20.00	72.67	85.03
Soybean biomass, Zn(II), $R(\%)$												
5	4.18	4.20	29.86	30.00	70.60	72.32	129.22	129.00	18.04	20.00	58.79	69.39
Soybean waste biomass, Zn(II), $R(\%)$												
6	4.18	4.20	27.3	30.00	69.93	72.32	118.59	119.00	33.02	30.00	59.03	69.97

Table 14. *Cont.*

No.	x_1 (pH) calc	x_1 (pH) exp	x_2 (DS) calc	x_2 (DS) exp	x_3 (c_0) calc	x_3 (c_0) exp	x_4 (t_c) calc	x_4 (t_c) exp	x_5 (T) calc	x_5 (T) exp	q_{exp} (mg/g)	q_{calc} (mg/g)
(b)												
Soybean biomass, Pb(II), q(mg/g)												
7	4.50	4.50	15.14	15.00	297.55	291.51	129.23	129.00	36.96	30.00	19.86	25.04
Soybean waste biomass, Pb(II), q(mg/g)												
8	4.29	4.30	15.16	15.00	299.15	291.51	73.55	74.00	36.96	30.00	20.12	28.36
Soybean biomass, Cd(II), q(mg/g)												
9	4.44	4.50	15.14	15.00	166.41	165.08	129.29	129.00	36.96	30.00	15.91	24.08
Soybean waste biomass, Cd(II), q(mg/g)												
10	4.19	4.20	19.18	19.00	166.03	165.08	55.71	56.00	36.94	30.00	15.93	28.18
Soybean biomass, Zn(II), q(mg/g)												
11	3.87	4.00	16.76	17.00	150.2	146.17	115.39	115.00	23.68	20.00	9.08	10.97
Soybean waste biomass, Zn(II), q(mg/g)												
12	3.83	4.00	15.46	15.50	143.90	146.17	98.83	99.00	36.78	30.00	10.25	15.98

Table 15. Experimental values of $R(\%)$ during the validation of the optimal solutions in Table 12.

No.	x_1 (pH) calc	x_1 (pH) exp	x_2 (DS) calc	x_2 (DS) exp	x_3 (c_0) calc	x_3 (c_0) exp	x_4 (t_c) calc	x_4 (t_c) exp	x_5 (T) calc	x_5 (T) exp	R_{exp} (%)	R_{calc} (%)
Soybean biomass, Pb(II), R(%)												
1	4.03	4.00	15.14	15.00	299.15	291.51	129.29	129.00	19.77	20.00	62.07	57.78
Soybean waste biomass, Pb(II), R(%)												
2	4.18	4.20	15.14	15.00	299.15	291.51	129.29	129.00	29.25	30.00	65.18	57.21
Soybean biomass, Cd(II), R(%)												
3	4.03	4.00	15.14	15.00	166.31	165.08	129.29	129.00	34.12	30.00	59.91	66.49
Soybean waste biomass, Cd(II), R(%)												
4	4.00	4.00	15.18	15.00	167.52	165.08	129.29	129.00	36.96	30.00	63.17	67.08
Soybean biomass, Zn(II), R(%)												
5	4.05	4.00	15.72	16.00	152.40	146.17	129.29	129.00	35.07	30.00	42.46	49.76
Soybean waste biomass, Zn(II), R(%)												
6	4.19	4.20	15.14	15.00	152.40	146.17	118.38	119.00	18.67	20.00	53.07	51.00

5. Conclusions

The sustainable use of natural materials and their waste for the removal of heavy metal ions from aqueous effluents via biosorption requires the knowledge of the most favorable combinations of independent variables that determine the performance of the biosorption process to maximize the biosorption efficiency and biosorption capacity of biosorbents used. To this end, in this paper, the empirical mathematical modeling of the biosorption process of Pb(II), Cd(II), and Zn(II) ions on soybean-based biosorbents was performed by applying the Response Surface Methodology, followed by the optimization of the biosorption process, as prerequisites for scale-up. Two categories of natural biosorbents were used: soybean biomass and soybean waste biomass resulting from oil the extraction process of soy seeds.

The purpose of the modeling was to find the most probable mathematical relationships between the dependent variables (biosorption efficiency, $R(\%)$, and biosorption capacity,

$q(mg/g)$, and the residual concentration of metal ion in the solution) and process parameters (pH; sorbent dose, DS (g/L); initial concentration of metal ion in the solution, c_0 (mg/L); contact time, t_c (min); temperature, T (°C)) in order to find sets of solutions to maximize biosorption efficiency and biosorption capacity by applying specific optimization techniques. The models were validated using methods specific to multiple regression analysis and the results of the optimization process were validated experimentally.

An ANOVA confirmed the adequacy of the developed models ((Prob > F) < 0.05). The standard deviations are small and the values of the coefficient of determination (R^2) are usually over 0.9. The high values of $R^2{}_{adj}$ show that the total variations of R and q can be described by the elaborated models. Additionally, the values $R^2{}_{pred}$ are in good agreement with $R^2{}_{adj}$, which demonstrates the correctness of the models in that they are well selected (and not as a consequence of the large number of variables). The sensitivity analysis of the models shows that the change in pH and the initial concentration of the metal in the solution significantly influences the biosorption process.

All this information showed that both soybean biomass and soybean waste biomass can be used effectively as biosorbents to remove heavy metal ions from aqueous effluents under certain operating conditions. Finally, a comparison was performed between the performances of the two biosorbents in terms of biosorption efficiency and biosorption capacity, which demonstrated, based on confirmed experimental data and through modeling and optimization, that soybean biomass waste performs similarly to soybean biomass.

Response Surface Methodology (RSM) was used as a mathematical and statistical tool to evaluate the effects of some factors on the performance of the process, namely, biosorption capacity and biosorption efficiency, and has proven to be a powerful tool for measuring optimal regional responses by analyzing a sequence of designed experiments. RSM was particularly useful because it reduced the number of tests required to find the optimal conditions, which were then validated experimentally.

The experimental program fulfilled in this paper demonstrates that biomass waste from the biofuel production industry can be used as a very efficient, low-cost biosorbent with which to remove heavy metals from water effluents in a sustainable way. The waste biomass can be exploited in environmental cleaning processes in accordance with the principles of the circular economy. These results require further research by conducting a cost–benefit analysis of optimized alternatives in order to scale-up the biosorption process to a larger scale.

Supplementary Materials: The following supporting information can be downloaded at: https://www.mdpi.com/article/10.3390/pr10030523/s1, Table S1. Equations of mathematical model for metal ion biosorption efficiency $R(\%)$ (a) and biosorption capacity $q(mg/g)$ in coded (a) and real (b) coordinates. Table S2. Results of tests for model adequacy. Table S3. ANOVA for the polynomial (quadratic) model for $R(\%)$ in case of Pb(II) retention on soybean biomass and soybean waste biomass, respectively. Table S4. ANOVA for the polynomial (quadratic) model for $q(mg/g)$ in case of Pb(II) retention on soybean biomass and soybean waste biomass, respectively.

Author Contributions: Conceptualization, L.B. and M.G.; methodology, L.B. and D.I.F.; formal analysis, L.B.; investigation, D.I.F. and L.B.; resources, L.B. and D.I.F.; data curation, L.B. and D.I.F.; writing—original draft preparation, L.B. and M.G.; writing—review and editing, L.B. and M.G.; visualization, D.I.F. and L.B.; supervision, L.B. and M.G.; project administration M.G.; funding acquisition, M.G. All authors have read and agreed to the published version of the manuscript.

Funding: This work was supported by a grant of the Romanian Ministry of Education and Research, CCCDI—UEFISCDI, project number PN-III-P2-2.1-PED-2019-5239, Contract no. 269PED/2020, within PNCDI III.

Institutional Review Board Statement: Not applicable.

Informed Consent Statement: Not applicable.

Data Availability Statement: Not applicable.

Conflicts of Interest: The authors declare no conflict of interest.

References

1. Minelgaitė, A.; Liobikienė, G. Waste problem in European Union and its influence on waste management behaviours. *Sci. Total Environ.* **2019**, *667*, 86–93. [CrossRef] [PubMed]
2. Cappellaro, F.; Fantin, V.; Barberio, G.; Cutaia, L. Circular economy good practices supporting waste prevention: The case of Emilia-Romagna region. *Environ. Eng. Manag. J.* **2020**, *19*, 1701–1710. [CrossRef]
3. Ewijk, S.; Stegemann, J.A. Recognising waste use potential to achieve a circular economy. *Waste Manag.* **2020**, *105*, 1–7. [CrossRef]
4. Ma, B.; Li, X.; Jiang, Z.; Jiang, J. Recycle more, waste more? When recycling efforts increase resource consumption. *J. Clean. Prod.* **2018**, *206*, 870–877. [CrossRef]
5. Gavrilescu, M.; Fortuna, M.E.; Simion, I.M.; Ghinea, C.; Petraru, M.; Cozma, P.; Apostol, L.C.; Hlihor, R.M.; Fertu, D.T. Analysis and management of specific processes from environmental engineering and protection based on sustainability indicators. *Environ. Eng. Manag. J.* **2012**, *11*, 333–350. [CrossRef]
6. Bian, Y.; Song, K.; Bai, J. Market segmentation, resource misallocation and environmental pollution. *J. Clean. Prod.* **2019**, *228*, 376–387. [CrossRef]
7. Premarathna, K.; Rajapaksha, A.U.; Sarkar, B.; Kwon, E.E.; Bhatnagar, A.; Ok, Y.S.; Vithanage, M. Biochar-based engineered composites for sorptive decontamination of water: A review. *Chem. Eng. J.* **2019**, *372*, 536–550. [CrossRef]
8. Dai, Y.; Sun, Q.; Wang, W.; Lu, L.; Liu, M.; Li, J.; Yang, S.; Sun, Y.; Zhang, K.; Xu, J.; et al. Utilizations of agricultural waste as adsorbent for the removal of contaminants: A review. *Chemosphere* **2018**, *211*, 235–253. [CrossRef]
9. Filote, C.; Roșca, M.; Hlihor, R.M.; Cozma, P.; Simion, I.M.; Apostol, M.; Gavrilescu, M. Sustainable Application of Biosorption and Bioaccumulation of Persistent Pollutants in Wastewater Treatment: Current Practice. *Processes* **2021**, *9*, 1696. [CrossRef]
10. Vareda, J.P.; Valente, A.J.M.; Durães, L. Assessment of heavy metal pollution from anthropogenic activities and remediation strategies: A review. *J. Environ. Manag.* **2019**, *246*, 101–118. [CrossRef]
11. Zhang, L.; Zhu, G.; Ge, X.; Xu, G.; Guan, Y. Novel insights into heavy metal pollution of farmland based on reactive heavy metals (RHMs): Pollution characteristics, predictive models, and quantitative source apportionment. *J. Hazard. Mater.* **2018**, *360*, 32–42. [CrossRef] [PubMed]
12. Jacob, J.M.; Karthik, C.; Saratale, R.G.; Kumar, S.S.; Prabakar, D.; Kadirvelu, K.; Pugazhendhi, A. Biological approaches to tackle heavy metal pollution: A survey of literature. *J. Environ. Manag.* **2018**, *217*, 56–70. [CrossRef] [PubMed]
13. Zhang, L.; Zhao, B.; Xu, G.; Guan, Y. Characterizing fluvial heavy metal pollutions under different rainfall conditions: Implication for aquatic environment protection. *Sci. Total Environ.* **2018**, *635*, 1495–1506. [CrossRef] [PubMed]
14. EL Hammoudani, Y.; Dimane, F.; El Ouarghi, H. Removal efficiency of heavy metals by a biological wastewater treatment plant and their potential risks to human health. *Environ. Eng. Manag. J.* **2021**, *20*, 995–1002. [CrossRef]
15. Gavrilescu, M. Microbial recovery of critical metals from secondary sources. *Bioresour. Technol.* **2021**, *344*, 126208. [CrossRef]
16. Nakagawa, K.; Imura, T.; Berndtsson, R. Distribution of heavy metals and related health risks through soil ingestion in rural areas of western Japan. *Chemosphere* **2021**, *290*, 133316. [CrossRef] [PubMed]
17. Hu, G.; Bakhtavar, E.; Hewage, K.; Mohseni, M.; Sadiq, R. Heavy metals risk assessment in drinking water: An integrated probabilistic-fuzzy approach. *J. Environ. Manag.* **2019**, *250*, 109514. [CrossRef]
18. Ahluwalia, S.S.; Goyal, D. Microbial and plant derived biomass for removal of heavy metal from wastewater. *Bioresour. Technol.* **2007**, *98*, 2243–2257. [CrossRef]
19. Povar, I.; Zinicovscaia, I.; Ubaldini, S.; Spinu, O.; Pintilie, B.; Lupascu, T.; Duca, G. Thermodynamic analysis of heavy metals precipitation for their recovery from industrial wastewaters. *Environ. Eng. Manag. J.* **2020**, *19*, 281–288. [CrossRef]
20. Rebello, S.; Sivaprasad, M.; Anoopkumar, A.; Jayakrishnan, L.; Aneesh, E.M.; Narisetty, V.; Sindhu, R.; Binod, P.; Pugazhendhi, A.; Pandey, A. Cleaner technologies to combat heavy metal toxicity. *J. Environ. Manag.* **2021**, *296*, 113231. [CrossRef]
21. Thasneema, K.; Dipin, T.; Thayyil, M.S.; Sahu, P.K.; Messali, M.; Rosalin, T.; Elyas, K.; Saharuba, P.; Anjitha, T.; Ben Hadda, T. Removal of toxic heavy metals, phenolic compounds and textile dyes from industrial waste water using phosphonium based ionic liquids. *J. Mol. Liq.* **2020**, *323*, 114645. [CrossRef]
22. Choumane, R.; Peulon, S. Development of an efficient electrochemical process for removing and separating soluble Pb(II) in aqueous solutions in presence of other heavy metals: Studies of key parameters. *Chem. Eng. J.* **2021**, *423*, 130161. [CrossRef]
23. Dąbrowski, A.; Hubicki, Z.; Podkoscielny, P.; Robens, E. Selective removal of the heavy metal ions from waters and industrial wastewaters by ion-exchange method. *Chemosphere* **2004**, *56*, 91–106. [CrossRef] [PubMed]
24. Rajivgandhi, G.; Gnanamangai, B.M.; Ramachandran, G.; Chackaravarthy, G.; Chelliah, C.K.; Maruthupandy, M.; Alharbi, N.S.; Kadaikunnan, S.; Li, W.-J. Effective removal of heavy metals in industrial wastewater with novel bioactive catalyst enabling hybrid approach. *Environ. Res.* **2021**, *204*, 112337. [CrossRef] [PubMed]
25. Bilal, M.; Ihsanullah, I.; Younas, M.; Shah, M.U.H. Recent advances in applications of low-cost adsorbents for the removal of heavy metals from water: A critical review. *Sep. Purif. Technol.* **2021**, *278*, 119510. [CrossRef]
26. Shrestha, R.; Ban, S.; Devkota, S.; Sharma, S.; Joshi, R.; Tiwari, A.P.; Kim, H.Y.; Joshi, M.K. Technological trends in heavy metals removal from industrial wastewater: A review. *J. Environ. Chem. Eng.* **2021**, *9*, 105688. [CrossRef]
27. Bhatnagar, A.; Sillanpää, M. Utilization of agro-industrial and municipal waste materials as potential adsorbents for water treatment—A review. *Chem. Eng. J.* **2010**, *157*, 277–296. [CrossRef]

28. Tudorache, D.I.F.; Gavrilescu, M. Application of natural zeolites as sorbents in the clean-up of aqueous streams. *Environ. Eng. Manag. J.* **2012**, *11*, 867–878.
29. Gautam, R.K.; Mudhoo, A.; Lofrano, G.; Chattopadhyaya, M.C. Biomass-derived biosorbents for metal ions sequestration: Adsorbent modification and activation methods and adsorbent regeneration. *J. Environ. Chem. Eng.* **2014**, *2*, 239–259. [CrossRef]
30. Zhou, Z.; Sun, Y.; Wang, Y.; Yu, F.; Ma, J. Adsorption behavior of Cu(II) and Cr(VI) on aged microplastics in antibiotics-heavy metals coexisting system. *Chemosphere* **2021**, *291*, 132794. [CrossRef]
31. Apostol, L.C.; Gavrilescu, M. Application of natural materials as sorbents for persistent organic pollutants. *Environ. Eng. Manag. J.* **2009**, *8*, 243–252. [CrossRef]
32. Shahrokhi-Shahraki, R.; Benally, C.; El-Din, M.G.; Park, J. High efficiency removal of heavy metals using tire-derived activated carbon vs commercial activated carbon: Insights into the adsorption mechanisms. *Chemosphere* **2020**, *264*, 128455. [CrossRef] [PubMed]
33. Cao, F.; Lian, C.; Yu, J.; Yang, H.; Lin, S. Study on the adsorption performance and competitive mechanism for heavy metal contaminants removal using novel multi-pore activated carbons derived from recyclable long-root *Eichhornia crassipes*. *Bioresour. Technol.* **2019**, *276*, 211–218. [CrossRef]
34. Ali, I.; Asim, M.; Khan, T.A. Low cost adsorbents for the removal of organic pollutants from wastewater. *J. Environ. Manag.* **2012**, *113*, 170–183. [CrossRef] [PubMed]
35. De Gisi, S.; Lofrano, G.; Grassi, M.; Notarnicola, M. Characteristics and adsorption capacities of low-cost sorbents for wastewater treatment: A review. *Sustain. Mater. Technol.* **2016**, *9*, 10–40. [CrossRef]
36. Esfandiar, N.; Suri, R.; McKenzie, E.R. Competitive sorption of Cd, Cr, Cu, Ni, Pb and Zn from stormwater runoff by five low-cost sorbents; Effects of co-contaminants, humic acid, salinity and pH. *J. Hazard. Mater.* **2021**, *423*, 126938. [CrossRef] [PubMed]
37. Godage, N.H.; Gionfriddo, E. Use of natural sorbents as alternative and green extractive materials: A critical review. *Anal. Chim. Acta* **2020**, *1125*, 187–200. [CrossRef] [PubMed]
38. Ata, A.; Nalcaci, O.O.; Ovez, B. Macro algae *Gracilaria verrucosa* as a biosorbent: A study of sorption mechanisms. *Algal Res.* **2012**, *1*, 194–204. [CrossRef]
39. Demirbas, A. Heavy metal adsorption onto agro-based waste materials: A review. *J. Hazard. Mater.* **2008**, *157*, 220–229. [CrossRef] [PubMed]
40. Ullrich, A.H.; Smith, M.W. The biosorption process of sewage and waste treatment. *Sew. Ind. Wastes* **1951**, *23*, 1248–1253.
41. Kratochvil, D.; Volesky, B. Advances in the biosorption of heavy metals. *Trends Biotechnol.* **1998**, *16*, 291–300. [CrossRef]
42. Lesmana, S.O.; Febriana, N.; Soetaredjo, V.; Sunarso, J.; Ismadji, S. Studies on potential applications of biomass for the separation of heavy metals from water and wastewater. *Biochem. Eng. J.* **2009**, *44*, 19–41. [CrossRef]
43. Demey, H.; Melkior, T.; Chatroux, A.; Attar, K.; Thiery, S.; Miller, H.; Grateau, M.; Sastre, A.M.; Marchand, M. Evaluation of torrefied poplar-biomass as a low-cost sorbent for lead and terbium removal from aqueous solutions and energy co-generation. *Chem. Eng. J.* **2018**, *361*, 839–852. [CrossRef]
44. Guérin, T.; Ghinet, A.; Hossart, M.; Waterlot, C. Wheat and ryegrass biomass ashes as effective sorbents for metallic and organic pollutants from contaminated water in lab-engineered cartridge filtration system. *Bioresour. Technol.* **2020**, *318*, 124044. [CrossRef]
45. Bulgariu, L.; Ferţu, D.I.; Cara, I.G.; Gavrilescu, M. Efficacy of Alkaline-Treated Soy Waste Biomass for the Removal of Heavy-Metal Ions and Opportunities for Their Recovery. *Materials* **2021**, *14*, 7413. [CrossRef]
46. Mobasherpour, I.; Salahi, E.; Pazouki, M. Comparative of the removal of Pb(II), Cd(II) and Ni(II) by nano crystallite hydroxyapatite from aqueous solutions: Adsorption isotherm study. *Arab. J. Chem.* **2012**, *5*, 439–446. [CrossRef]
47. Al Hamouz, O.C.S.; Ali, S.A. Removal of Zinc and Cadmium ions using a cross-linked polyaminophosphonate. *J. Macromol. Sci. Part A* **2013**, *50*, 375–384. [CrossRef]
48. Bashir, A.; Manzoor, T.; Malik, L.A.; Qureashi, A.; Pandith, A.H. Enhanced and Selective Adsorption of Zn(II), Pb(II), Cd(II), and Hg(II) Ions by a Dumbbell- and Flower-Shaped Potato Starch Phosphate Polymer: A Combined Experimental and DFT Calculation Study. *ACS Omega* **2020**, *5*, 4853–4867. [CrossRef]
49. He, J.; Chen, P.A. A comprehensive review on biosorption of heavy metals by algal biomass: Materials, performances, chemistry, and modeling simulation tools. *Bioresour. Technol.* **2014**, *160*, 67–78. [CrossRef]
50. Suzaki, P.Y.R.; Munaro, M.T.; Triques, C.C.; Kleinübing, S.J.; Klen, M.R.F.; Bergamasco, R.; Jorge, L.M.D.M. Phenomenological mathematical modeling of heavy metal biosorption in fixed-bed columns. *Chem. Eng. J.* **2017**, *326*, 389–400. [CrossRef]
51. Laurent, J.; Casellas, M.; Dagot, C. Heavy metals biosorption on disintegrated activated sludge: Description of a new equilibrium model. *Chem. Eng. J.* **2010**, *164*, 63–69. [CrossRef]
52. Özen, R.; Sayar, N.A.; Durmaz-Sam, S.; Sayar, A.A. A sigmoidal model for biosorption of heavy metal cations from aqueous media. *Math. Biosci.* **2015**, *265*, 40–46. [CrossRef] [PubMed]
53. Moreira, V.R.; Lebron, Y.A.R.; Santos, L.V.D.S. Predicting the biosorption capacity of copper by dried Chlorella pyrenoidosa through response surface methodology and artificial neural network models. *Chem. Eng. J. Adv.* **2020**, *4*, 100041. [CrossRef]
54. Blagojev, N.; Vasić, V.; Kukić, D.; Šćiban, M.; Prodanović, J.; Bera, O. Modelling and efficiency evaluation of the continuous biosorption of Cu(II) and Cr(VI) from water by agricultural waste materials. *J. Environ. Manag.* **2021**, *281*, 111876. [CrossRef] [PubMed]

55. Sarubbo, L.A.; Filho, A.A.P.S.; Nascimento, L.A.D.; Rufino, R.D.; de Luna, J.M.; Brasileiro, P.P.F.; da Silva, C.F.S.; Benachour, M.; dos Santos, V.A. Study of an oily water treatment process in a pilot hybrid system combining air flotation and a constructed wetland: Data analysis, efficiency optimization and scale-up. *Environ. Eng. Manag. J.* **2021**, *20*, 247–256. [CrossRef]
56. Jaafari, J.; Yaghmaeian, K. Optimization of heavy metal biosorption onto freshwater algae (Chlorella coloniales) using response surface methodology (RSM). *Chemosphere* **2018**, *217*, 447–455. [CrossRef]
57. Khan, A.; Khan, M.S.; Hadi, F.; Saddiq, G.; Khan, A.N. Energy-Dispersive X-ray (EDX) fluorescence based analysis of heavy metals in marble powder, paddy soil and rice (*Oryza sativa* L.) with potential health risks in District Malakand, Khyber Pakhtunkhwa, Pakistan. *Environ. Pollut. Bioavailab.* **2021**, *33*, 301–316. [CrossRef]
58. Lu, Z.; Hu, X.; Lu, Y. Particle Morphology Analysis of Biomass Material Based on Improved Image Processing Method. *Int. J. Anal. Chem.* **2017**, *2017*, 5840690. [CrossRef]
59. UK College of Engineering. *Sample Preparation for Electron Microscopy*; University of Kentucky: Lexington, KY, USA. Available online: http://emc.engr.uky.edu/equipment/sample-preparation-electron-microscopy (accessed on 20 January 2022).
60. Box, G.E.P.; Wilson, K.B. On the experimental attainment of optimum condition. *J.R. Stat. Soc. Ser. B. Methods* **1951**, *13*, 1–45. [CrossRef]
61. Witek-Krowiak, A.; Chojnacka, K.; Podstawczyk, D.; Dawiec, A.; Pokomeda, K. Application of response surface methodology and artificial neural network methods in modelling and optimization of biosorption process. *Bioresour. Technol.* **2014**, *160*, 150–160. [CrossRef]
62. Leupin, O.X.; Hug, S.J. Oxidation and removal of arsenic (III) from aerated groundwater by filtration through sand and zero-valent iron. *Water Res.* **2005**, *39*, 1729–1740. [CrossRef] [PubMed]
63. Sheikh, Z.; Amin, M.; Khan, N.; Khan, M.N.; Sami, S.K.; Khan, S.B.; Hafeez, I.; Khan, S.A.; Bakhsh, E.M.; Cheng, C.K. Potential application of *Allium Cepa* seeds as a novel biosorbent for efficient biosorption of heavy metals ions from aqueous solution. *Chemosphere* **2021**, *279*, 130545. [CrossRef] [PubMed]
64. Zhang, C.; Ren, H.X.; Zhong, C.Q.; Wu, D. Biosorption of Cr(VI) by immobilized waste biomass from polyglutamic acid production. *Sci. Rep.* **2020**, *10*, 3705. [CrossRef] [PubMed]
65. Sheng, P.X.; Ting, Y.P.; Chen, J.P.; Hong, L. Sorption of lead, copper, cadmium, zinc, and nickel by marine algal biomass: Characterization of biosorptive capacity and investigation of mechanisms. *J. Colloid Interface Sci.* **2004**, *275*, 131–141. [CrossRef] [PubMed]
66. Jalali, R.; Ghafourian, H.; Asef, Y.; Davarpanah, S.; Sepehr, S. Removal and recovery of lead using nonliving biomass of marine algae. *J. Hazard. Mater.* **2002**, *92*, 253–262. [CrossRef]
67. Isam, M.; Baloo, L.; Kutty, S.R.M.; Yavari, S. Optimisation and Modelling of Pb(II) and Cu (II) Biosorption onto Red Algae (*Gracilaria changii*) by Using Response Surface Methodology. *Water* **2019**, *11*, 2325. [CrossRef]
68. Gavrilescu, M. Removal of Heavy Metals from the Environment by Biosorption. *Eng. Life Sci.* **2004**, *4*, 219–232. [CrossRef]
69. Gavrilescu, M. Biosorption in environmental remediation. In *Bioremediation Technology*; Fulekar, M.H., Ed.; Springer: Dordrecht, The Netherlands, 2010; pp. 35–99.
70. Beni, A.A.; Esmaeili, A. Biosorption, an efficient method for removing heavy metals from industrial effluents: A Review. *Environ. Technol. Innov.* **2019**, *17*, 100503. [CrossRef]
71. Sibi, G. Biosorption of chromium from electroplating and galvanizing industrial effluents under extreme conditions using Chlorella vulgaris. *Green Energy Environ.* **2016**, *1*, 172–177. [CrossRef]
72. Dhanarani, S.; Viswanathan, E.; Piruthiviraj, P.; Arivalagan, P.; Kaliannan, T. Comparative study on the biosorption of aluminum by free and immobilized cells of *Bacillus safensis* KTSMBNL 26 isolated from explosive contaminated soil. *J. Taiwan Inst. Chem. Eng.* **2016**, *69*, 61–67. [CrossRef]
73. Cholico-González, D.; Lara, N.O.; Macedo, A.M.F.; Salas, J.C. Adsorption Behavior of Pb(II), Cd(II), and Zn(II) onto Agave Bagasse, Characterization, and Mechanism. *ACS Omega* **2020**, *5*, 3302–3314. [CrossRef] [PubMed]
74. Salem, N.D.; Awwad, A.; Al-Dujaili, A. Biosorption of Pb(II), Zn(II), and Cd(II) from aqueous solutions by (*Eriobotrya japonica*) loquat bark. *Int. J. Environ. Prot.* **2002**, *2*, 1–7.
75. Lezcano, J.M.; Gonzalez, F.; Ballester, A.; Blázquez, M.L.; Muñoz, J.A.; García-Balboa, C. Biosorption of Cd(II), Cu(II), Ni(II), Pb(II) and Zn(II) using different residual biomass. *Chem. Ecol.* **2010**, *26*, 1–17. [CrossRef]
76. Amer, M.; Ahmad, R.A.; Awwad, A.M. Biosorption of Cu(II), Ni(II), Zn(II) and Pb(II) ions from aqueous solution by *Sophora japonica* pods powder. *Int. J. Ind. Chem.* **2015**, *6*, 67–75. [CrossRef]
77. Kamar, F.H.; Nechifor, A.C.; Nechifor, G.; Al-Musawi, T.J.; Mohammed, A.H. Aqueous phase biosorption of Pb(II), Cu(II), and Cd(II) onto cabbage leaves powder. *Int. J. Chem. React. Eng.* **2016**, *15*, 20150178. [CrossRef]
78. Sulaymon, A.; Mohammed, A.A.; Al-Musawi, T.J. Removal of lead, cadmium, copper, and arsenic ions using biosorption: Equilibrium and kinetic studies. *Desalination Water Treat.* **2013**, *51*, 4424–4434. [CrossRef]
79. Anwar, J.; Shafique, U.; Zaman, W.U.; Salman, M.; Dar, A.; Anwar, S. Removal of Pb(II) and Cd(II) from water by adsorption on peels of banana. *Bioresour. Technol.* **2010**, *101*, 1752–1755. [CrossRef]
80. Ezeonuegbu, B.A.; Machido, D.A.; Whong, C.M.; Japhet, W.S.; Alexiou, A.; Elazab, S.T.; Qusty, N.; Yaro, C.A.; Batiha, G.E.-S. Agricultural waste of sugarcane bagasse as efficient adsorbent for lead and nickel removal from untreated wastewater: Biosorption, equilibrium isotherms, kinetics and desorption studies. *Biotechnol. Rep.* **2021**, *30*, e00614. [CrossRef]

81. Ozdemir, S.; Kılınç, E.; Acer, Ö.; Soylak, M. Simultaneous preconcentrations of Cu(II), Ni(II), and Pb(II) by SPE using E. profundum loaded onto Amberlite XAD-4. *Microchem. J.* **2021**, *171*, 106758. [CrossRef]
82. MacFarland, T.W. *Two-Way Analysis of Variance: Statistical Tests and Graphics Using R*; Springer Nature: Cham, Switzerland, 2012.
83. Myers, R.H.; Montgomery, D.C.; Anderson-Cook, C.M. *Response Surface Methodology: Process and Product Optimization Using Designed Experiments*, 4th ed.; Wiley: Hoboken, NJ, USA, 2016; ISBN 9781118916025.
84. Chattoraj, N.K.; Mondal, B.; Das, P.; Roy, B.; Sadhukan, B. Biosorption of carbaryl from aqueous solution onto *Pistia stratiotes* biomass. *Appl. Water Sci.* **2014**, *4*, 79–88. [CrossRef]
85. Sadhukhan, B.; Mondal, N.K.; Chattoraj, S. Optimization using central composite design (CCD) and the desirability function for sorption of methylene blue from aqueous solutions onto *Lemna major*. *Karbala Int. J. Mod. Sci.* **2016**, *2*, 145–155. [CrossRef]

processes

MDPI

Article

Surface Hydrophobic Modification of Biochar by Silane Coupling Agent KH-570

Muxi Zhang [1,2], Hongxiang Zhu [3], Beidou Xi [1,2,4], Yuxin Tian [1,2], Xiaojie Sun [1,2,*], Hongxia Zhang [1,2,*] and Beibei Wu [1,2]

1 Guangxi Key Laboratory of Environmental Pollution Control Theory and Technology, Guilin University of Technology, Guilin 541004, China; zhangmuxi2022@163.com (M.Z.); xibeidou@263.net (B.X.); ttianyyxx@163.com (Y.T.); beibei_wu@webmail.hzau.edu.cn (B.W.)

2 Guangxi Collaborative Innovation Center for Water Pollution Control and Water Safety in Karst Area, Guilin University of Technology, Guilin 541004, China

3 College of Light Industry and Food Engineering, Guangxi University, Nanning 530004, China; zhx@gxu.edu.cn

4 State Environmental Protection Key Laboratory of Simulation and Control of Groundwater Pollution, Chinese Research Academy of Environmental Sciences, Beijing 100012, China

* Correspondence: sunxiaojie@glut.edu.cn (X.S.); zhx75@glut.edu.cn (H.Z.); Tel.: +86-15078329789 (X.S.)

Abstract: Biochar-amended soil cover (BSC) in landfills can improve the oxidation of methane. However, adding biochar can cause a larger amount of rainwater to enter the soil cover and landfill because it increases the permeability of the soil cover, which increases leachate production. Improving the hydrophobicity and waterproof ability of BSC is expected to reduce rainwater that goes into landfills. Silane coupling agent KH-570 is used to modify biochar to improve its hydrophobicity and waterproof ability after being added to the soil cover. The waterproofness of hydrophobic biochar-amended soil cover (HBSC) was studied by conducting a precipitation simulation test. Results showed that the optimum hydrophobicity of the surface-modified biochar was obtained when the mass fraction of KH-570 was 7%, the biochar dosage was 7 g, and the modification temperature was 60 °C. In these conditions, the contact angle was 143.99° and the moisture absorption rate was 0.10%. The analysis results of thermogravimetric, X-ray diffractometer and scanning electron microscopy before and after the biochar modification showed that KH-570 formed a hydrophobic organic coating layer on the biochar surface, indicating that the surface hydrophobic modification of biochar was successfully carried out by silane coupling agent. The waterproof ability of HBSC was significantly better than that of BSC in the simulated precipitation test.

Keywords: silane coupling agents; biochar; hydrophobic modification; landfill cover

Citation: Zhang, M.; Zhu, H.; Xi, B.; Tian, Y.; Sun, X.; Zhang, H.; Wu, B. Surface Hydrophobic Modification of Biochar by Silane Coupling Agent KH-570. *Processes* **2022**, *10*, 301. https://doi.org/10.3390/pr10020301

Academic Editor: Carmen Branca

Received: 27 December 2021
Accepted: 31 January 2022
Published: 2 February 2022

1. Introduction

Landfill soil cover has been used to oxidize methane to reduce methane emission [1,2]. The CH_4 emission reduction of landfill depends on the choice of soil cover material. In contrast to traditional soil cover, the biocover in landfill have been extensively studied to increase the aerobic oxidation of methane [3]. Many kinds of biocover materials include compost [4–6], mineralized refuse [7], and solidified sludge [8]. However, CH_4 oxidation capacity of traditional landfill soil cover material was generally insufficient [9,10]. Biomass-based carbon materials have attracted attention due to their economical, sustainable, and environmentally friendly features [11]. Biochar can be considered as a suitable and alternate material as a final cover system for landfills because its high specific surface area and high porosity can improve the permeability of the biocover and promote the growth and reproduction of methanotrophs [1,12–14]. Prior studies noted that average CH_4 removal efficiencies were up to 85.2% for biochar-amended soil cover (BSC) [6].

However, a number of hydrophilic groups, such as carboxyl and hydroxyl groups, are on the surface of biochar (C > 60%) [15]. As the content of biochar added to the soil cover increases, so does the permeability coefficient of the cover. Studies have shown that the permeability coefficient of the soil with 10% biochar added has been more than 10^{-7} cm/s [16]. The increase of permeability coefficient leads to the increase of rainwater into the methane oxide layer and landfill. Water content has been found to be an important environmental factor that controls methane oxidation and adsorption capacity of landfill cover soil [17,18]. The increase of rainwater entering the methane oxide layer decreases methane oxidation and adsorption efficiency [19]. In addition, the increase of rainwater entering the landfill leads to the increase of leachate production, thereby increasing the cost of leachate treatment. Therefore, a problem to be solved is how to reduce rainwater entering the soil cover after adding biochar.

KH-570 is an environmentally friendly hydrophobic modifier that has been studied and applied in environmental pollution control [20]. For example, Chen [21] used four kinds of silane coupling agents to modify the interfaces of composites and found that KH-570 modified composite possessed the best mechanical properties and water resistance. Therefore, in this study, silane coupling agent KH-570 was selected as the modifier to conduct an experiment on hydrophobic modification of biochar, thereby improving its hydrophobicity and reducing rainwater that enters the soil cover and landfill. The hydrophobic properties of biochar were tested under various concentrations of modifier, dosages, and modification temperatures to determine the optimal hydrophobic modification conditions. A simulated landfill cover precipitation experiment was conducted to study the waterproof ability of modified biochar. The hydrophobic modification of biochar was conducted to develop materials to improve the hydrophobic permeability of the landfill cover layer, providing technical support for methane emission reduction in landfills.

2. Experimental

2.1. Materials

Biochar was purchased from Desheng Carbon Industry Co. Ltd., silane coupling agent KH-570 from Sinopharm Chemical Reagent Co. Ltd., ethanol absolute (analytically pure) from Cologne Chemicals Co. Ltd., and acetic acid (analytically pure) from Sinopharm Chemical Reagent Co. Ltd. (all manufacturers are based in China).

2.2. Preparation of Biochar

The biochar used was produced by pyrolysis of discarded rice straw in a completely anoxic environment through pyrolysis at 500 °C [15]. Total pore volume of adsorption was 0.07 cm^3/g. The biochar used in the test was alkaline, with high C content (64.2%) and low P (0.16%) and K (0.33%) contents. The main component of rice straw was cellulose, and its main structures were sieve tube and conduit, which mainly account for the large specific surface area of rice straw biochar [15].

2.3. Surface Modification of Biochar with KH-570

First, biochar was added to alcohol-water solution to hydrolyze it sufficiently. The volume proportion of ethanol absolute and water was 1:1, and 0.1 mol/L acetic acid was used to adjust pH to 4. Then, the mixture was stirred in a water bath at 30 °C–70 °C for 30 min. Second, the prepared silane coupling agent KH-570 was added into the above solution and stirred for 2 h. Third, the reaction products were filtered and washed with ethanol absolute several times. Finally, the surface hydrophobic modified biochar was obtained by drying in an electric blast drying oven at 50 °C for 6 h.

2.4. Hydrophobic Modification Mechanism of KH-570

Silane coupling agent KH-570 represents a type of important difunctional group modifiers that can react with adsorbed water on organic molecules and inorganic surfaces to form a strong bond [22]. Figure 1 shows the surface hydrophobic modification reaction mechanism of KH-570. This silane coupling agent undergoes hydrolysis, dehydration, and condensation to form oligomers. The dehydration of oligomers and hydroxyl groups on the surface of the biochar formed partial covalent bonds, which coated the biochar surface with coupling agent and increased the hydrophobicity [15].

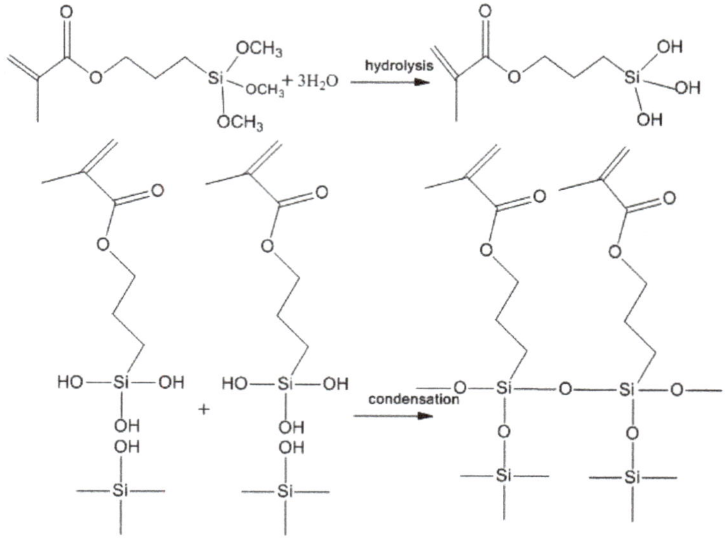

Figure 1. Hydrophobic modification mechanism of KH-570.

2.5. Characterization Analysis

The changes of functional groups of biochar before and after modification were characterized by Fourier transform infrared spectroscopy (FTIR) (IS10, Thermo Fisher Scientific, Madison, WI, USA). The morphology and structure of biochar before and after modification were characterized by scanning electron microscopy (SEM) (JSM-7900F Plus, Nidec Corporation, Kyoto, Japan). The thermogravimetric (TGA) curves of biochar before and after modification were measured by synchronous thermal analyzer to characterize its thermal stability (SDT-Q600, Waters, Milford, MA, USA, USA). The phase of biochar before and after modification was analyzed by X'Pert3 Power multifunctional X-ray diffractometer (XRD) (PANalytical B.V., Alemlo, The Netherlands). Surface contact angle was an important parameter for evaluating hydrophobicity or hydrophilicity. JC2000D1 contact angle tester (Shanghai Zhongchen Digital Technology Equipment Co. Ltd., Shanghai, China) was used to determine the wettability of biochar before and after modification. The contact angle was less than or equal to 90° for hydrophilicity and greater than 90° for hydrophobicity [23]. Hygroscopicity refers to the ability of a solid material to absorb water from air. The lower the moisture content, the better its hydrophobicity [23,24]. Then, 4 g of dried to constant weight original biochar and modified biochar were evenly spread on glass sheets and placed at room temperature for 24 h. Their mass changes were measured and their moisture absorption rates were calculated. The moisture absorption rates can be calculated as follows:

$$\text{moisture absorption rates} = [(m_2 - m_1) / m_1] \times 100\% \tag{1}$$

where m_1 is the drying mass, g; and m_2 is the mass after 24 h of hygroscopicity, g.

2.6. Waterproofing Experiment

In this study, a simple model was designed to simulate the actual landfill soil cover (Figure 2) to compare the waterproof performance of biochar before and after modification. Biochar-amended soil and hydrophobic biochar-amended soil, in which the volume proportion of biochar or modified biochar and soil was 1:5 [16], were, respectively, added into the model. The material was repeatedly compacted, with one side flush with the water guide plate. The slope of the top was adjusted to 7% according to the Technical Specification for Sealing of Sanitary Landfill Sites of Domestic Waste (China, 2017). The initial water content of the cover material was adjusted to approximately 22%. Then, 2.5 L water was evenly sprayed by simulated precipitation at the top of the compacted slope, and the water flowed out of the water guide plate into the collection container. The volume of outflow water was measured.

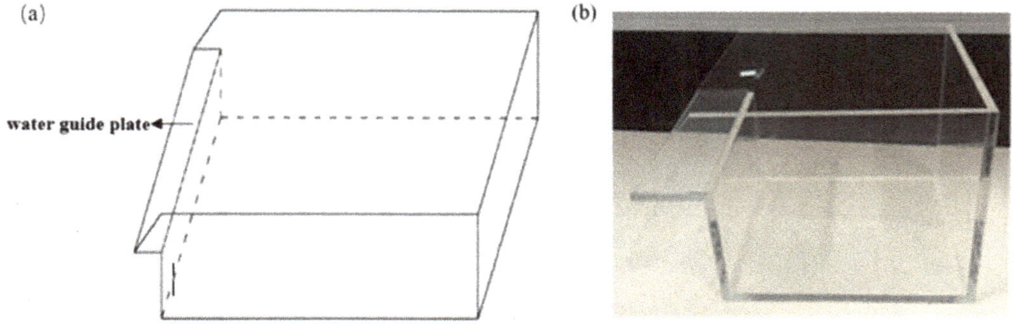

Figure 2. Schematic and physical map of a simple landfill soil cover model.

3. Results and Discussion

3.1. Effect of Mass Fraction of KH-570 on Surface Hydrophobic Properties of Modified Biochar

Under the biochar dosage of 5 g and modification temperature of 70 °C, the effects of various mass fractions of KH-570 on the surface hydrophobicity of biochar were investigated. As shown in Figure 3, the contact angle of biochar was 7.04° and the water absorption rate was 0.22% without the addition of KH-570. With the increase of the mass fraction of KH-570, the contact angle of the modified biochar increased initially, and then showed a trend of fluctuation and decline. The reason was that when the mass fraction of KH-570 continued increasing (<7%), the hydrophobic performance improved as more hydrophilic groups on the biochar surface were replaced by hydrophobic groups. However, when the mass fraction of KH-570 exceeded 7%, the siloxane anion generated by the hydrolysis of KH-570 formed a bridge, leading to the flocculation of the powder, thereby affecting the hydrophobic modification effect of the biochar [25]. The contact angle of modified biochar reached 131.99° and the water absorption rate was 0.04% when the mass fraction of KH-570 was 7%. Therefore, the optimal mass fraction of KH-570 was 7% when the biochar dosage was 5 g and the modification temperature was 70 °C.

3.2. Effect of Biochar Addition on Surface Hydrophobic Properties

The effect of the additional amount of biochar on the surface hydrophobic performance was studied at 70 °C modification temperature and 7% mass fraction of KH-570. As shown in Figure 4, the contact angle reached a maximum value of 141.99°, and the moisture absorption rate correspondingly had a minimum value of 0.08% when the biochar dosage was 7 g, indicating that the hydrophilic groups on the surface of biochar were basically replaced by the hydrophobic groups of KH-570, forming a hydrophobic organic cover on the surface of biochar in the form of chemical bonding, which resulted in the transformation of biochar from hydrophilic surface to hydrophobic surface [25]. However, when the biochar dosage increased from 7 g to 13 g, the contact angle decreased from 141.99° to 125.49°, and the moisture absorption rate increased from 0.08% to 0.10%, indicating that the hydrophilic group on the surface of biochar could not be completely replaced, and the hydrophobic effect worsened. Therefore, the optimal dosage of biochar was 7 g when the mass fraction of KH-570 was 7% and the modification temperature was 70 °C.

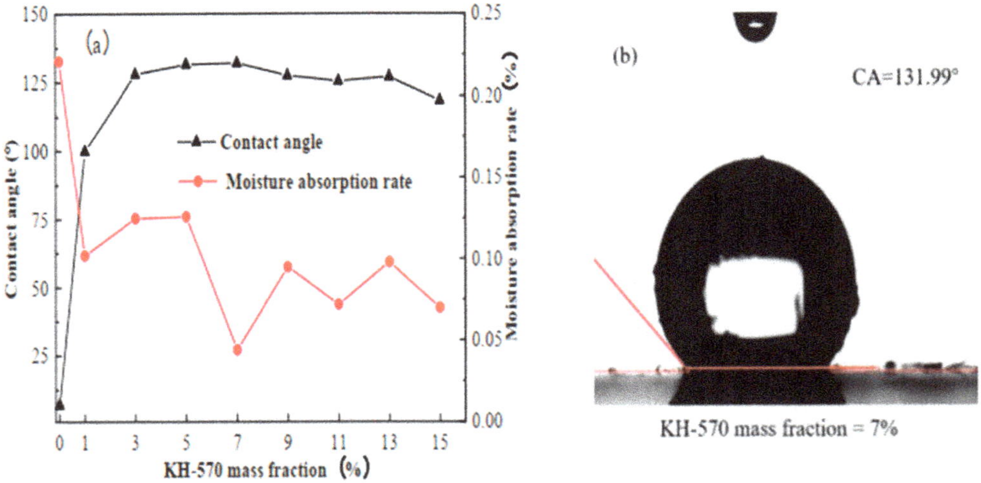

Figure 3. (**a**) Contact angles and moisture absorption rates of biochar modified by KH-570, and (**b**) contact angle of modified biochar when KH-570 mass fraction was 7%.

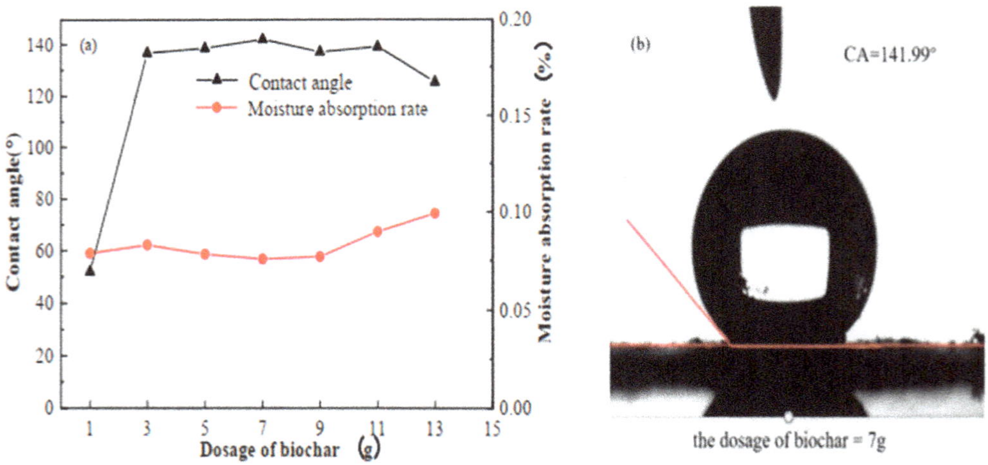

Figure 4. (**a**) Contact angles and moisture absorption rates of modified biochar with different dosage, and (**b**) contact angle of modified biochar when the dosage of biochar was 7 g.

3.3. Effect of Reaction Temperature on Surface Hydrophobicity of Modified Biochar

The effect of the reaction temperature on the surface hydrophobicity of biochar was studied when the mass fraction of KH-570 was 7% and the biochar dosage was 7 g. The test results of the contact angle and moisture absorption rates under 30, 40, 50, 60, and 70 °C are shown in Figure 5. The hygroscopicity decreased greatly when the reaction temperature reached 50 °C, and modified biochar changed from hydrophilic surface to hydrophobic surface. The maximum contact angle of modified biochar was 143.99°, and the minimum moisture absorption rate was 0.10% until the reaction temperature reached 60 °C. When the reaction temperature was increased, the contact angle decreased and the moisture absorption rate increased. This result may be due to the fracture and release of a large amount of hydrophilic oxygen-containing functional groups in biomass [24]. Therefore, the optimal modification temperature was 60 °C when the mass fraction of KH-570 was 7% and the biochar dosage was 7 g.

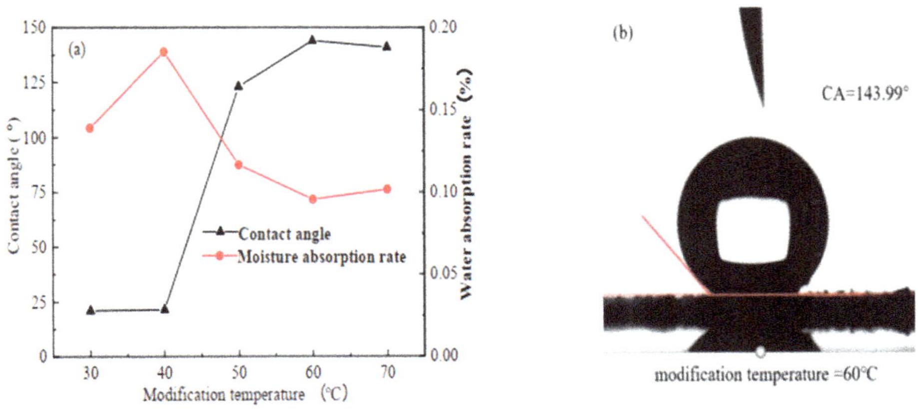

Figure 5. (**a**) Contact angles and moisture absorption rate of modified biochar at different reaction temperatures, and (**b**) contact angle of modified biochar when temperature was 60°C.

In summary, the hydrophobic modified biochar was prepared under the following conditions: 7% mass fraction of KH-570, 7 g biochar dosage, and 60 °C modification temperature. In these conditions, the contact angle was 143.99° and the moisture absorption rate was 0.10%. Compared with the original biochar, the contact angle of the modified biochar increased by 136.95° and the moisture absorption rate decreased by 0.12%.

3.4. FT-IR and TGA Analysis

The surface functional groups of the original and modified biochar were detected using FT-IR spectroscopy [26]. As shown in Figure 6a, the spectra of the biochar before and after modification were basically consistent, indicating that the biochar was the same substance before and after modification, but the microstructure changed. Some proper functional groups on the surface of biochar were effectively identified such that the intensity of the wide adsorption band around 3400 cm^{-1} could be attributed to O-H stretching of acid or to alcohol structures, that around 2368–2381 cm^{-1} could indicate the presence of carbon dioxide (O=C=O), and around 1590–1850 cm^{-1} may indicate C=C stretching from alkenes, around 1360 cm^{-1} may indicate the presence of C-O, C=O and probably C-H bending modes. The stretching vibration absorption peak around 1050 cm^{-1} of the modified hydrophobic biochar was significantly stronger and wider than that of the original biochar, indicating that the modified biochar was grafted with the modifier and generated a large amount of C-O-Si and Si-O [20,26–31]. Hydrophobic groups in KH-570 effectively bound to the surface of the biochar, forming a hydrophobic organic coating [20].

The thermal stabilities of the biochar before and after powder modification in an atmosphere of N_2 were measured using TGA [9]. As shown in Figure 6b, the first weight loss occurred at 100 °C, which was attributed to the removal of adsorbed water [32]. Before 200 °C, the weight loss rate of original biochar was 5.21%. However, the weight loss rate of the modified biochar was 2.11%, which was significantly reduced, indicating that the surface hydrophobicity of the modified biochar was enhanced and the moisture absorption rate decreased. The second weight loss occurred at 400–600 °C due to the combustion of free carbon and ash in biochar. The decreasing trend of the modified biochar mass was smoother than that of the original biochar, indicating that the silane coupling agent coated on the surface of biochar delayed the combustion of carbon and ash. The results showed that the modified biochar exhibited better thermal stability than the original biochar.

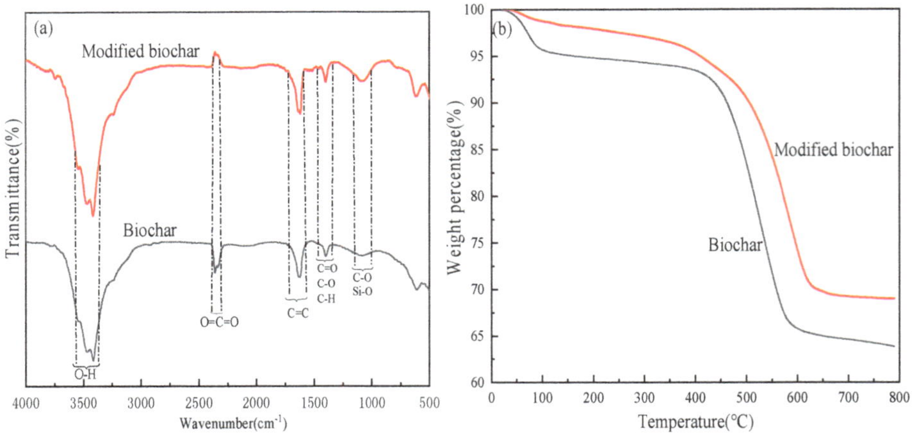

Figure 6. (**a**) Fourier transform infrared spectroscopy spectra and (**b**) thermogravimetric curves of biochar before and after modification.

3.5. XRD Analysis

The effects of silane coupling agent modification on the biochar were explored by characterizing them through XRD analysis. As shown in Figure 7, biochar before and after modification basically coincided, indicating that the surface hydrophobicity modification of the biochar by KH-570 had no significant effect on the phase composition of the biochar [21]. The XRD pattern of the biochar before and after modification was dominated by crystalline structures with a signal centered around 27° corresponding to the (002) reflection as the most predominant [33]. High peak density was observed in the modified biochar, indicating that higher crystalline structures were formed [34]. The impurity characteristic peak of modified biochar decreased compared with that of original biochar, indicating that the modifier KH-570 had been coated on the surface of biochar.

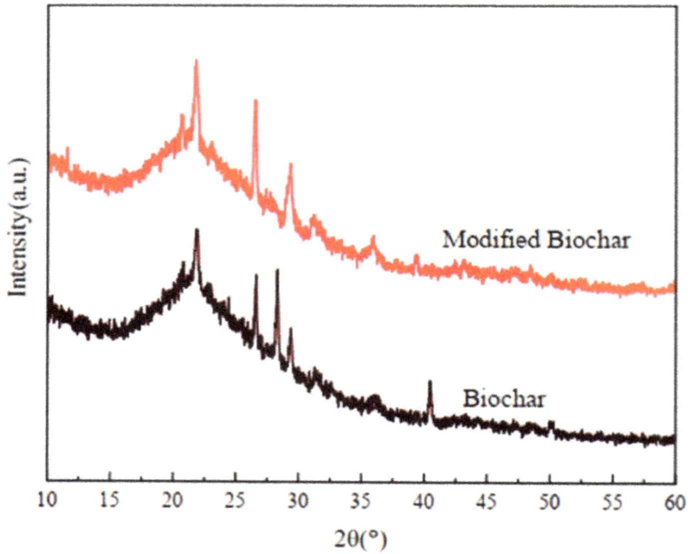

Figure 7. X-ray diffractometer images of biochar before and after modification.

3.6. SEM Analysis

Figure 8 shows the surface morphology of biochar before and after modification. As shown in Figure 8a, rice biochar was tubular porous with a honeycomb-like structure and a rough surface [34]. Figure 8b shows that the porous structure of the modified biochar does not change due to modification, and the surface of the modified biochar has an obvious covering layer [20], which was consistent with the previous characterization results.

Figure 8. (**a**) Scanning electron microscope images of biochar and (**b**) modified biochar.

3.7. Waterproof Performance

The measured infiltration water of the biochar-amended soil was 200 mL, and that in the hydrophobic biochar-amended soil was 20 mL, which was significantly less than that in the biochar-amended soil. This result indicates that the waterproofing ability of the hydrophobic biochar-amended soil was significantly better than that of the biochar-amended soil.

4. Conclusions

In this study, the biochar was modified by silane coupling agent KH-570 to improve its hydrophobicity. The optimum conditions for preparation of surface hydrophobic biochar were as follows: mass fraction of KH-570 was 7%, biochar dosage was 7 g, and modification temperature was 60 °C. The contact angle and hygroscopicity tests confirmed that the hydrophobicity of the surface-modified biochar was greatly increased compared with that of the original biochar. The contact angle of the modified biochar was 143.99°, which was 136.95° higher than that of the original biochar. The moisture absorption rate was 0.10%, which was 0.12% lower than that of the original biochar. The analysis results of TGA, XRD, and SEM before and after the biochar modification showed that the hydrophobic organic coating was formed on the surface of the modified biochar, resulting in the biochar turning from hydrophilic to hydrophobic surface. The simulated precipitation test indicated that the waterproof ability of the HBSC was significantly better than that of BSC.

Author Contributions: Conceptualization, software, formal analysis, writing—original draft, M.Z.; conceptualization, writing—review and editing, H.Z. (Hongxiang Zhu); supervision, writing—review and editing, B.X.; writing—review and editing, Y.T.; funding acquisition, project administration, supervision, writing—review and editing, X.S.; conceptualization, writing—review and editing, H.Z. (Hongxia Zhang); methodology, data curation, formal analysis, B.W. All authors have read and agreed to the published version of the manuscript.

Funding: This work was financially supported by National Natural Science Foundation of China (No. 51668014); Natural Science Foundation of Guangxi (No. 2018GXNSFGA281001); Science and Technology Major Project of Guangxi (GuikeAA18118013).

Institutional Review Board Statement: Not applicable.

Informed Consent Statement: Not applicable.

Data Availability Statement: Not applicable.

Conflicts of Interest: The authors declare no conflict of interest.

References

1. Sadasivam, B.Y.; Reddy, K.R. Landfill methane oxidation in soil and bio-based cover systems: A review. *Rev. Environ. Sci. Bio/Technol.* **2014**, *13*, 79–107. [CrossRef]
2. Huang, D.; Xu, W.; Wang, Q.; Xu, Q. Impact of hydrogen sulfide on biochar in stimulating the methane oxidation capacity and microbial communities of landfill cover soil. *Chemosphere* **2022**, *286*, 131650. [CrossRef] [PubMed]
3. Parsaeifard, N.; Sattler, M.; Nasirian, B.; Chen, V.C.P. Enhancing anaerobic oxidation of methane in municipal solid waste landfill cover soil. *Waste Manag.* **2020**, *106*, 44–54. [CrossRef] [PubMed]
4. Bukh, P.G.; Charlotte, S.; Peter, K. Availability and properties of materials for the Fakse Landfill biocover. *Waste Manag.* **2011**, *31*, 884–894. [CrossRef]
5. Charlotte, S.; Alessio, P.; Bukh, P.G.; Peter, K. Evaluation of respiration in compost landfill biocovers intended for methane oxidation. *Waste Manag.* **2011**, *31*, 895–902. [CrossRef]
6. Huang, D.; Yang, L.; Xu, W.; Chen, Q.; Ko, J.H.; Xu, Q. Enhancement of the methane removal efficiency via aeration for biochar-amended landfill soil cover. *Environ. Pollut.* **2020**, *263*, 114413. [CrossRef]
7. Zhang, Y.; Zhang, H.; Jia, B.; Wang, W.; Zhu, W.; Huang, T.; Kong, X. Landfill CH_4 oxidation by mineralized refuse: Effects of NH_4^+-N incubation, water content and temperature. *Sci. Total Environ.* **2012**, *426*, 406–413. [CrossRef]
8. He, J.; Feng, X.Y.; Zhou, L.R.; Zhang, L. The effect of leachate seepage on the mechanical properties and microstructure of solidified sludge when used as a landfill temporary cover material. *Waste Manag.* **2021**, *130*, 127–135. [CrossRef]
9. Liu, Y.; Gao, C.; Wang, Y.; He, L.; Lu, H.; Yang, S. Vermiculite modification increases carbon retention and stability of rice straw biochar at different carbonization temperatures. *J. Clean. Prod.* **2020**, *254*, 120111. [CrossRef]
10. Yargicoglu, E.N.; Reddy, K.R. Biochar-Amended Soil Cover for Microbial Methane Oxidation: Effect of Biochar Amendment Ratio and Cover Profile. *J. Geotech. Geoenviron. Eng.* **2017**, *144*, 1845. [CrossRef]
11. Gopinath, P.; Vo, D.; Gnana Prakash, D.; Adithya Joseph, A.; Viswanathan, S.; Arun, J. Environmental applications of carbon-based materials: A review. *Environ. Chem. Lett.* **2021**, *19*, 557–582. [CrossRef]
12. Srivatsav, P.; Bhargav, S.; Shanmugasundaram, V.; Arun, J.; Gopinath, P.; Bhatnagar, A. Biochar as an Eco-Friendly and Economical Adsorbent for the Removal of Colorants (Dyes) from Aqueous Environment: A Review. *Water* **2020**, *12*, 3561. [CrossRef]
13. Reddy, K.R.; Yargicoglu, E.N.; Yue, D.; Yaghoubi, P. Enhanced Microbial Methane Oxidation in Landfill Cover Soil Amended with Biochar. *J. Geotech. Geoenviron. Eng.* **2014**, *140*, 1148. [CrossRef]
14. Zhang, X.; Xia, J.; Pu, J.; Cai, C.; Tyson, G.W.; Yuan, Z.; Hu, S. Biochar-Mediated Anaerobic Oxidation of Methane. *Environ. Sci. Technol.* **2019**, *53*, 6660–6668. [CrossRef]
15. Wu, B.; Xi, B.; He, X.; Sun, X.; Li, Q.; Ouche, Q.; Zhang, H.; Xue, C. Methane Emission Reduction Enhanced by Hydrophobic Biochar-Modified Soil Cover. *Processes* **2020**, *8*, 162. [CrossRef]
16. Poupak, Y. *Development of Biochar-amended Landfill Cover for Landfill Gas Mitigation, Chicago: University of Illinois at Chicago*; United States of America: Chicago, IL, USA, 2011.
17. Scheutz, C.; Kjeldsen, P.; Bogner, J.E.; De Visscher, A.; Gebert, J.; Hilger, H.A.; Huber-Humer, M.; Spokas, K. Microbial methane oxidation processes and technologies for mitigation of landfill gas emissions. *Waste Manag. Res.* **2009**, *27*, 409–455. [CrossRef]
18. Majdinasab, A.; Yuan, Q. Performance of the biotic systems for reducing methane emissions from landfill sites: A review. *Ecol. Eng.* **2017**, *104*, 116–130. [CrossRef]
19. Sadasivam, B.Y.; Reddy, K.R. Adsorption and transport of methane in landfill cover soil amended with waste-wood biochars. *J. Environ. Manag.* **2015**, *158*, 11–23. [CrossRef]
20. Sun, X.; Qin, Y.; Wu, B.; Li, J.; Xue, C. Optimization of hydrophobic properties of biochar modified by silane coupling agent. *Environ. Sci. Technol.* **2019**, *42*, 68–73. (In Chinese)
21. Chen, K.; Li, P.; Li, X.; Liao, C.; Li, X.; Zuo, Y. Effect of silane coupling agent on compatibility interface and properties of wheat straw/polylactic acid composites. *Int. J. Biol. Macromol.* **2021**, *182*, 2108–2116. [CrossRef]
22. Fatemeh, A.; Navarchian, A.H. Recent advances in chemical surface modification of metal oxide nanoparticles with silane coupling agents: A review. *Adv. Colloid Interface Sci.* **2020**, *286*, 102298. [CrossRef]
23. Li, G.; Yue, J.; Guo, C.; Ji, Y. Influences of modified nanoparticles on hydrophobicity of concrete with organic film coating. *Constr. Build. Mater.* **2018**, *169*, 1–7. [CrossRef]
24. Jiang, H.; Ye, Y.; Lu, P.; Zhao, M.; Xu, G.; Chen, D.; Song, T. Effects of torrefaction conditions on the hygroscopicity of biochars. *J. Energy Inst.* **2021**, *96*, 260–268. [CrossRef]
25. He, L.H.; Li, L.; Zhou, C.; Li, W.H. Hydrophobic surface modification of diatomit with silane coupling agent KH-570. *Mod. Chem. Ind.* **2014**, *34*, 93–97. (In Chinese)

26. Yang, Q.; Sun, Y.; Sun, W.; Qin, Z.; Liu, H.; Ma, Y.; Wang, X. Cellulose derived biochar: Preparation, characterization, and Benzo[a]pyrene adsorption capacity. *Grain Oil Sci. Technol.* **2021**, *4*, 182–190. [CrossRef]
27. Wang, L.; Zhang, S.; Wu, S.; Long, Y.; Li, L.; Zheng, Z.; Hei, Y.; Zhou, L.; Luo, L.; Jiang, F. Controlling wettability of AgI/$BiVO_4$ composite photocatalyst and its effect on photocatalytic performance. *J. Alloys Compd.* **2020**, *835*, 155367. [CrossRef]
28. Manfrin, J.; Gonçalves, A.C., Jr.; Schwantes, D.; Conradi, E., Jr.; Zimmermann, J.; Ziemer, G.L. Development of biochar and activated carbon from cigarettes wastes and their applications in Pb^{2+} adsorption. *Environ. Chem. Eng.* **2021**, *9*, 104980. [CrossRef]
29. Tran, H.N.; Tomul, F.; Thi Hoang Ha, N.; Nguyen, D.T.; Lima, E.C.; Le, G.T.; Chang, C.-T.; Masindi, V.; Woo, S.H. Innovative spherical biochar for pharmaceutical removal from water: Insight into adsorption mechanism. *J. Hazard. Mater.* **2020**, *394*, 122255. [CrossRef]
30. Stylianou, M.; Christou, A.; Dalias, P.; Polycarpou, P.; Michael, C.; Agapiou, A.; Papanastasiou, P.; Fatta-Kassinos, D. Physicochemical and structural characterization of biochar derived from the pyrolysis of biosolids, cattle manure and spent coffee grounds. *J. Energy Inst.* **2020**, *93*, 2063–2073. [CrossRef]
31. Khoshnood Motlagh, E.; Asasian-Kolur, N.; Sharifian, S.; Ebrahimian Pirbazari, A. Sustainable rice straw conversion into activated carbon and nano-silica using carbonization-extraction process. *Biomass Bioenergy* **2021**, *144*, 105917. [CrossRef]
32. Chen, R.; Zhao, X.; Jiao, J.; Li, Y.; Wei, M. Surface-Modified Biochar with Polydentate Binding Sites for the Removal of Cadmium. *Int. J. Mol. Sci.* **2019**, *20*, 1775. [CrossRef] [PubMed]
33. Clurman, A.M.; Rodríguez-Narvaez, O.M.; Jayarathne, A.; Silva, G.D.; Ranasinghe, M.I.; Goonetilleke, A.; Bandala, E.R. Influence of surface hydrophobicity/hydrophilicity of biochar on the removal of emerging contaminants. *Chem. Eng. J.* **2020**, *402*, 126277. [CrossRef]
34. Hong, N.; Cheng, Q.; Goonetilleke, A.; Bandala, E.R.; Liu, A. Assessing the effect of surface hydrophobicity/hydrophilicity on pollutant leaching potential of biochar in water treatment. *J. Ind. Eng. Chem.* **2020**, *89*, 222–232. [CrossRef]

processes MDPI

Article

Use of the Solid By-Product of Anaerobic Digestion of Biomass to Remove Anthropogenic Organic Pollutants with Endocrine Disruptive Activity

Elisabetta Loffredo *, Claudia Carnimeo, Roccangelo Silletti and Carmine Summo

Dipartimento di Scienze del Suolo, della Pianta e degli Alimenti, Università degli Studi di Bari Aldo Moro, 70126 Bari, Italy; claudia.carnimeo@uniba.it (C.C.); roccangelo.silletti@uniba.it (R.S.); carmine.summo@uniba.it (C.S.)

* Correspondence: elisabetta.loffredo@uniba.it; Tel.: +39-080-5442282

Abstract: Anaerobic digestion of biomass has increasing implementation for bioenergy production. The solid by-product of this technology, i.e., the digestate, has relevant potential in agricultural and environmental applications. This study explored the capacity of a digestate from mixed feedstock to remove from water four endocrine-disrupting chemicals, namely the pesticides metribuzin (MET) and boscalid (BOS) and the xenoestrogens bisphenol A (BPA) and 4-tert-octylphenol (OP). The surface micromorphology and functional groups of the digestate were investigated using scanning electron microscopy (SEM) and Fourier-transform infrared (FTIR) spectroscopy, respectively. Results of sorption kinetics showed that all compounds reached the steady state in a few hours according to a pseudo-first-order model in the cases of MET and OP, a pseudo-second-order model for BOS and both models in the case of BPA. Data of adsorption isotherms were fitted to the Henry, Freundlich, Langmuir and Temkin equations. The adsorption of MET preferentially followed the non-linear Freundlich model, whereas the adsorption of the other compounds was properly described by both the linear and Freundlich models. The organic carbon partition coefficients, K_{OC}, were 170, 1066, 256 and 2180 L kg^{-1} for MET, BOS, BPA and OP, respectively. The desorption of BOS, BPA and OP was slow and incomplete, indicating a phenomenon of hysteresis. In conclusion, the digestate showed a remarkable efficiency in the removal of the compounds, especially those with high hydrophobicity, thus behaving as a promising biosorbent for environmental remediation.

Keywords: digestate; endocrine-disrupting chemical; metribuzin; boscalid; bisphenol A; octylphenol; xenoestrogen; sorption kinetics; sorption isotherm; desorption

check for updates

Citation: Loffredo, E.; Carnimeo, C.; Silletti, R.; Summo, C. Use of the Solid By-Product of Anaerobic Digestion of Biomass to Remove Anthropogenic Organic Pollutants with Endocrine Disruptive Activity. *Processes* **2021**, *9*, 2018. https://doi.org/10.3390/pr9112018

Academic Editors: Andrea Petrella, Marco Race and Danilo Spasiano

Received: 26 September 2021
Accepted: 9 November 2021
Published: 11 November 2021

1. Introduction

The current practice of processing waste biomass to produce bioenergy is increasingly adopted all over the world, representing a virtuous alternative to the consumption of fossil fuels. This approach can effectively cope with more than one emergency, such as the ever-increasing global demand for energy, the need to dispose of the huge mass of solid and liquid organic wastes from agro-zootechnical, industrial and urban activities and the need to implement a circular economy.

Several traditional and innovative conversion technologies, such as pyrolysis, microgasification, hydrothermal carbonization and anaerobic digestion (AD), have been developed by scientists and specialists to obtain combustible biogas and biofuels from organic wastes or dedicated crops [1]. These processes can also be combined to improve bioenergy production and usable remains [2,3]. The resulting by-products from these technologies are carbon-rich materials suitable for agricultural and environmental applications [4]. In agricultural practice, these materials can be used to improve soil fertility, as they are able to compensate for the widespread progressive decline of soil organic matter and also act as modulators of the bioavailability of phytonutrients and contaminants [5]. In environmental remediation strategies, these by-products can be exploited as biosorbents of various types of pollutants [4].

The AD process is a biological conversion of plant and animal wastes operated by various bacterial and archaeal populations [6]. Starting from different organic feedstocks, AD produces biogas and a concentrated semi-solid organic/inorganic mixture that, after a separation phase, gives a solid digestate (DIG) and a separate clarified liquid also called liquid digestate [3]. The chemical and physical properties of the DIG depend on the type of biomass used and the production parameters adopted and significantly influence the best utilization of the DIG [7]. During the AD process, the easily degradable organic compounds are readily converted into biogas, while the remaining more recalcitrant lignocellulosic components have remarkable retention properties towards inorganic and organic compounds. Some properties of DIG, such as surface reactive functional groups, a microstructure with some porosity and a large surface area, make this material a good biosorbent for organic compounds. Recently, DIG has been used successfully in mixtures with other carbon-rich substrates to prepare biofilters and biobeds [8].

Practices of conventional agriculture, such as the use of agrochemicals and the application to the soil of wastewater and sewage sludge not thoroughly decontaminated, have contributed to the release into the environment of anthropogenic organic pollutants (AOPs). AOPs include different classes of compounds, such as synthetic biocides, dyes, pharmaceuticals, personal care products, surfactants, wood preservatives and industrial products and by-products [9]. Many AOPs are biopersistent pollutants, and several of them have been recognized as endocrine-disrupting chemicals (EDCs) [10,11]. Even at very low levels, EDCs are capable of altering the normal hormonal functionality of humans and animals, especially aquatic animals, causing a variety of health effects, such as dysfunctions and pathologies of the reproductive apparatus and the cardiovascular system [12,13].

Among the AOPs with EDC activity, there are many plant protection products used intensively around the world to control crop diseases and ensure food production. Inappropriate and repeated use of these compounds can exceed the self-depollution capacity of the soil and accumulate in plants, soil and natural waters, with high risk for humans and ecosystems [14]. Metribuzin (4-amino-6-tert-butyl-3-(methylsulfanyl)-1,2,4-triazin-5(4H)-on) (MET) is a triazinone herbicide widely used to control broadleaf weeds in various crops. Due to its high water solubility, MET is considered to be one of the pesticides with the greatest potential for moving to surface and groundwater [15]. Another largely used pesticide is boscalid (2-chloro-N-(4′-chlorobiphenyl-2-yl)-nicotinamide) (BOS), which is a broad-spectrum carboxamide fungicide applied to protect fruit and horticultural plants. BOS is persistent in soil and especially in aquatic systems [16]. Recently, the European Commission has included both MET and BOS in the list of suspected EDCs [10].

Well-known EDCs are the xenoestrogens bisphenol A (2,2-Bis(4-hydroxyphenyl) propane) (BPA) and 4-tert-octylphenol (OP). BPA is widely used for the industrial production of polycarbonate and epoxy resins, flame retardants, food and beverage packaging, bottle caps and water supply systems [17]. The OP molecule is generated by the microbial degradation of octylphenol polyethoxylates which are non-ionic surfactants used in the production of paints, detergents and pesticides [18]. OP is largely present in the effluents of sewage sludge treatment plants and is biopersistent in the environment due to recalcitrance [18]. All these compounds can be detected in soil and natural waters where they pose serious environmental problems and health risks [19,20].

New sustainable strategies for removing EDCs from environmental matrices have recently been explored, including adsorption techniques using low-cost carbon-rich adsorbents obtained from processed biowaste, such as DIG. The sorption process consists of the accumulation of an adsorbate at the interface between the adsorbent phase and the solution phase. It is favourably regarded by researchers and operators for its easy-to-make features, low energy consumption and considerable efficiency.

In soil, the organic components, in particular the humic fraction, play a prominent role in the retention of pollutants. The low level of organic matter in some soils can therefore seriously compromise the self-depollution capacity of the soil and, consequently, the quality and fertility of the soil. Furthermore, the adsorption/desorption process controls the entire dynamic of contaminants in the soil–plant system, including movement and persistence. Therefore, the addition to the soil of organic amendments, such as DIG, can replace, at least in part, the native organic matter, with multiple benefits for the environment. The sorption of pollutants on DIG can prevent the leaching of these compounds, especially the more polar ones and control their bioavailability for plants and microorganisms. This is extremely important in the case of biopersistent molecules [21]. The removal of contaminants from water and wastewater is also very important. However, the use of expensive synthetic adsorbents, such as commercial activated carbon, and sophisticated technologies have become environmentally and economically unsustainable. Therefore, recent research has focused on the possible exploitation of biosorbents from biomass recycling and by-products of bioenergy production [4,22].

The interaction between pesticides and/or EDCs and carbon-rich materials has been extensively studied [8,23,24]. However, most studies concern the adsorption of individual AOPs on biosorbents such as compost and biochar [25–28], while little information exists on the use of DIG to remove AOPs [8]. Furthermore, few studies have evaluated the removal of AOPs from multi-contaminated matrices [29]. In real environmental systems, such as soil, contaminants with different properties and hydrophobicity coexist and interact simultaneously with solid and dissolved components, especially organic ones. The same happens in wastewater which is usually contaminated with many types of inorganic and organic compounds. Considering all this, the objective of this study was to investigate the quantitative aspects of the adsorption/desorption process of four EDCs with contrasting physicochemical properties, namely MET, BOS, BPA and OP, on a DIG sample obtained from a mixed plant and animal biomass.

2. Materials and Methods

2.1. Chemicals and Digestate

Metribuzin (MET) at a purity of ≥98.0%, boscalid (BOS) at 99.0% purity, bisphenol A (BPA) at 99.0% purity and 4-tert-octylphenol (OP) at 99.5% purity were purchased from Sigma-Aldrich S.r.l., Milano, Italy. Structural formula and some chemical properties of the compounds are reported in Table 1. All other chemicals of extra pure grade were obtained from commercial sources and used without further purification. Aqueous mixtures of MET, BOS, BPA and OP were prepared by diluting appropriate aliquots of individual methanol (HPLC grade) solutions of the compounds at a concentration of 2000 mg L^{-1} with double-distilled water. At the maximum concentration of the compounds used in the experiments (2 mg L^{-1}), the percentage of methanol in the final mixture was 0.4%.

The digestate (DIG) sample was obtained through anaerobic digestion from a mixture of oat silage, manure, slurry, cereal by-products, and two-phase olive pomace. The DIG sample was provided by F.lli Caione Azienda Agricola La Quercia Soc. Coop., Foggia, Italy. Before use, the DIG was air-dried and then characterized according to conventional methods. Some characteristics of air-dried DIG are reported in Table 2.

2.2. Scanning Electron Microscopy Analysis

Scanning electron microscopy (SEM) analysis of the DIG was performed to evaluate the surface micromorphology. The sample was fixed with an adhesive carbon tape, metallized with graphite, and analyzed with a high-resolution field emission scanning electron microscope VP FE-SEM ΣIGMA 300 (ZEISS, Oberkochen, Germany) equipped with an energy dispersive X-ray (EDX) elemental analyzer. The SEM micrographs of the DIG were captured at both 2500× and 13,000× magnifications using a 5 kV acceleration potential.

Table 1. Some properties of the compounds.

Compound	Chemical Structure	Molecular Weight (g/mol)	Water Solubility (mg L^{-1})	Log Kow
Metribuzin	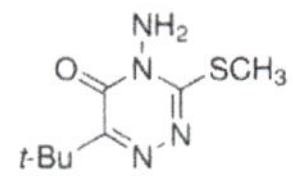	214.29	1200	1.70
Boscalid	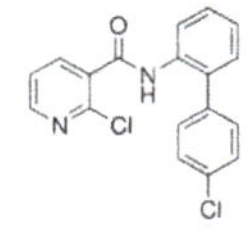	343.21	4.6	2.96
Bisphenol A	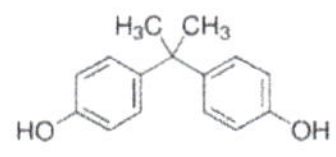	228.29	300	3.32
4-tert-octylphenol	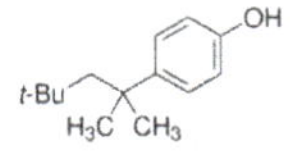	206.32	3.1	5.50

Data from PubChem [30].

Table 2. Some properties of the air-dried digestate.

Parameter	Value
pH [a]	8.73 ± 0.08 [b]
EC (dS m^{-1}) [a]	1.36 ± 0.04
Dry matter (%)	87.41 ± 0.63
Volatile solids (% d.m.)	87.15 ± 0.47
Total organic carbon (% d.m.)	50.55 ± 0.27
Ash (% d.m)	12.85 ± 0.54

[a] DIG/H_2O, 1:10 *w*/*v*. [b] All values are the mean ± SD (n = 3).

2.3. Fourier-Transform Infrared Analysis

Fourier-transform infrared (FTIR) analyses were performed to investigate the surface functional groups of the DIG sample before and after the 24-h interaction with an aqueous solution of the four compounds at the individual concentration of 2 mg L^{-1}. A mixture of 2 mg of dried DIG (before and after interaction) and 400 mg of dried KBr (FTIR grade) was homogenized using an agate mortar and pestle. A pellet was obtained from the mixture by pressing under vacuum at a pressure of 6000 kg cm^{-1} for 10 min. Fourier-transform infrared (FTIR) spectra were recorded using a Thermo Nicolet iS50 FTIR spectrophotometer in the range 4000–400 cm^{-1} with a resolution of 2 cm^{-1} and 16 scans min^{-1} for each acquisition.

2.4. Preliminary Adsorption Experiments

Preliminary adsorption experiments were conducted in batch mode to evaluate the efficiency of different solution/DIG ratios in removing the compounds. For the purpose, aliquots of 40, 50, 100 and 200 mg of DIG were made to interact with a volume of 20 mL of a mixture of MET, BOS, BPA and OP at the individual concentration of 2 mg L^{-1}, thus obtaining solution/substrate ratios equal to 500, 400, 200 and 100. All samples were mechanically shaken at 350× *g* for 24 h at a temperature of 20 ± 1 °C to reach sorption equilibrium. Previous experiments showed that, for each compound, no further adsorption occurred after 24 h. Successively, the suspensions were centrifuged at 10,000× *g* for 10 min. A volume of 18 mL of supernatant solution was collected from each sample and the equilibrium concentration of each compound was measured by high performance liquid chromatography (HPLC) as described in Section 2.6. All experiments were triplicated. The amount of compound sorbed on the DIG unit after 24 h, q_t (mg kg^{-1}), was calculated from the equation:

$$q_t = (C_0 - C_t)\ V/m \qquad (1)$$

where C_0 (mg L^{-1}) is the initial concentration of the compound in solution, C_t (mg L^{-1}) is the concentration of the compound at time t (24 h in these experiments), V (L) is the volume of the solution and m (kg) is the mass of the substrate.

The experimental sorption data for each compound (triplicated values) were statistically analyzed using one-way analysis of variance (ANOVA) followed by the Duncan's new multiple range test at $P \leq 0.01$.

2.5. Adsorption and Desorption Experiments

Sorption kinetics were performed to evaluate the adsorption rates of MET, BOS, BPA and OP onto the DIG sample and to establish the time required to reach equilibrium. Based on the results obtained from the preliminary trials, in the subsequent experiments, a solution/substrate ratio of 500 was adopted, that is the ratio that allowed the maximum adsorption efficiency of the DIG. Hence, volumes of 10 mL of an aqueous mixture of the four compounds at individual concentration of 2 mg L^{-1} and pH of 8.09 were added to 20 mg of DIG in glass centrifuge tubes. The suspensions were then stirred for 0.25, 0.5, 1, 4, 8, 16 and 24 h in the dark at 20 $\pm$ 1 °C. After each time, the suspensions were centrifuged at 10,000$\times$ g for 10 min and the supernatants were analyzed by HPLC to determine the residual concentration of each compound in solution (see Section 2.6). All experiments were triplicated. The amount of compound sorbed on the substrate unit at each time t, q_t (mg kg^{-1}) was calculated from Equation (1). To determine the sorption equilibrium time, the Student's *t*-test was used to compare, two-by-two, the quantities of compound adsorbed at any time ($P \leq 0.05$). The equilibrium time was established when there was no significant difference between the values at two successive times. Then, ANOVA and the Duncan's new multiple range test ($P \leq 0.01$) were performed to evaluate significant differences among the adsorbed quantities at equilibrium.

Adsorption isotherms of the four compounds onto the DIG were performed using the batch equilibration method. Aliquots of 20 mg of DIG were added with 10 mL of aqueous mixtures of MET, BOS, BPA and OP at individual concentrations of 0.2, 0.4, 0.5, 1 and 2 mg L^{-1} in glass centrifuge tubes. Based on the adsorption kinetics results, the suspensions were stirred for 24 h at 20 $\pm$ 1 °C in the dark to reach equilibrium. Then, the suspensions were centrifuged at 10,000$\times$ g for 10 min and the equilibrium concentration of each compound in the supernatants was measured by HPLC (see Section 2.6). All experiments were triplicated.

Desorption experiments were started immediately after adsorption using the DIG sample added with the maximum initial concentration of each compound (2 mg L^{-1}). For all four desorption cycles, an aliquot of 8 mL of equilibrium supernatant solution was replaced with the same volume of distilled water. The sample was stirred again for 24 h at 20 $\pm$ 1 °C and centrifuged in the conditions reported above. Then, the supernatant solution was collected, and the compounds were analyzed using the HPLC procedure described in the next section.

2.6. Chromatographic Analysis

The concentration of each compound in solution was measured using an HPLC apparatus equipped with a Spectra System^{TM} pump (Thermo Electron Corporation, San Jose, CA, USA), a Rheodyne® 7125 injector with a 20 μL loop and a Supelcosil^{TM} LC-18 chromatographic column (250 mm $\times$ 4.6 mm $\times$ 5 μm). The mobile phase was a mixture of water (A) and acetonitrile (B) flowing at 0.8 mL min^{-1}. The gradient elution adopted was: 0–4 min, 50% B; 4–8 min, from 50 to 70% B; 8–15 min, from 70 to 90% B. Retention times of MET, BOS, BPA and OP were 4.0, 6.2, 9.2 and 13.6 min, respectively. MET and BOS were detected using a Spectra System UV6000LP™ diode array detector at wavelengths of 294 nm and 207 nm, respectively. BPA and OP were quantified by a Spectra System FL3000 fluorescence detector operating at wavelengths of 240-nm excitation and 310-nm emission. All compounds were quantified using the external standard method.

2.7. Sorption and Desorption Models

Two theoretical models were used to interpret the sorption kinetics data, determine the kinetic constants and investigate the sorption mechanisms of the compounds onto the DIG. The pseudo-first-order model of Lagergren [31] is based on sorbent capacity. The non-linear form of the pseudo-first-order kinetic model is given by the equation [32]:

$$q_t = q_e\,(1 - \exp^{-k1t}) \tag{2}$$

where q_e and q_t are the amount of the compound adsorbed per mass unit of adsorbent (mg kg^{-1}) at equilibrium and at time t, respectively, and k_1 (h^{-1}) is the rate constant of the pseudo-first-order kinetics. The non-linearized form of the pseudo-second-order kinetic model, based on equilibrium sorption, is expressed as [33]:

$$q_t = \frac{q_e^2 k_2 t}{1 + k_2 q_e t} \tag{3}$$

In this equation, q_t and q_e have already been described for the previous model and k_2 (kg mg^{-1} h^{-1}) is the second-order adsorption rate constant. Then, using the solver add-in component of Microsoft Excel, a trial-and-error procedure was adopted to estimate the pseudo-first-order and pseudo-second-order kinetic parameters with the non-linear regression method [33]. To quantify the fit of the two equations to the experimental data, the widely used error function coefficient of correlation, r, was adopted:

$$r = \sqrt{\frac{\Sigma(q_t m - \overline{q_t})^2}{\Sigma(q_t m - \overline{q_t})^2 + \Sigma(q_t m - q_t)^2}} \tag{4}$$

where $q_t m$ is the amount of solute adsorbed per mass unit of sorbent (mg kg^{-1}) at time t according to the kinetic model considered, q_t is the experimental amount of solute adsorbed per mass unit of sorbent (mg kg^{-1}) and $\overline{q_t}$ is the mean of q_t (mg kg^{-1}).

The adsorption isotherms data for each compound were interpreted with four sorption models, the two-parameter non-linear empirical equations of Freundlich, Langmuir and Temkin and the linear Henry equation. The Freundlich isotherm model assumes a multilayer adsorption of the sorbate on the substrate and is described by the equation:

$$q_e = K_F\, C_e^{1/n} \tag{5}$$

where q_e (mg kg^{-1}) is the amount of compound adsorbed per unit of substrate, C_e (mg L^{-1}) is the equilibrium concentration of the sorbate in solution, $1/n$ indicates the degree of nonlinearity between the concentration of the compound in solution and that adsorbed, the reciprocal n expresses the sorption intensity and K_F (better indicated as $K_{F\text{-}ads}$ and $K_{F\text{-}des}$ for adsorption and desorption, respectively) is the measure of the sorption capacity of the adsorbent. The Langmuir model is based on a monolayer adsorption and is expressed by:

$$q_e = (K_L C_e b)/(1 + K_L C_e) \tag{6}$$

where q_e and C_e are defined as in Equation (5), b (mg kg^{-1}) is the maximum adsorption capacity of the adsorbent, i.e., the amount of compound forming a monolayer on the adsorbent and K_L (L mg^{-1}) is the Langmuir constant that expresses the energy of adsorption, that is the affinity between the adsorbent and the sorbate. The Temkin isotherm predicts a logarithmic reduction of adsorptive site and energy and is expressed by:

$$q_e = B\ln(A_T C_e) \tag{7}$$

where q_e and C_e have the same meaning as the previous Equations (5) and (6), A_T is the Temkin isotherm equilibrium binding constant (L mg^{-1}) and B (J mol^{-1}) is a parameter expressing the enthalpy of the adsorption. In this equation, B = RT/b_T, where b_T is a constant related to the heat of adsorption, T is the absolute temperature (K) and R is the universal gas constant (8.314 J mol^{-1} K^{-1}). The adsorption parameters K_F and 1/n of the Freundlich equation (Equation (5)), b and K_L of the Langmuir equation (Equation (6)) and B and A_T of the Temkin equation (Equation (7)) were estimated through the non-linear regression method using the solver add-in component of Microsoft Excel and a trial-and-error procedure which minimized the sum of squared residuals (SSR) between observed and simulated concentrations [8]. The fit of the three equations to the experimental data was quantified using the error function coefficient of correlation, r, as described for the calculation of the kinetic parameters (Equation (4)).

Finally, the Henry linear equation:

$$q_e = K_d\ C_e \quad (8)$$

assumes a constant proportion between the concentration of the sorbate in solution and the concentration of the sorbate on the adsorbent over the concentration range studied. Equation (7) allowed the calculation of the distribution coefficient, K_d (L kg^{-1}), from the slope. The amount of adsorbed compound per unit of organic C (OC) of substrate, i.e., the organic-carbon-partition coefficient K_{OC}, was also calculated according to: K_{OC} = (K_d × 100)/(%OC)) [23].

Using the same calculation procedure adopted to obtain the sorption parameters, the desorption data, i.e., the quantity of each compound remained adsorbed at each desorption step and the corresponding equilibrium concentration, were fitted in the Freundlich equation to calculate the parameters $K_{F\text{-}des}$ and 1/n_{des} and in the Henry equation to calculate $K_{d\text{-}des}$ and $K_{OC\text{-}des}$. Finally, the hysteresis coefficient, H, for the adsorption–desorption isotherm was calculated from the ratio H = (1/n_{des})/(1/n_{ads}) [34]. In general, a H value < 1 is indicative of a hysteretic condition, i.e., difficulty for the sorbate to be desorbed from the substrate.

3. Results and Discussion

3.1. SEM Analysis

The SEM technique identifies the micromorphological aspects of the material surface with information on the distribution and allocation of the pores. The surface features of the DIG sample were evaluated using SEM coupled with an EDX elemental analyzer. Images obtained at 2500× and 13,000× magnifications are shown in Figure 1. The surface of the DIG sample appeared rough and presented numerous irregularly shaped ridges, sharp edges, microparticles, channels and cavities of mostly less than 10 μm (Figure 1). The pores, which generally originate from cell walls and vascular tissues, were not so evident. Porosity and a large surface area are extremely important properties when the material is used as an adsorbent for decontamination purposes. The EDX spectrum obtained for the DIG sample evidenced the presence on the surface of various elements typical of plant-based materials (Figure 1).

3.2. FTIR Analysis

The surface functional groups of the DIG and their modification after adsorption of the four compounds were investigated using Fourier-transform infrared (FTIR) spectroscopy (Figure 2). The main features of the FTIR spectrum of the DIG were the following: (a) a strong band centered at 3410 cm^{-1}, typical of O–H and N–H stretching, also hydrogen bonded; (b) two twin peaks of medium intensity at 2925 and 2853 cm^{-1} attributable to aliphatic C–H stretching; (c) a medium-strong peak at 1638 cm^{-1} with a shoulder at 1598 cm^{-1} that can be ascribed to various vibrations, including aromatic C=C stretching, C=O stretching of amide groups (amide I band) and ketonic C=O and COO^- symmetric stretching; (d) a weak peak at 1510 cm^{-1} feasibly due to aromatic C=C stretching and N–H deformation and C=N stretching of amide groups (amide II band); (e) two weak peaks at 1456 and 1423 cm^{-1} possibly due to aliphatic C–H deformation; (f) an intense peak at 1385 cm^{-1} that can be attributed to various vibrations, including O–H deformation and C–O stretching of phenolic groups, C–H deformation of CH_2 and CH_3 groups and COO^- asymmetric stretching; (g) a weak peak with a shoulder at 1230 cm^{-1} possibly due to C–O stretching, O–H deformation of COOH and C–O stretching of aryl ethers and phenols; (h) a shoulder at 1153 cm^{-1} attributable to aliphatic C–OH stretching; (i) a medium-strong band at 1040 cm^{-1} with a shoulder at 1107 cm^{-1} typical of C–O stretching of polysaccharide-like substances and Si–O silicate impurities in the digestate (Figure 2) [23].

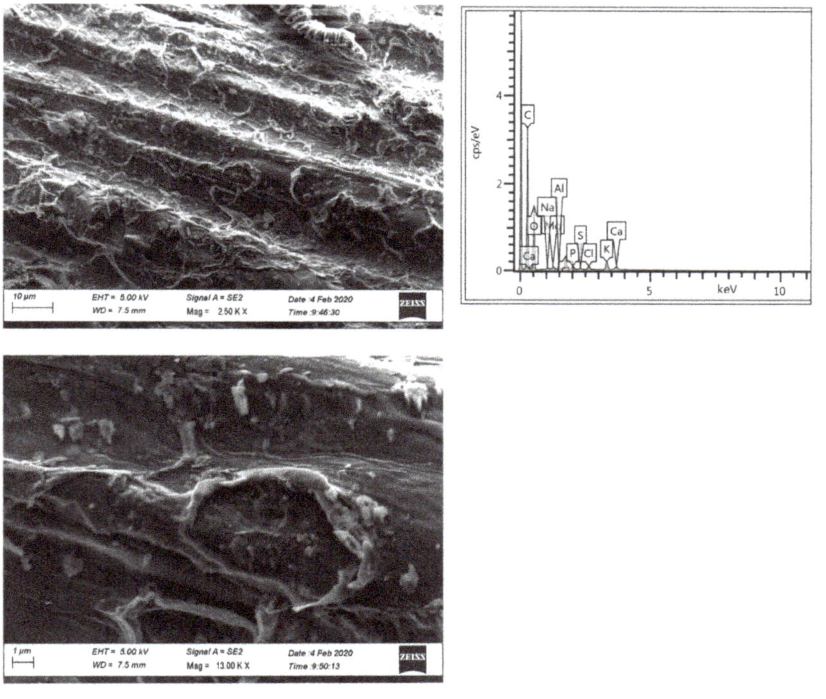

Figure 1. Scanning electron microscopy (SEM) images and energy-dispersive X-ray (EDX) spectrum of the digestate sample at magnifications of 2500 (**top**) and 13,000 (**bottom**) times. Images were taken with secondary electrons.

The FTIR spectrum of the DIG after interaction with the four molecules appeared not to be informative as it showed no evident modifications in the bands/peaks wavenumbers and relative intensity compared to that of the original DIG, except for the strong peak at 1385 cm^{-1} possibly due to the presence of phenolic O–H and C–O in the four molecules. However, this result is not surprising given the low concentrations of the molecules used in the interaction and the richness and strong intensity of the bands/peaks of the DIG.

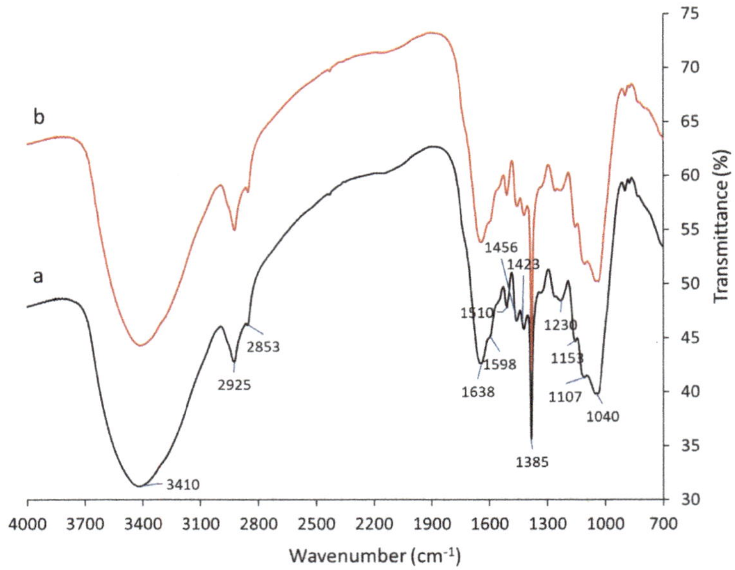

Figure 2. FTIR spectra of the DIG before (**a**) and after (**b**) adsorption of the compounds.

3.3. Preliminary Adsorption Experiments

The amounts of each compound adsorbed on DIG at equilibrium are reported in Figure 3 and Table 3. The different solution/adsorbent ratios adopted are quite common and comparable with those used in similar studies [8,27]. By increasing the solution/DIG ratio from 100 to 500, the adsorbed quantity of BOS, BPA and OP increased significantly with each subsequent ratio tested, whereas that of MET only increased to the solution/DIG ratio of 500 when the adsorption doubled, compared to that measured at the ratios of 100 and 200 (Table 3). It is reasonable to assume that, at low adsorbent dosage, more sorption sites were available, including the innermost and less accessible ones, for the molecules, especially the more hydrophobic BOS and OP, and, at high adsorbent dosage, fewer sites were available. In addition, in the case of MET, a relevant increase of the solution/DIG ratio favored the sorption extent, probably due to the reduced competition from water molecules. At the highest ratio (500), the percentages of MET, BPA, BOS and OP adsorbed on the DIG after 24-h interaction were approximately 11, 45, 24 and 66% of the initial compound added (40 μg), respectively. Recently, Yao et al. [35], investigating the capacity of a DIG to remove various dyes from a wastewater, reported an increase in the adsorption capacity as the solution/DIG ratio increased. The authors commented that this could be due to the large number of sites not occupied by the dye when the DIG was at a high dosage, thus reducing the adsorption capacity of the latter [35]. Considering the physicochemical properties of the four molecules, as expected, the affinity for the DIG (MET < BPA < BOS < OP) was inversely related to their solubility in water. In fact, it is well known that an organic solute is more adsorbable by a substrate the weaker its interaction with the solvent is. A negative correlation was already demonstrated between the sorption efficiency of carbon-rich materials and the water solubility of some EDCs and pesticides [24]. Based on these results, the highest ratio was chosen for the subsequent sorption experiments.

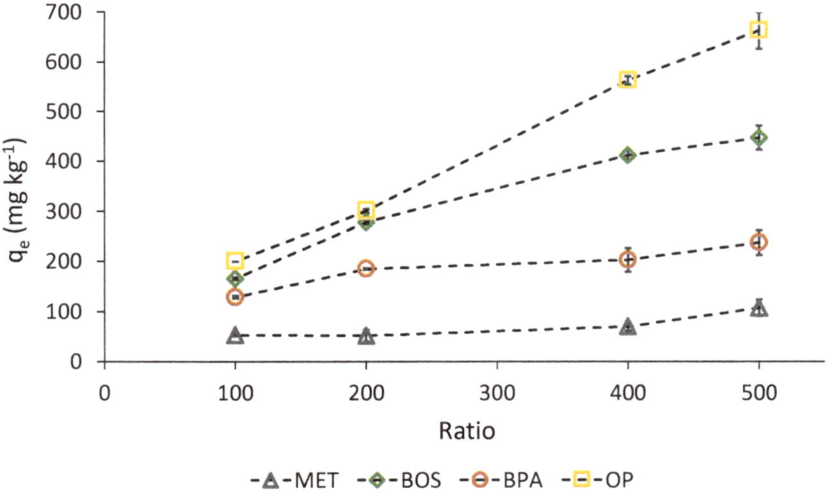

Figure 3. Adsorption of metribuzin (MET), boscalid (BOS), bisphenol A (BPA) and 4-tert-octylphenol (OP) from water on the digestate at equilibrium concentration using different solution/adsorbent ratios.

Table 3. Amount of compound adsorbed (mg kg^{-1}) as a function of the solution/digestate ratio adopted.

Compound	100	200	400	500
MET	53.11 ± 2.65 B	51.56 ± 11.88 B	69.83 ± 8.42 B	107.78 ± 16.42 A
BOS	165.99 ± 2.23 D	278.19 ± 4.19 C	412.97 ± 8.72 B	448.74 ± 24.12 A
BPA	128.88 ± 2.48 D	184.99 ± 1.25 C	203.08 ± 24.04 B	238.20 ± 25.04 A
OP	200.00 ± 0.12 D	301.11 ± 3.65 C	563.65 ± 8.15 B	664.51 ± 36.52 A

Note: data were analyzed by ANOVA and means were separated by the Duncan's new multiple range test at $P \leq 0.01$ (n = 3).

Although, as expected, the extent of DIG sorption in this study was somehow lower than that of biochar for MET [26,27,36], BOS [8], BPA [37] and OP [24], it can be considered remarkable and suggests a valuable use of this material as biosorbent. To the best of our knowledge, the literature does not report information on the removal of the four molecules studied by the DIG alone. The only comparable study found in the literature is that of Mukherjee et al. [8], who used soil/DIG biomixtures to remove some pesticides, including BOS.

3.4. Adsorption Kinetics

The study of adsorption kinetics allows the evaluation of the retention rate of the solute on the substrate in a certain period and provides useful information on the type of interaction. Based on the kinetics curves of adsorption obtained, it was evident that each molecule reached the steady state in a relatively short time. The Student's *t*-test applied to the quantities of compound adsorbed by the DIG at each time stated that the equilibrium times were 4 h for MET, BPA and OP and 8 h for BOS (Figure 4). The longer equilibrium time shown by BOS, compared to the other three molecules, might be due to its larger size. Then, subsequent adsorption isotherms of all compounds were performed using a contact time of 16 h between the substrate and the solution. The shapes of the kinetics curves indicate that the adsorption was a multi-step process, comprising a first rapid adsorption of the compounds on the most available external sites of the material followed by a slower sorption on the innermost active sites. Upon reaching the equilibrium condition, the maximum sorption concentration of 106.20 ± 4.77, 591.61 ± 11.89, 208.87 ± 11.49 and 712.80 ± 2.47 mg kg^{-1} were measured for MET, BOS, BPA and OP, respectively (Figure 4 and Table 4). Statistical treatment of these data (ANOVA and Duncan's test) showed a high significant difference ($P \leq 0.01$) among the amounts of compounds adsorbed at equilibrium (Table 4). The two less water-soluble BOS and OP showed the greatest affinity for the DIG sample, their concentration on the material being about 6 and 7 times higher, respectively, than that of MET. The lower adsorption of MET on organic substrates, compared to other pesticides and ECDs, is generally responsible for the high mobility of MET in soil and sediments and the consequent dangerous release into natural waters [15].

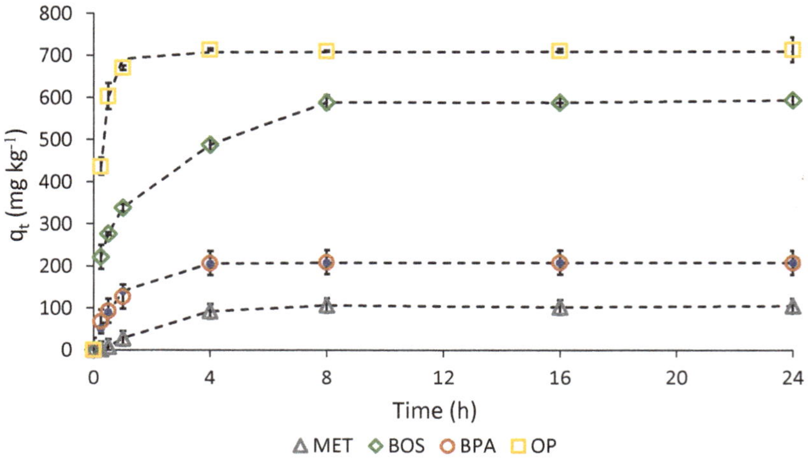

Figure 4. Adsorption kinetics data and plots of predicted pseudo-first-order kinetics of metribuzin (MET), bisphenol A (BPA) and 4-tert-octylphenol (OP) and predicted pseudo-second-order kinetics of boscalid (BOS) onto the digestate. Standard error is reported as vertical bar on each point (n = 3).

Table 4. Kinetic pseudo-first-order and pseudo-second-order parameters obtained through the non-linear method for the adsorption of the compounds onto the digestate.

Compound	Pseudo-First-Order					Pseudo-Second-Order			
	$q_{e,experimental}$ (mg kg^{-1})	r	SSR	$q_{e,1}$ (mg kg^{-1})	k_1 (h^{-1})	r	SSR	$q_{e,2}$ (mg kg^{-1})	k_2 (kg mg^{-1} h^{-1})
MET	106.20 D	0.985	339	107.17	0.372	0.965	793	124.78	0.003
BOS	591.61 C	0.958	16,199	561.81	1.217	0.983	5525	601.64	0.003
BPA	208.87 B	0.991	485	207.53	1.125	0.990	472	221.21	0.007
OP	712.80 A	0.996	513	706.35	3.801	0.971	3450	728.61	0.010

Note: experimental data at equilibrium were analyzed by ANOVA and the means were separated by the Duncan's new multiple range test at $P \leq 0.01$ (n = 3).

To investigate the adsorption mechanisms of the compounds onto the DIG, the kinetic data were analyzed using the non-linearized forms of the pseudo-first-order (Equation (2)) and pseudo-second-order (Equation (3)) models. Both models are widely applied in the study of AOPs' adsorption on organic and mineral substrates [32,38,39]. Table 4 shows the values of the kinetics parameters of the molecules according to the two models along with the correlation coefficients, r, and the sum of squared residuals, SSR. High r values and low SSR values indicate the adequacy of the model to the experimental data. Based on the r values, MET and OP preferentially followed the pseudo-first-order model and BOS followed the pseudo-second-order model, whereas BPA fitted both models well (Table 4). This was confirmed by the SSR values. Figure 4 shows the sorption kinetics data and plots of the predicted pseudo-first-order kinetics of MET, BPA and OP and the predicted pseudo-second-order kinetics of BOS.

The pseudo-first-order kinetics suggest a prevailing role of the adsorbent capacity [32]. The pseudo-second-order kinetics is based on sorption equilibrium capacity and assumes that the rate-limiting step is a chemisorption process involving valency forces through the sharing or exchange of electrons between the adsorbent and the sorbate [38,39]. The lignocellulosic fraction of the DIG may have interacted with the different molecules through various mechanisms, such as weak physisorption through van der Waals forces and hydrogen bonding and chemisorption through strong ionic or covalent bonds. As far as we know, there are no other studies on the sorption kinetics of the molecules examined here onto DIG. A recent work reported that the adsorption of MET on different chars was better described by the pseudo-second-order model [27]. This suggests an important role of the sorbent properties on the type of interaction prevailing between a solute and the substrate. Essandoh et al. [36], studying the adsorption of MET on biochar, concluded that the adsorption mechanism of MET could be explained mainly by hydrogen bonds and Coulombic forces and a minor contribution of van der Waals, dipole–dipole and π–π interactions.

Although, in these experiments, no significant relationship was calculated between the amounts of the molecules adsorbed at equilibrium and the corresponding log Kow or water solubility, it was evident that the greatest removals by the DIG were obtained with the most hydrophobic OP and BOS.

3.5. Adsorption and Desorption Isotherms

The adsorption isotherm describes quantitatively the interaction between an adsorbent and a sorbate at a fixed temperature. It allows the evaluation of the adsorption parameters, such as the adsorption constant and the maximum adsorption, and provides indications on the allocation of the sorbate on sorbent sites. Different sorption models were adopted to fit the experimental adsorption and desorption data of the four compounds, such as the Freundlich, Langmuir, Temkin and Henry equations. Modeling of isotherm data is useful for predicting the adsorption mechanism. The Freundlich model can account for reversible adsorption on a heterogeneous substrate surface and is not limited to monolayer adsorption. Differently, the Langmuir equation is a good fit when the sorbent surface is homogeneous, the molecular interaction among the adsorbed species is negligible and the adsorption occurs as a monolayer on the adsorbent. The Temkin isotherm predicts a logarithmic reduction of available sites and sorptive energy and is best applied at intermediate concentrations.

The experimental sorption data obtained, along with the plots of the predicted Freundlich, Langmuir, Temkin and Henry equations, are shown in Figure 5. The sorption parameters calculated by fitting the equilibrium data in all four models and the desorption parameters calculated by fitting the experimental data in the Henry and Freundlich models are given in Tables 5 and 6, respectively. In addition to the values of the correlation coefficients (r), these tables show the values of the sum of squared residuals (SSR) that are an indicator of the matching of experimental data with the theoretical model (Tables 5 and 6). Higher r values and lower SSR values indicate a better fit of the experimental data in the model.

Based on the calculated r values over the whole concentration range tested, the sorption of MET was better described by the Freundlich model, whereas the sorption of the other three compounds better matched the Henry equation (Table 5). However, considering the SSR values, it was evident that, for all four compounds, the Freundlich model was the best fit because it reduced errors (lowest SSR), even if the differences among the SSR values of the four models were not always relevant (Table 5). Hence, the adsorption of each compound occurred through the multilayer formation on the heterogeneous surface of the DIG. Moreover, based on the exponent of the Freundlich equation, $1/n_{ads}$, and according to Giles et al. [40], the isotherm of MET was L-shaped ($1/n < 1$), those of BPA and OP were C-type ($1/n \sim 1$) and that of BOS was S-shaped ($1/n > 1$) (Table 5). A non-linear L-shaped

Freundlich isotherm indicates that the compound has a high affinity for the sorbent in the initial stage of adsorption, while successively it decreases as adsorption sites are filled, and the process never reaches saturation. The L-shaped isotherm is typical of a solute of low hydrophobicity, such as MET (log Kow = 1.70), onto heterogeneous substrates. A C-type isotherm indicates a constant partitioning of the sorbate between the solution and the sorbent, without reaching saturation in the concentration range adopted. Finally, S-type isotherms are usually observed at low concentrations and indicate an increasing adsorption rate as the concentration of the sorbate in solution increases. The S-type isotherm is typical of adsorbents having high affinity for the solvent, e.g., water competes with the solute for adsorption sites [40]. The Freundlich exponent (1/n) is related to the strength and feasibility of the adsorption. The reciprocal n is the heterogeneity factor. Based on 1/n value, the process can be considered mainly physical when $1/n < 1$, chemical when $1/n > 1$ and linear if 1/n is equal to 1 [41]. Thus, the low 1/n value obtained for MET suggests that the adsorption on the DIG was predominantly physical (Table 5). Differently, the high 1/n value for BOS suggests a chemisorption process. The 1/n values of the two phenolic molecules, BPA and OP, were not very different from the unit (Table 5). These findings agree with the results obtained in the kinetic study. Based on the structure and functional groups of the molecules used in the mixture (very weak organic acids or bases) in this study, no interactions between/among them can be expected. AOPs may interact with organic matter (OM) in several ways which affect both the retention capacity of the adsorbent and the rate of desorption. In general, AOPs can be adsorbed to OM through specific physical and chemical binding mechanisms and forces with varying degrees and strengths, which include ionic, hydrogen and covalent bonding, charge-transfer or electron donor-acceptor mechanisms, dipole–dipole and van der Waals forces, ligand exchange and cation and water bridging [42]. The capacity of DIG to retain organic compounds is essentially due to the numerous hydrophobic and hydrophilic surface sites and chemically reactive functional groups, such as carboxylic and phenolic OH, alcoholic OH, ketonic C=O, amine groups and so on, where the molecules can be linked with bonds of different type and strength. For example, DIG sites containing O and N- can form hydrogen bonds with molecules containing suitable complementary groups. This can happen for MET that, at the pH of this study, is almost all unprotonated. The adsorption of MET onto different biosorbents has been mainly ascribed to H bonds and Coulombic forces and, to a lesser extent, to van der Waals and dipole–dipole interactions [36]. It is likely that the non-polar BOS molecule interacted with DIG through hydrophobic interaction [8]. Phenolic groups of BPA and OP can interact through electron donor-acceptor mechanisms or charge-transfer, by binding with complementary groups of DIG. However, adsorption of low-polar AOPs can also occur through non-specific hydrophobic or partitioning processes between water and the OM phase, in particular on hydrophobic active sites of OM, such as aliphatic side chains and aromatic structures [42]. OP has been proven to bind to organic materials through covalent and H bonds [24]. The presence of strong and weak binding of various phenolic EDCs to organic matter has been reported [42]. Finally, the mechanisms and the extent of adsorption certainly depend on the solution/adsorbent ratio, the physical and chemical properties of both the solute and the adsorbent and the condition of the medium, such as pH and ionic strength.

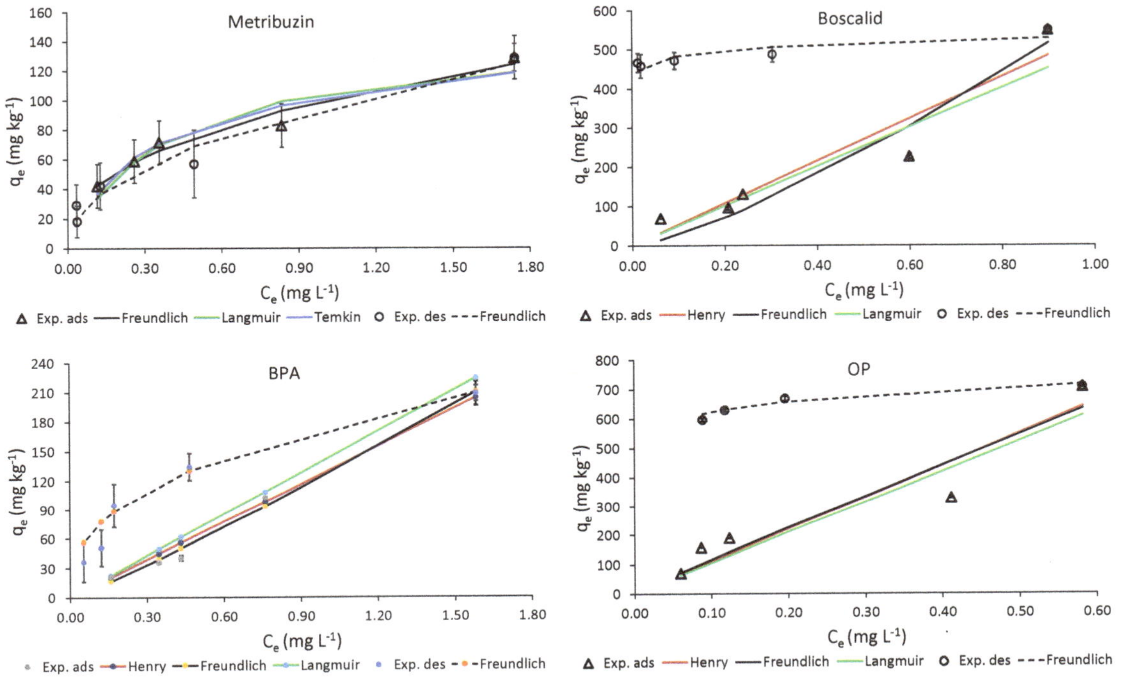

Figure 5. Adsorption (experimental points and solid lines representing plots of Henry, Freundlich, Langmuir and Temkin models) and desorption (experimental points and dashed lines representing plot of Freundlich model) isotherms of the compounds on the digestate. Standard error is reported as vertical bar on each point (n = 3).

As far as we know, there are no studies concerning the adsorption of the four compounds used here on the DIG, and therefore it is not possible to compare our results with those of the literature. The adsorption of MET on various plant-based chars was well interpreted by L-shaped Freundlich isotherms [27]. In a recent work, BOS adsorption on different sediments followed a Freundlich model [43], and the same model was the best fit for OP on an aquifer material [44] and various carbon-rich substrates [24].

The Henry isotherm (Equation (8)) allowed the calculation of the distribution coefficient K_d that is a reliable parameter to express the sorption efficiency of a substrate. The adsorption parameters ($K_{d\ ads}$, $K_{F\ ads}$ and $1/n_{\ ads}$) and the coefficient $K_{OC\ ads}$ for the four compounds are referred to in Table 5. The adsorption constants $K_{d\ ads}$, $K_{F\ ads}$ and $K_{OC\ ads}$ followed the same order: MET < BPA < BOS < OP (Table 5). The $K_{F\ ads}$ value of the more hydrophobic OP was approximately 11 times higher than that of the more water-soluble MET (Table 5). The values of the normalized organic carbon, $K_{OC\ ads}$, of the four compounds varied widely from 170 for the least hydrophobic MET to 2180 for the most hydrophobic OP, which are comparable to the values found in the literature [8,27,36,43,44]. The interpretation of the adsorption data with Langmuir's equation allowed the calculation of the maximum adsorption, b. On the basis of the b values, the adsorption trend was the following: BOS > OP > BPA > MET. The Temkin model is applied for an intermediate concentration range and takes into account the interaction between the sorbate and the adsorbent. It assumes that the heat of adsorption of the sorbate decreases linearly with the surface coverage. From the Temkin plot, the parameters estimated were A_T, B and b_T (Table 5). When the Temkin equation was used to fit the experiment data, a good match was obtained only for MET with r = 0.962, though this value was lower than in the Freundlich model (Table 5). The plot of the Temkin model for MET is depicted in Figure 5. Much lower r values were obtained for the other compounds, indicating that the Temkin model was not appropriate for BPA, OP and, especially, BOS. The parameter B gives an indication of the heat of adsorption; for each molecule, especially the most hydrophobic ones, it was quite high, suggesting an exothermic sorption process of the molecules on the DIG [41].

Unfortunately, there is no information in the literature on the extent and modeling of adsorption of these molecules on DIG, and this does not make a comparison possible. In a recent study, Gaullier et al. [43] reported $K_{F\ ads}$ values for BOS on sediments ranging from 2.8 to 13.5 mg kg^{-1}. The $K_{OC\ ads}$ value calculated in this study for MET on DIG (170 L kg^{-1}) was almost twice and 73%, respectively, of those found averagely for the adsorption of MET on vermicompost (93 L kg^{-1}) and hydrochar (232 L kg^{-1}) [27]. The $K_{OC\ ads}$ value of BOS calculated here (1066 L kg^{-1}) was much higher than that reported for the adsorption of BOS on a biomixture of 30% of DIG and soil [8]. Ying et al. [44] studied the adsorption of various EDCs on a sediment and reported $K_{F\ ads}$ values of 3.89 and 90.9 L kg^{-1} for BPA and OP, respectively.

Table 5. Adsorption parameters of the compounds on the digestate.

Compound	Henry Model				Freundlich Model				Langmuir Model				Temkin Model				
	r	SSR	K_d (L kg^{-1})	K_{OC} (L kg^{-1})	r	SSR	$K_{F\ ads}$ (L kg^{-1})	$1/n_{\ ads}$	r	SSR	b (mg kg^{-1})	K_L (L mg^{-1})	r	SSR	A_T (L kg^{-1})	B (J mol^{-1})	b_T
MET	0.930	4588	86.19	170.50	0.983	147	99.88	0.40	0.954	445	145	2.587	0.962	322	29.46	30.12	80.87
BOS	0.980	15,049	538.73	1065.74	0.968	11872	593.88	1.32	0.948	15,130	126,113	0.004	0.815	51,869	14.57	153.57	15.86
BPA	0.997	354	129.20	255.59	0.996	205	126.31	1.11	0.992	361	47,424	0.003	0.925	3459	5.44	83.15	29.30
OP	0.980	26,465	1102.09	2180.20	0.951	26,322	1067.82	0.96	0.953	26,528	117,129	0.009	0.903	46,769	20.54	229.41	10.62

Table 6. Desorption parameters of the compounds from the digestate.

Compound	Henry Model				Freundlich Model				H
	r	SSR	$K_{d\ des}$ (L kg^{-1})	$K_{OC\ des}$ (L kg^{-1})	r	SSR	$K_{F\ des}$ (L kg^{-1})	$1/n_{\ des}$	
MET	0.947	2369	78.80	155.88	0.982	287	96.35	0.47	1.175
BOS	0.673	651,881	769.65	1522.55	0.874	1287	531.02	0.04	0.030
BPA	0.904	14,170	151.70	300.10	0.994	170	175.59	0.39	0.351
OP	0.753	889,423	1709.20	3381.21	0.963	507	749.07	0.08	0.083

The experimental desorption data and the Freundlich plots for the four compounds are shown in Figure 5, while the desorption coefficients of Henry ($K_{d\ des}$ and $K_{OC\ des}$) and Freundlich ($K_{F\ des}$ and $1/n_{\ des}$) are given in Table 6. When desorption data were fitted to both Henry and Freundlich equations, based on both r and SSR indicators, the Freundlich model was the best fit for all compounds (Table 6). In all cases, the desorption isotherm was L-type ($1/n_{\ des} < 1$). After four desorption steps, approximately 86, 17, 74 and 16% of the initially retained MET, BOS, BPA and OP were desorbed from the DIG, respectively. The values of $K_{F\ des}$ for BOS and OP were lower than the corresponding values of $K_{F\ ads}$, suggesting a partial reversibility of the process. Comparing the values of $1/n_{\ des}$ and $1/n_{\ ads}$, we found that for MET alone the desorption rate was slightly higher than the adsorption rate ($1/n_{\ des} > 1/n_{\ ads}$), whereas the desorption of all the other compounds occurred very slowly and was incomplete (Table 6). This behavior is clearly shown in Figure 5. In other words, the DIG demonstrated a good capacity to retain MET, but also an easy release (negative hysteresis, $H > 1$) as soon as the liquid phase was diluted. This was easily predictable considering the physicochemical properties of MET, such as the low log Kow and high water solubility. Differently, the other three molecules were difficult to desorb from the DIG, denoting a strong sequestration and a significant positive hysteresis ($H < 1$) (Table 6). The formation of covalent bonds could be the reason for the low desorption rate and the hysteresis phenomenon observed for these molecules. These results suggest that the type of interaction between MET and the DIG was weaker than that formed between the other three compounds and the DIG. We can assume that the more hydrophobic molecules were retained through high-energy bonds which hindered their reversibility. The marked retention of BOS, BPA and OP by the DIG and the low desorption, especially for BOS and OP, are reasonably due to their chemical structures and hydrophobic character. In particular, the large size and the low water solubility of BOS can explain the very low release observed. The desorption of BOS from a 30% DIG/soil mixture was almost negligible after three desorption steps [8]. Studying the adsorption of BOS on sediments, Gaullier et al. [43] reported a scarce reversibility of this compound and explained this behavior with the formation of irreversible bonds between BOS and the organic fraction of the sediment.

Possible relationships between the sorption/desorption constants ($K_{F\,ads}$, $K_{d\,ads}$, $K_{F\,des}$ and $K_{d\,des}$) and the corresponding physicochemical properties of the compounds, including water solubility and log Kow, were assessed through linear regressions. No significant correlations were obtained between each constant and water solubility, whereas all constants were sufficiently correlated with log Kow (Figure 6). This confirmed the crucial role of hydrophobicity in the interaction of AOPs with organic materials.

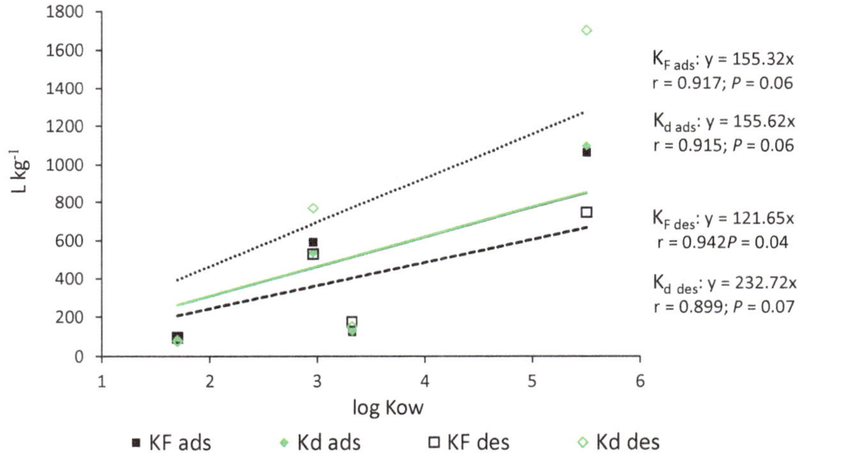

Figure 6. Plots of the correlations between the values of the adsorption and desorption constants, $K_{F\,ads}$, $K_{d\,ads}$, $K_{F\,des}$ and $K_{d\,des}$, of the four compounds and the corresponding log Kow.

4. Conclusions

Digestate (DIG), a by-product of the anaerobic biological conversion of waste biomass, is produced in increasing quantities and is a promising material for both agricultural and environmental applications. This study evaluated for the first time the capacity of DIG to remove from water four pollutants with endocrine-disrupting activity, namely two pesticides and two xenoestrogens. The DIG used showed a remarkable efficiency in adsorbing all compounds, especially the more hydrophobic OP and BOS. The sorption kinetic data of MET and OP preferentially fitted a pseudo-first-order kinetic equation and those of BOS followed a pseudo-second-order kinetic equation, whereas those of BPA fitted both models well. The extent of adsorption of the compounds on the DIG followed the order: OP > BOS > BPA > MET. The adsorption of BOS, BPA and OP were adequately described by both the Henry and Freundlich isotherms, whereas the adsorption of MET was better interpreted by the Freundlich model. The desorption rates of BOS, BPA and OP were lower than the adsorption rates, indicating strong retention of the compounds on the DIG and the occurrence of hysteretic conditions. Conversely, MET was easily and completely released from the DIG, denoting a slight negative hysteresis. The sorption/desorption constants K_F and K_d were sufficiently correlated with the hydrophobicity of the molecules. The overall results of this study evidenced the good potential of this material as a biosorbent of organic pollutants. Therefore, the incorporation of DIG into the soil can yield more than one benefit, such as reducing the bioavailability of pollutants for plants and microorganisms, counteracting the leaching of these compounds into groundwater and preventing their entry into the food chain.

Author Contributions: Conceptualization, E.L. and C.C.; methodology, C.C. and R.S.; software, E.L. and C.C.; validation, E.L., C.C. and R.S.; formal analysis, E.L. and C.C.; investigation, C.C., C.S. and R.S.; resources, E.L. and C.S.; data curation, C.C. and R.S.; writing—original draft preparation, E.L.; writing—review and editing, E.L.; visualization, E.L. and C.C.; supervision, E.L. and C.S.; project administration, E.L. and C.C.; funding acquisition, E.L. and C.S. All authors have read and agreed to the published version of the manuscript.

Funding: This research was funded by Ordinary Fund for Scientific Research of the University of Bari Aldo Moro, Italy.

Acknowledgments: The authors thank F.lli Caione Azienda Agricola La Quercia Società Coop., Foggia, Italy, for providing the digestate sample used in this study. The authors are grateful to the anonymous reviewers for their valuable comments and suggestions.

Conflicts of Interest: The authors declare no conflict of interest.

References

1. Singh, L.; Kalia, V.C. *Waste Biomass Management—A Holistic Approach*; Springer International Publishing AG: Cham, Switzerland, 2017. [CrossRef]
2. Fabbri, D.; Torri, C. Linking pyrolysis and anaerobic digestion (Py-AD) for the conversion of lignocellulosic biomass. *Curr. Opin. Biotechnol.* **2016**, *38*, 167–173. [CrossRef] [PubMed]
3. Wang, W.; Lee, D.-J. Valorization of anaerobic digestion digestate: A prospect review. *Bioresour. Technol.* **2021**, *323*, 124626. [CrossRef]
4. Zhang, Z.; Zhu, Z.; Shen, B.; Liu, L. Insights into biochar and hydrochar production and applications: A review. *Energy* **2019**, *171*, 581–598. [CrossRef]
5. Parlavecchia, M.; Carnimeo, C.; Loffredo, E. Soil amendment with biochar, hydrochar and compost mitigates the accumulation of emerging pollutants in rocket salad plants. *Water Air Soil Pollut.* **2020**, *231*, 554. [CrossRef]
6. Braguglia, C.M.; Gallipoli, A.; Gianico, A.; Pagliaccia, P. Anaerobic bioconversion of food waste into energy: A critical review. *Bioresour. Technol.* **2018**, *248*, 37–56. [CrossRef]
7. Peng, W.; Pivato, A. Sustainable management of digestate from the organic fraction of municipal solid waste and food waste under the concepts of back to earth alternatives and circular economy. *Waste Biomass Valor.* **2019**, *10*, 465–481. [CrossRef]
8. Mukherjee, S.; Weihermüller, L.; Tappe, W.; Hofmann, D.; Köppchen, S.; Laabs, V.; Vereecken, H.; Burauel, P. Sorption-desorption behaviour of bentazone, boscalid and pyrimenthanil in biochar and digestate based soil mixtures for biopurification systems. *Sci. Total Environ.* **2016**, *559*, 63–73. [CrossRef]
9. López-Pacheco, I.Y.; Silva-Núñez, A.; Salinas-Salazar, C.; Arevalo-Gallegos, A.; Lizarazo-Holguin, L.A.; Barcelo, D.; Iqbal, H.M.N.; Parra-Saldívar, R. Anthropogenic contaminants of high concern: Existence in water resources and their adverse effects. *Sci. Total Environ.* **2019**, *690*, 1068–1088. [CrossRef]
10. European Commission (EC) 2016. Defining Criteria for Identifying Endocrine Disruptors in the Context of the Implementation of the Plant Protection Products Regulation and Biocidal Products Regulation. Available online: https://ec.europa.eu/transparency/regdoc/rep/10102/2016/EN/SWD-2016-211-F1-EN-MAIN-PART-6.PDF (accessed on 13 July 2021).
11. European Commission (EC) 2020. Endocrine Disruptors. Available online: https://ec.europa.eu/environment/chemicals/endocrine/index_en.htm (accessed on 13 July 2021).
12. Calafat, A.M.; Ye, X.; Wong, L.Y.; Reidy, J.A.; Needham, L.L. Exposure of the U.S. population to bisphenol A and 4-tertiary-octylphenol: 2003–2004. *Environ. Health Perspect.* **2008**, *116*, 39–44. [CrossRef]
13. Diamanti-Kandarakis, E.; Bourguignon, J.-P.; Giudice, L.C.; Hauser, R.; Prins, G.S.; Soto, A.M.; Zoeller, R.T.; Gore, A.C. Endocrine-disrupting chemicals: An Endocrine Society scientific statement. *Endocr. Rev.* **2009**, *30*, 293–342. [CrossRef]
14. Silva, V.; Mol, H.G.J.; Zomer, P.; Tienstra, M.; Ritsema, C.J.; Geissena, V. Pesticide residues in European agricultural soils—A hidden reality unfolded. *Sci. Total Environ.* **2019**, *653*, 1532–1545. [CrossRef]
15. USEPA Office of Water Report 2003. Candidate Contaminant List Regulatory Determination Support Document for Metribuzin. Available online: https://www.epa.gov/sites/production/files/2014-09/documents/support_cc1_metribuzin_ccl_regdet.pdf (accessed on 13 July 2021).
16. USEPA Office of Pesticide Programs 2010. Environmental Fate and Ecological Risk Assessment for Boscalid New Use on Rapeseed, Including Canola (Seed Treatment). Available online: https://www3.epa.gov/pesticides/chem_search/cleared_reviews/csr_PC-128008_23-Dec-10_a.pdf (accessed on 20 July 2021).

17. Geens, T.; Aerts, D.; Berthod, C.; Bourguignon, J.-P.; Goeyens, L.; Lecomte, P.; Maghuin-Rogister, G.; Pironnet, A.-M.; Pussemier, L.; Scippo, M.-L.; et al. A review of dietary and non-dietary exposure to bisphenol-A. *Food Chem. Toxicol.* **2012**, *50*, 3725–3740. [CrossRef] [PubMed]
18. Olaniyan, L.W.B.; Okoh, O.O.; Mkwetshana, N.T.; Akoh, A.I. Environmental water pollution, endocrine interference and ecotoxicity of 4-*tert*-octylphenol: A review. *Rev. Arch. Environ. Contam. Toxicol.* **2018**, *248*, 81–109. [CrossRef]
19. Campbell, C.G.; Borglin, S.E.; Green, F.B.; Grayson, A.; Wozei, E.; Stringfellow, W.T. Biologically directed environmental monitoring, fate, and transport of estrogenic endocrine disrupting compounds in water: A review. *Chemosphere* **2006**, *65*, 1265–1280. [CrossRef] [PubMed]
20. Linhart, C.; Panzacchi, S.; Belpoggi, F.; Clausing, P.; Zaller, J.G.; Hertoge, K. Year-round pesticide contamination of public sites near intensively managed agricultural areas in South Tyrol. *Environ. Sci. Eur.* **2021**, *33*, 1. [CrossRef]
21. Fernandes, M.C.; Cox, L.; Hermosín, M.C.; Cornejo, J. Organic amendments affecting sorption, leaching and dissipation of fungicides in soils. *Pest. Manag. Sci.* **2006**, *62*, 1207–1215. [CrossRef] [PubMed]
22. Mojiri, A.; Zhou, J.L.; Robinson, B.; Ohashi, A.; Ozaki, N.; Kindaichi, T.; Farraji, H.; Vakili, M. Pesticides in aquatic environments and their removal by adsorption methods. *Chemosphere* **2020**, *253*, 126646. [CrossRef]
23. Senesi, N.; Loffredo, E.; D'Orazio, V.; Brunetti, G.; Miano, T.M.; La Cava, P. Adsorption of pesticides by humic acids from organic amendments and soils. In *Humic Substances and Chemical Contaminants*; Clapp, C.E., Hayes, M.H.B., Senesi, N., Bloom, P.R., Jardine, P.M., Eds.; John Wiley & Sons, Inc.: Madison, WI, USA, 2015; pp. 129–153. [CrossRef]
24. Loffredo, E.; Taskin, E. Adsorptive removal of ascertained and suspected endocrine disruptors from aqueous solution using plant-derived materials. *Environ. Sci. Pollut. Res.* **2017**, *24*, 19159–19166. [CrossRef]
25. Mohan, D.; Sarswat, A.; Ok, Y.S.; Pittman, C.U., Jr. Organic and inorganic contaminants removal from water with biochar, a renewable, low cost and sustainable adsorbent—A critical review. *Bioresour. Technol.* **2014**, *160*, 191–202. [CrossRef]
26. Ferreira Mendes, K.; Nogueira de Sousa, R.; Takeshita, V.; Gimenes Alonso, F.; Justiniano Régo, A.P.; Tornisielo, V.L. Cow bone char as a sorbent to increase sorption and decrease mobility of hexazinone, metribuzin, and quinclorac in soil. *Geoderma* **2019**, *343*, 40–49. [CrossRef]
27. Loffredo, E.; Parlavecchia, M.; Perri, G.; Gattullo, R. Comparative assessment of metribuzin sorption efficiency of biochar, hydrochar and vermicompost. *J. Environ. Sci. Health B* **2019**, *54*, 728–735. [CrossRef] [PubMed]
28. Yu, J.; Zhu, Z.; Zhang, H.; Guanglan, D.; Qiu, Y.; Yin, D.; Wang, S. Hydrochars from pinewood for adsorption and nonradical catalysis of bisphenols. *J. Hazard. Mat.* **2020**, *385*, 121548. [CrossRef]
29. Altenburger, R.; Scholze, M.; Busch, W.; Escher, B.I.; Jakobs, G.; Krauss, M.; Krüger, J.; Neale, P.; Ait-Aissa, S.; Almeida, A.; et al. Mixture effects in samples of multiple contaminants—An inter-laboratory study with manifold bioassays. *Environ. Int.* **2018**, *114*, 95–106. [CrossRef]
30. PubChem Open Chemistry Database at the National Institutes of Health (NIH); U.S. National Library of Medicine. Available online: https://pubchem.ncbi.nlm.nih.gov/compound/8814 (accessed on 20 July 2021).
31. Lagergren, S. Zur theorie der sogenannten adsorption gelöster stoffe, Kungliga Svenska Vetenskapsakademiens. *Handlingar* **1898**, *24*, 1–39. [CrossRef]
32. Kumar, K.V. Linear and non-linear regression analysis for the sorption kinetics of methylene blue onto activated carbon. *J. Hazard. Mater.* **2006**, *B137*, 1538–1544. [CrossRef]
33. Ho, Y.S. Second-order kinetic model for the sorption of cadmium onto tree fern: A comparison of linear and non-linear methods. *Water Res.* **2006**, *40*, 119–125. [CrossRef]
34. Barriuso, E.; Laird, D.A.; Koskinen, W.C.; Dowdy, R.H. Atrazine desorption from smectites. *Soil Sci. Soc. Am. J.* **1994**, *58*, 1632–1638. [CrossRef]
35. Sicong Yao, S.; Fabbricino, M.; Race, M.; Ferraro, A.; Pontoni, L.; Aimone, O.; Chen, Y. Study of the digestate as an innovative and low-cost adsorbent for the removal of dyes in wastewater. *Processes* **2020**, *8*, 852. [CrossRef]
36. Essandoh, M.; Wolgemuth, D.; Pittman, C.U., Jr.; Mohan, D.; Mlsna, T. Adsorption of metribuzin from aqueous solution using magnetic and nonmagnetic sustainable low-cost biochar adsorbents. *Environ. Sci. Pollut. Res.* **2017**, *24*, 4577–4590. [CrossRef]
37. Zhao, J.; Zhou, D.; Zhang, J.; Li, F.; Chu, G.; Wu, M.; Pan, B.; Steinberg, C.E.W. The contrasting role of minerals in biochars in bisphenol A and sulfamethoxazole sorption. *Chemosphere* **2021**, *264*, 128490. [CrossRef]
38. Aly, Z.; Graulet, A.; Scales, N.; Hanley, T. Removal of aluminium from aqueous solutions using PAN-based adsorbents: Characterisation, kinetics, equilibrium and thermodynamic studies. *Environ. Sci. Pollut. Res.* **2014**, *21*, 3972–3986. [CrossRef] [PubMed]
39. Ho, Y.S.; Mckay, G. Pseudo-second order model for sorption processes. *Process. Biochem.* **1999**, *34*, 451–465. [CrossRef]
40. Giles, C.H.; MacEwan, T.H.; Nakhwa, S.N.; Smith, D. Studies in adsorption. Part XI. A system of classification of solution adsorption isotherms, and its use in diagnosis of adsorption mechanisms and in measurement of specific surface areas of solids. *J. Chem. Soc.* **1960**, *111*, 3973–3993. [CrossRef]
41. Prasannamedha, G.; Senthil Kumar, P.; Mehala, R.; Sharumitha, T.J.; Surendhar, D. Enhanced adsorptive removal of sulfamethoxazole from water using biochar derived from hydrothermal carbonization of sugarcane bagasse. *J. Hazard. Mat.* **2021**, *407*, 124825. [CrossRef]

42. Loffredo, E.; Senesi, N. The role of humic substances in the fate of anthropogenic organic pollutants in soil with emphasis on endocrine disruptor compounds. In *Soil and Water Pollution Monitoring, Protection and Remediation;* NATO Science Series IV-Earth and Environmental Sciences; Springer International Publishing AG: Cham, Switzerland, 2006; Volume 69, pp. 69–92. [CrossRef]
43. Gaullier, C.; Dousset, S.; Billet, D.; Baran, N. Is pesticide sorption by constructed wetland sediments governed by water level and water dynamics? *Environ. Sci. Pollut. Res.* **2018**, *25*, 14324–14335. [CrossRef] [PubMed]
44. Ying, G.-G.; Kookana, R.S.; Dillon, P. Sorption and degradation of selected five endocrine disrupting chemicals in aquifer material. *Water Res.* **2003**, *37*, 3785–3791. [CrossRef]

processes MDPI

Article

To Advance Industrial Green Technology via Environmental Governance—Evidence from China's Industrial Sector

Junwei Shi and Yingjing Yu *

School of Business Administration, Zhongnan University of Economics and Law, Wuhan 430074, China; jw.shi@163.com
* Correspondence: yuyingjing163@163.com

Abstract: Industrial green technology progress is an effective way to realize high-quality economic development in China. The different competitive incentives of local governments make a difference in the intensity of environmental regulation between regions. The intensity of environmental regulation is different in different areas of the same industry, leading to the inter-regional transfer of pollution enterprises. The regional distribution of industries is different, which determines the different industries' needs to coordinate different regions. Only when various industries realize the coordination of regional governance can they jointly promote the progress of industrial green technology. Based on data from 33 industrial sectors in China from 2001 to 2015 and considering the institutional evolution of governance synergy, this study comprehensively investigated the influence mechanism of local government environmental regulation on the industrial green technology progress using the mediating effect model. We found that environmental regulation promoted industrial technological progress through governance synergy and a low degree of inter-regional regulation coordination hindered the industries' green technology progress. With the change in inter-regional governance synergy levels, we further discovered that the impact of environmental regulation on industrial green technology progress changed substantially. At a low level of governance synergy, environmental regulation inhibited industrial green technology progress. At a high level of governance synergy, environmental regulation promoted industrial green technology progress. While strengthening environmental regulation, we should promote inter-regional cooperation at the industry level. Only by collectively enforcing pollution regulations in industrial level can industrial green technology progress be promoted.

Keywords: environmental regulation; governance synergy; industrial green technology progress; mediating effect

Citation: Shi, J.; Yu, Y. To Advance Industrial Green Technology via Environmental Governance—Evidence from China's Industrial Sector. *Processes* **2021**, *9*, 1797. https://doi.org/10.3390/pr9101797

Academic Editor: Andrea Petrella

Received: 28 July 2021
Accepted: 27 September 2021
Published: 11 October 2021

1. Introduction

The extensive development mode with high energy consumption and high emissions in the early Chinese industrial movement created huge economic dividends. However, it also brought severe challenges to environmental governance. In the 19th National Congress of the Communist Party of China (CPC), the party proposed to promote green development and solve environmental problems to realize high-quality economic development. In the long run, whether the existing environmental regulation policy can help to realize the development of the environment and economy depends on whether it can promote the progress of green technology.

The relationship between environmental regulation and green technology progress has been the focus of academic debate for a long time. In China, the research on the relationship has not yet reached a unanimous conclusion. The reason for this lies in the fact that all of this research ignored the governance synergy of local government in the specific national conditions of China. The central government promulgates environmental protection policy to promote green technology based on the regulation of the industrial sector. However, local governments execute this policy at their discretion. Under the administrative decentralization system in China, the different competition incentives of local governments cause the different intensities of inter-regional environmental regulations. The intensity of environmental regulation is different in different areas of the same industry, it easily leads to polluting enterprises avoiding the high cost of local innovation by moving the enterprises to the lax supervision areas. While the regional distribution of different industries is different, it determines that different industries need different areas of co-governance. If local governments do their own thing to manage environmental pollution, not much progress will be made in green technology in the entire industrial sector. Because of the great differences in the industrial structure between each region, there will be great differences in the key areas where each industry needs to be jointly governed if the environmental regulation is carried out at the level of industry. The relevant environmental policies of governance synergy focus on the problem of air pollution and the key cooperation areas of implementing regulations at the industry level are also based on the inherent economic circle of air pollution prevention and control. Without governance synergy, the effect of the environmental policy that is promulgated by the central government will be greatly undermined at the stage of implementation by local governments. It is important to construct industry-level governance synergy to effectively promote industrial green technology and solve China's environmental problems. However, there is a lack of theoretical and empirical research on the mechanism of how environmental regulation affects green technology progress through governance synergy.

The main contributions of this study are as follows: (1) It constructed the mechanism of how environmental regulation promotes industrial green technology progress through governance synergy at the industry level. This study is different from the previous research on air pollution in urban agglomerations or inter-regional joint prevention. There are regional differences in the distribution of industries, and each industry needs different provinces for co-governance. It is significant whether the co-operation of different local governments in a specific industry can realize the industrial green technology progress. This is highly complementary to existing research. (2) A new measurement of governance synergy is constructed. By using the two-digit industries in a province of industrial gross output value panel data to calculate the weights, two-digit industries in a province of environmental regulation intensity were calculated, and then the inter-regional governance synergy degree of each industry was measured. Different from the previous studies using virtual variable measurement, this study constructed a continuous variable that reflected the collaborative governance level of different regions within a specific industry. (3) Based on the heterogeneity of governance synergy, it was found that the impact of environmental regulation on industrial green technology progress will change substantially at the different levels of governance synergy.

The rest of this paper is arranged as follows: the second part is the literature review, the third part provides the theoretical mechanism and hypothesis, the fourth part provides the research design, the fifth part shows the empirical results and discussions, and the final part provides the summary and policy enlightenment.

2. Review of the Literature

The existing studies were mainly undertaken from the following two perspectives: one was against the Porter hypothesis, arguing that environmental regulation aggravates the production burden of enterprises. It holds that environmental regulation fails to stimulate green technology innovation of enterprises and hinders industrial green technology progress [1–3]. Greenstone et al. [4] found that stringent air pollution regulations reduced the green total factor productivity of polluting firms in regulated areas based on data of the U.S. manufacturing plant survey. To avoid the restriction of environmental policies or reduce environmental costs, polluting enterprises transfer production due to regional differences in environmental standards or regulations, leading to the pollution shelter effect [5,6]. Yuan and Xiang [7] employed panel data on Chinese 28 manufacturing industries from

2003 to 2014 to examine the effects of environmental regulation on industrial innovation and green development. They found that the impacts of environmental regulations on green total factor productivity were insignificant over both the short and long term. Some scholars hold that the validity of the Porter hypothesis depends not only on the intensity of environmental regulation but also on the type of environmental regulation [8,9]. The other perspective involved supporting the Porter hypothesis, i.e., strict and appropriate environmental regulation will encourage enterprises to engage in green technology innovation activities and reduce the cost of environmental governance by improving the technological level [10–12]. Acemoglu et al. [13] divided the production sector into clean and non-clean sectors, analyzed the impact of environmental policy incentives on technological innovation, and deduced the endogenous process of technological progress by constructing the model of technological progress direction. Ulucak [14] found that environment-related technologies positively contribute to green growth. The strictness of the environmental regulations positively impacts the green innovation of the companies [15]. Some scholars hold that there is a U-shaped threshold effect between environmental regulation and industrial green technology progress [16,17].

Appropriate environmental policies stimulate technological innovation, but the translocation of polluting industries will worsen the environmental quality of the destination, resulting in bottom-up competition between local governments. Wu et al. [18] constructed a spatial Durbin dynamic threshold panel model with provincial-level data to study the nonlinear relationship between environmental regulation and local decentralization. The results showed that there was a significant U-shaped relationship between environmental regulation and green total factor productivity, and a high level of local decentralization inhibited the green technology progress. From the perspective of local decentralization, different competition incentives of local governments lead to different intensities of environmental regulation in different areas. As environmental regulation becomes tighter in one place, polluting enterprises may relocate to nearby regions where environmental regulation is weak [19]. The non-synchronous stimulation of environmental policy among regions will weaken the effect of environmental regulation. It is difficult for environmental policy to stimulate the innovation of industrial green technology.

If local environmental regulation has an impact on green technology progress in neighboring regions [20], will environmental regulation promote green technology progress through inter-regional industrial governance collaboration? Only a few relevant studies have been done on the relationship between governance synergy and environmental pollution. Li et al. [21] constructed a comprehensive index of environmental regulation and the degree of environmental co-governance at the enterprise level. The analysis of the influencing mechanism shows that environmental co-governance can reduce the probability of enterprise migration, inhibit the transfer of pollution to nearby areas, and improve the efficiency of environmental governance. Based on the theory of collective action, Hu et al. [22] put forward the optimal regional control scheme for the governance synergy of air pollution in China and considered that the establishment of inter-regional joint organizations will effectively promote the degree of governance synergy.

There are abundant studies on the influence of environmental regulation on green technology progress. The research on the relationship between governance synergy and environmental pollution has been paid more attention to. However, few studies have combined environmental regulation, governance synergy, and industrial green technology progress for empirical analysis. This study was based on the practical evidence that there were different intensities of environmental regulations in different areas, as well as the unique national conditions of China. It examined how environmental regulation promotes industrial green technology progress through inter-regional governance synergy. Furthermore, if the level of governance synergy is different, how does environmental regulation affect industrial green technology progress? These questions are related to which level of governance synergy is necessary to ensure that environmental policies have a positive effect on green technology progress. The study constructed a governance synergy index, considered how environmental regulation affects industrial green technology progress through governance synergy with the mediating effect model, and investigated the heterogeneity characteristics of the governance synergy degree.

3. System Evolution and Theory Analysis

As early as 1996, the Water Pollution Control Act of China incorporated the water pollution prevention and control planning system for key river basins into the legal framework for the first time, which was essentially the embryonic form of regional joint prevention and control. In May 2015, the State Council issued the Action Plan on Water Pollution Prevention and Control. It suggested that ten key industries, including papermaking and coking, should be cleansed. Furthermore, it proposed the establishment of a governance synergy mechanism for regional water pollution prevention in the Beijing–Tianjin–Hebei region, the Yangtze River Delta, and the Pearl River Delta. In 2017, the Water Pollution Prevention and Control Act was amended with new regulations in the form of a law to ensure the governance synergy mechanism for water pollution prevention in major rivers and lakes. Compared with the water pollution prevention cooperation mechanism, the establishment of the air pollution prevention cooperation mechanism was relatively late.

In 1998, the Ministry of Environmental Protection proposed the Two Control Area Divisions Program for acid rain and SO_2. In May 2010, the Ministry of Environmental Protection and nine other departments issued the Guiding Opinions on Promoting Joint Prevention of Air Pollution to Improve Regional Air Quality. It put forward a solution to the problem of regional air pollution with the idea of joint prevention for the first time and strengthened the promotion of cleaner production technology in key industries, such as thermal power, iron, and steel. In September 2013, the State Council issued the Action Plan on Air Pollution Prevention, which proposed the establishment of a regional cooperation mechanism of air pollution prevention in the Beijing–Tianjin–Hebei and Yangtze River delta regions, especially for the waste gas pollution in key industries. The following year, the Working Plan on strengthening air pollution prevention in the energy sector was formulated. It proposed that the local governments of ten provinces and cities, including Beijing, Shanghai, and Guangzhou, were responsible for implementing the tasks of controlling the total amount of energy and coal consumption. A long-term mechanism for joint prevention from the central government to the local authorities should be established. At the same time, the Ministry of Environmental Protection formulated the measures of air pollution prevention in key industries, such as power, steel, cement, and flat glass, for the Yangtze River Delta Economic Zone, the Beijing–Tianjin–Hebei region, and the surrounding areas. In 2015, the air pollution prevention law was amended again to set up new content for the Joint Prevention of Air Pollution in Key Regions, which ensured the joint prevention mechanism of air pollution in key regions from the national legal level. In 2018, the State Council issued the three-year action plan for winning the battle to defend the Blue Sky, which identified 28 cities in the Beijing–Tianjin–Hebei region and its surrounding areas as key areas and reorganized the coordination group into a leading group on air pollution prevention.

No matter the key river basins of water pollution prevention or the three economic circles of air pollution prevention, there is a lack of institutional planning and regulatory rules that consider the coordinated actions of different local governments and governments from industry. This is the defect of the current policy system of governance synergy. At present, the coordination areas of industrial pollution prevention depend on the existing air prevention areas. If such prevention is not implemented in the inter-regional coordination action within specific industrial sectors, environmental regulation will not upgrade the level of green technology of the overall industry, and the environmental policy pollution effect will be greatly compromised. The establishment of the pollution prevention system from the central government to the local governments shows that governance synergy is imminent. Based on the above analysis, hypothesis 1 was proposed.

Hypothesis 1. *Environmental regulation promotes industrial green technology progress through governance synergy to some extent.*

The improvement of governance synergy will help enterprises to improve their innovation level and achieve the policy objectives [23]. The higher the level of governance synergy, the smaller the differences in environmental regulation intensity between different provinces in the industry, and the smaller the space for pollution enterprises to reduce the regulation cost via inter-regional transfer. When sustainable economic development is faced with tight environmental regulation constraints, the high level of governance coordination becomes an important way to solve the dilemma of maintaining growth and promoting carbon emissions reduction. However, the current governance synergy lacks intrinsic motivation. There are three reasons why it is difficult for local governments to manage environmental problems in coordination: The first reason is the fiscal incentive. Fiscal decentralization causes local governments to pursue economic development at the expense of environmental governance, resulting in a bottom-up competition effect. In contrast, to compete for the essential resources that favor a high-quality environment, local governments are competing with each other in spending on pollution prevention and forming top-down competition. The second reason is the promotion incentives for officials. Distortions in local government efforts are inevitable because green technology progress indicators are not easy to quantify. Driven by their political achievements, officials may ignore the long-term effects of rapid economic growth, especially those related to environmental pollution, which are not easily assessed, resulting in a bottom-up competition effect. In contrast, to implement the Scientific Outlook on Development, the competition for local government environmental regulation has subsided. Officials have even pursued an achievement project that involves beautifying the environment during their term of office, resulting in a top-down competition effect. The third reason is catching up with and surpassing strategic incentives. Decentralization reform causes local governments to continue to give priority to the development of heavy industries to catch up with the advanced provinces. This lowers the standard of environmental regulation and leads to a bottom-up competition effect. After the reform and opening-up, the rapid economic growth of China expanded the regional differences in development. The differences in the competitive incentives of local governments are likely to lead to differences in behavior and intensity of environmental regulation.

However, the intervention of environmental regulation just makes up the deficiency of this intrinsic incentive and aggravates the regulation cost burden of incomplete implementation by local governments. It ensures that local governments effectively implement environmental protection policies that are related to governance synergy. In 2005, the State Council issued the Decision on Implementing the Scientific Outlook on Development and Strengthening Environmental Protection. It incorporated the performance of environmental protection into the assessment system as the basis of local official selection for the first time. In July 2015, the Environmental Protection Supervision Plan (Trial) was released. In 2016, the State Environmental Protection Supervision Office was set up and the reform on vertical environmental management system was piloted. In June 2020, the State Council General Office issued a reform plan on the division of financial powers and expenditure responsibilities between the central and local governments in the field of ecological environment. It established a fiscal relationship between the central and local governments with clear powers and responsibilities, coordinated financial resources, and balanced regional development. Therefore, a reasonable standard of environmental regulation is helpful for the construction of a regional joint prevention and control system, as well as driving the inter-regional governance coordination and industrial green technology progress to a high level. However, for those regions whose economic development is heavily dependent on a single industry, especially pollution-intensive manufacturing, the tightening of environmental regulation standards will make it difficult to raise the level of governance synergy in the short term. Moreover, it will magnify the costs of regulation such that the level of low-carbon technology in the region may decrease as the intensity of environmental regulation increases. In this case, it is inefficient to promote industrial green technology progress through environmental regulation. Therefore, according to the mechanism of governance synergy, the impact of environmental regulation on the progress of industrial green technology will change with the inter-regional governance synergy level in a specific industry. Moreover, the higher the level of inter-regional governance synergy, the more obvious the industrial green technology progress effect of environmental regulation will be, and vice versa. Based on the above analysis, we proposed hypothesis 2.

Hypothesis 2. *The impact of environmental regulation on the industrial green technology progress will change substantially with the change in the inter-regional governance synergy level. At a low level of governance synergy, environmental regulation is not good for industrial green technology progress, while at a high level of governance synergy, environmental regulation promotes industrial green technology progress.*

4. Research Design

4.1. Setting the Empirical Model

In this study, the mediating effect model was used to examine how environmental regulation affects industrial green technology progress through governance synergy. The mediating effect test is divided into three steps. First, the explanatory variable X has a significant effect on the explained variable Y. If the coefficient of X is significant, the mediating effect is examined. Second, the explanatory variable X has a significant effect on the mediating variable M. Third, M is added to the regression equation of the first step. While the coefficients of M and X are significant, M is considered a partial mediating effect. In this study, the explanatory variable X was environmental regulation, the explained variable Y was industrial green technology progress, and the mediating variable M was governance synergy, as shown in Figure 1. Based on the above analysis, the panel regression model was constructed as follows:

$$Y_{it} = \theta_0 + \theta_1 ER_{it} + \eta_1 C_{it} + \varepsilon_{it1} \quad (1)$$

$$CG_{it} = \gamma_0 + \gamma_1 ER_{it} + \eta_2 C_{it} + \varepsilon_{it2} \quad (2)$$

$$Y_{it} = \beta_0 + \beta_1 ER_{it} + \beta_2 CG_{it} + \eta_3 C_{it} + \varepsilon_{it3} \quad (3)$$

where i indicates the industry number, i = 1, 2, ..., 33. t indicates the year, t = 2001, 2002, ..., 2015. Y_{it} represents industrial green technological progress. ER_{it} represents environmental regulation. C_{it} represents the control variable, which incorporates the output rate of new products ($innov_{it}$), the level of capital management ($mngt_{it}$), and the level of urban wages ($wage_{it}$), the level of industry innovation (pat_{it}). ε_{it} represents the random perturbation term. The θ, γ, β, and η coefficients are values to be determined.

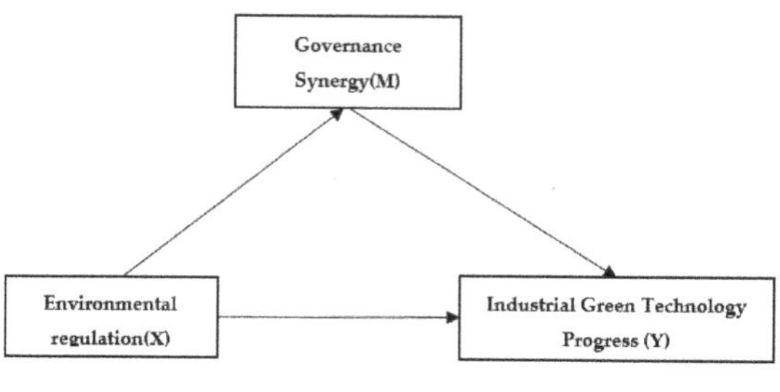

Figure 1. The mediating effect mechanism of environmental regulation, governance synergy, and industrial green technology progress.

The benchmark of the empirical model equation was the linear model equation of environmental regulation affecting industrial green technology progress. If θ_1 was significantly positive, then environmental regulation had a significant positive effect on the progress of industrial green technology. If γ_1 was significantly negative, then environmental regulation had a significant negative effect on governance synergy. The third step can be continued. Based on the benchmark model, the mediating variable CG_{it} was added. If β_2 was negative and significant and β_1 was significant and had decreased, then the environmental regulation had a partial mediating effect on industrial green technology progress through governance synergy.

4.2. Variable Calculation and Description

(1) Governance Synergy (M)

There are two main methods to measure the degree of governance synergy, including the existence of virtual variables of governance synergy, and the total number of policy joint publications multiplied by the policy intensity. Due to the lack of quantitative research on governance synergy degree, this study needed to create a measure of governance synergy from indicators of environmental regulations intensity. Different from the previous discrete variable measurement of air pollution prevention coordination, the indicator used continuous variables that represent the level of governance synergy of environmental regulation among different local governments in a given industry.

Before measuring the degree of governance synergy, it was necessary to reconstruct the intensity of environmental regulation in different provinces and industries. Industrial structure is an important factor that affects the environmental regulation of local government. Environmental policy is regulated on the basis of industries. Under certain local environmental regulations, due to the different industrial structures in different regions, the intensities of environmental regulations for industries are different in different areas. This results in different effects of environmental regulation. To analyze the intensity of environmental regulation corresponding to the industrial structure in different regions of China, this study reconstructed the environmental regulation intensity of two-digit industries in a province panel data, where ER_{NI} describes the intensity of environmental regulation of industry I in province N, as follows:

$$ER_{NI} = ER_N \times W_{NI}$$

where n is the province, $n = 1, 2, 3, \ldots, 30$; I is the industry, I = 1, 2, 3, ..., 33.

W_{NI} represents the proportion of industry I in province N and the adjustment coefficient of the two-digit industries in a province; it was measured using the proportion of industry gross output value O_{NI} of industry I in province N to the total industrial output value of the province O_N. It reflected the proportion of an industry in a specific province and was used to calculate the intensity of environmental regulation in that province. The adjustment factor was calculated as follows:

$$W_{NI} = O_{NI}/O_N$$

From the relation $W_{NI} = \frac{O_{IN}}{O_N} = \frac{ER_{IN}}{ER_N}$, we can see the rationality of the definition of environmental regulation.

ER_N represents the intensity of environmental regulation in each province, which was consistent with the measurement method of environmental regulation intensity in industries. It was measured using the proportion of the operating cost of the administrative facilities in each province to the total industrial output value, indicating the intensity of environmental regulation in province N. Given the lack of data on the operating cost of industrial solid waste treatment facilities in the provinces, the operating cost of treatment facilities included the operating cost of wastewater and waste gas treatment facilities.

If the technology level of a specific industry is constant in a given period, the proportion of each industry is different for a given province; therefore, the intensity of environmental regulation is different for different industries. Taking into account the relevance of data availability and environmental regulation while avoiding the estimation bias caused by the under-representation of indicators, this study adjusted the intensity and weight of environmental regulation in different provinces to reflect the different industries in the province corresponding to the intensity of environmental regulation differences. For example, if the ratio of the industrial output value of industry I in province N is very low, then the corresponding environmental regulation intensity of this industry ER_{NI} is relatively small. ER_{NI} is determined by the environmental regulation intensity of province N and the weight W_{NI}. If the proportion of industrial sectors in two provinces is equal and the intensity of environmental regulation is different, then the environmental regulation of the province with stronger environmental regulation is higher than that of the province with weaker environmental regulation. If the environmental regulation of the two provinces is equal and the proportion of industrial sectors is different, then the provinces with higher weights have higher environmental regulation ER_{NI} in this industry. If the weight is zero, then the province does not have industrial sector I, and the ER_{NI} is 0, that is, province N does not need to implement environmental regulation on industry I. If $ER_{NI} > ER_{MI}$ $(N \neq M)$, then $\Delta = ER_{NI} - ER_{MI} > 0$, indicating that different local governments have different intensities of environmental regulation in the industry. Other things being equal, enterprises in province N have the motive to move into province M to avoid the high environmental regulation cost in province N.

Finally, measuring the provinces of governance synergy (GS) within the industry. The calculation was as follows:

$$GS_I = \frac{1}{\ln[\text{SD} + 2]}$$

$\text{SD} = \sqrt{\sum_{N}^{30} (ER_{NI} - \overline{ER}_{NI})^2}$ represents the standard deviation of the environmental regulation intensity of all provinces in a certain industry, reflecting the degree of regional governance coordination. This measurement shows that the bigger the index is, the higher the governance coordination degree is, as well as the normal distribution of the data. The larger the GS_I, the higher the degree of governance synergy, and vice versa. The governance synergy degree of the 15 provinces with a large proportion of each industry (GS2) was tested as a substitution variable.

(2) Dependent Variables: Industrial Green Technology Progress (Y)

The measurement methods of industrial green technology progress are mainly divided into the Solow residual value method and the non-parametric DEA method. The former is measured using Solow residuals and the latter by decomposing technological progress from productivity.

Based on the input and output data of industries from 2000 to 2016, the non-parametric DEA method was used to measure the industrial green technology progress [24,25], and the green total factor productivity was calculated and decomposed into an ML index [26]. The ML index reflects the growth rate of industrial green technological progress. It assumes that the industrial green technological progress in 2001 was 1, then the ML index was multiplied by the industrial green technology progress from 2001 to 2015, reflecting the dynamic change. Given the availability of data, the research object of this study was the industrial enterprises above the scale. Given the lack of the statistical data of industry before 2001 in the Chinese Industrial Economy Statistical Yearbook and Chinese Environmental Statistical Yearbook, this study selected the input–output data of 33 industrial sectors from 2001 to 2015. The input index includes three items, namely labor input, capital input, and energy input. The expected output index is the total industrial output value of industry. Although the selection of the non-expected output index is controversial and has not been unified, considering the various emission of industrial pollutants, two indexes were selected as the non-expected output indexes, namely, the industrial CO_2 and SO_2 emissions.

The relevant indicators and data processing for inputs, expected outputs, and non-expected outputs are described below:

(1) Labor input: Labor hours provide a better measure than labor force when measuring labor input, but it is hard to obtain. We chose the average number of all employees in industrial enterprises above scale in sector to replace the number of labor hours. The relevant data was from the China Industrial Economic Statistical Yearbook.
(2) Capital investment: The total fixed assets of industrial enterprises above scale in sector were selected as the approximate estimation of the capital stock, and the fixed assets investment price index was converted into the constant price in 2001.
(3) Energy input: This study considered not only the capital input and labor input but also the energy input. Energy consumption is the main source of undesired output. The total energy consumption data of industrial enterprises above scale in sector were converted into 10,000 tons of standard coal according to the conversion coefficient of standard coal, where the conversion coefficient came from the appendix of the China Energy Statistics Yearbook.
(4) Gross industrial output value: By using the ex-factory price index provided by the China Industrial Economic Statistics yearbook, the total industrial output value of each industry was adjusted to the constant price in 2001.
(5) Industrial CO_2 emissions: According to the calculation method of carbon emissions in the guidelines of national greenhouse gas inventories, which was compiled by the Intergovernmental Panel on Climate Change (IPCC), CO_2 emissions were estimated according to the amount of fuel burned and the emission factors.
(6) Industrial SO_2 emissions: Considering that the large amount of industrial SO_2 emissions in industrial production is also one of the main sources of air pollution, we chose industrial SO_2 emissions as an undesired output index.

(3) Core independent variable: environmental regulation (X)

Scholars mainly measure environmental regulation from four perspectives, namely, the proportion of the total investment of industrial pollution governance in the industrial added value [27], the proportion of operating expenses of pollution facilities in the industrial output value [28], the comprehensive index of pollution emissions [29], and the number of environmental regulation policies or regulatory bodies inspecting polluting enterprises [30].

Considering the availability of industrial panel data, the proportion of the operating cost of each industry's pollution prevention in the industrial output value (*ER*) was chosen as the proxy variable of environmental regulation intensity. Due to the fact that the data of the governance operation costs of industrial solid waste in the annual report of China environmental statistics were not collected, the total operating costs of pollution treatment included the operating costs of wastewater and waste gas prevention.

(4) Control variables: The output rate of new product ($innov_{it}$), expressed as the proportion of new product sales to industry sales; the level of capital management ($mngt_{it}$), expressed as the proportion of main business income to total assets; the level of urban wages ($wage_{it}$), expressed as the average wages of employees in urban units in logarithms; the level of industry innovation (pat_{it}), expressed as the logarithm of the number of patent applications in industries.

4.3. Descriptive Statistics

The descriptive statistics of the variables are given in Table 1. The correlation coefficients of the variables are given in Table 2. According to the test results, the correlation coefficients between the variables were not large, which indicated that the variables had good independence and no serious multicollinearity problems were present.

4.4. Data Sources

Given the lack of environmental data of 33 industrial sectors in China in other years, this study used panel data of 33 industries in China from 2001 to 2015. The sample size was 495. The sample data came from the China Industrial Economic Statistical Yearbook, the China Environmental Statistical Yearbook, and the China Energy Statistical Yearbook. To eliminate the influence of the price factor, all the price-related data were in the form of a ratio. Due to the differences in the classification of manufacturing sectors in the 2002 and 2011 editions, we made the necessary divisions and combinations of data according to the principle of maximizing the use of data. Thus, 33 manufacturing subsectors were formed.

Table 1. Descriptive statistics.

Variable	Mean	Std.Dev.	Min	Max
Y	1.116	0.704	0	5.316
ER	0.233	0.293	0.005	1.782
CG	0.614	0.768	0	3.595
innov	11.681	10.418	0.004	62.941
mngt	1.22	0.466	0.352	3.11
wage	9.923	0.558	6.763	11.46
pat	7.336	2.03	1.099	11.547

Table 2. Matrix of correlations.

Variables	(1) Y	(2) ER	(3) CG	(4) Innov	(5) Mngt	(6) Wage	(7) Pat
(1) Y	1.000						
(2) ER	−0.295	1.000					
(3) CG	−0.245	0.286	1.000				
(4) innov	0.221	−0.334	−0.127	1.000			
(5) mngt	0.181	−0.414	−0.308	0.137	1.000		
(6) wage	0.513	−0.092	0.063	0.132	0.121	1.000	
(7) pat	0.461	−0.300	−0.046	0.460	0.282	0.548	1.000

The industry classification standards for industrial sectors referred to the China Industrial Economic Statistics Yearbook. Five subsectors were excluded because of missing data for some years. That is, the handicraft and other manufacturing industries, the waste resource and waste material recovery and processing industries, other mining industries, the gas production and supply industry, and the water production and supply industry. The missing data of other industry years were made up using the interpolation method. To keep the statistics consistent, the plastics industry and the rubber industry were merged into the plastic and rubber industry, the automobile manufacturing industry and the railway, ship, aerospace, and other transportation equipment manufacturing industries were merged into the transportation equipment manufacturing industry. After the above adjustments, 33 industrial sectors were formed.

5. Empirical Results and Discussion

5.1. Regression Analysis of the Mediating Effect Model

To avoid the problem of endogeneity among the variables, the lag phase of environmental regulation was used as an explanatory variable to estimate the dynamic model. First, we considered the baseline relationship between environmental regulation and industrial green technology progress and used industry-level fixed effects for the regression analysis. The estimated results are shown in Table 3. Model 1 was a regression model of the effect of the control variables on the industrial green technology progress. Model 2 added environmental regulation as an explanatory variable on the basis of model 1. The results showed that the coefficient of environmental regulation intensity was significantly positive, at least at the level of 5%, which indicated that environmental regulation promoted the industrial green technology progress. In model 3, the coefficient of environmental regulation intensity was significantly negative, at least at the level of 5%, which indicated that environmental regulation was becoming tighter and required higher inter-regional governance synergy. Therefore, if the assumption of environmental regulation influencing the green technology progress through governance synergy is established, we will observe that environmental regulation is more obvious under higher governance synergy.

Table 3. Model estimation and results.

	(1)	(2)	(3)	(4)	(5) (GSH)	(6) (GSL)
Variables	**Y**	**Y**	**GS**	**Y**	**Y**	**Y**
ER		0.300 **	0.0517 **	0.238 *	0.513 ***	0.171
		(0.130)	(0.0249)	(0.127)	(0.150)	(0.239)
GS				1.195 ***		
				(0.247)		
innov	0.00715 *	0.00796 **	−0.000266	0.00827 **	−0.00214	0.0270 ***
	(0.00393)	(0.00384)	(0.000737)	(0.00375)	(0.00392)	(0.00751)
mngt	0.677 ***	0.545 ***	−0.129 ***	0.699 ***	0.298 **	0.814 ***
	(0.122)	(0.127)	(0.0243)	(0.128)	(0.140)	(0.254)
wage	0.244 ***	0.368 ***	0.0205	0.343 ***	0.570 ***	0.269 *
	(0.0695)	(0.104)	(0.0199)	(0.101)	(0.148)	(0.155)
pat	0.121 ***	0.0744 **	−0.00358	0.0787 **	0.0518	0.0931
	(0.0290)	(0.0347)	(0.00666)	(0.0338)	(0.0414)	(0.0614)
Constant	−3.100 ***	−3.876 ***	0.995 ***	−5.065 ***	−5.243 ***	−3.577 ***
	(0.542)	(0.820)	(0.157)	(0.836)	(1.167)	(1.239)
Observations	495	462	462	462	249	213
R-squared	0.486	0.401	0.124	0.432	0.519	0.360
Number of ids	33	33	33	33	22	23

Note: *, **, and *** indicate the significance levels of 10, 5, and 1%, respectively.

Based on model 2, model 4 added the governance synergy as a mediating variable, improving the explanatory power of the model ($\Delta R^2 = 0.031$). The coefficient of environmental regulation was still positive but decreased. The coefficient of governance synergy was significantly negative at the level of 5%. This indicated that there was a partial mediating effect. Environmental regulation had a direct impact on the industrial green technology progress; meanwhile, it promoted industrial green technology progress through the mediating effect of governance synergy. This showed that the decrease in governance synergy weakened the influence of environmental regulation on the industrial green technology progress, which validated hypothesis 1.

To reduce their environmental governance costs, enterprises have two choices: one is to promote green technological progress through local technological innovation, while the other is to migrate to the other regions with lower environmental regulation intensity. For polluting enterprises, local innovation and inter-regional transfer have a substitution effect on reducing the cost of environmental treatment. On the one hand, to maintain and improve the original market advantage, enterprises will try technological innovation, eliminate backward production capacity for pollution prevention, and promote industrial green technology progress. The enterprises hope to eliminate pollution emissions and promote industrial green technology progress. On the other hand, the goal of an enterprise is its profit maximization. They lack environmental awareness because the pressure of environmental regulation will reduce its short-term profits. Even if the local government takes public welfare as its goal, the enterprises may evade the environmental regulation policy in disguise and migrate to the areas where the environmental regulation is weak. As mentioned above, it is not enough to simply rely on local strict and appropriate environmental regulations that encourage enterprises to innovate and promote industrial green technology progress. If environmental regulations in other areas are less stringent, enterprises will migrate to those other areas. The new site provides a refuge for polluters. If there is a small difference in the intensity of environmental regulation between the local governments, the degree of coordination governance is high. When environmental regulation is strengthened in one area, it also means that the intensity of environmental regulation is strengthened in neighboring areas. It is difficult for enterprises to reduce the regulation cost via migration, even if they bear the high cost of technology research and development. Environmental policy can encourage enterprises to carry out green technology innovation locally and promote industrial green technology progress. If the intensity of environmental regulation varies greatly between local governments, then the degree of governance coordination is low. When environmental regulation is strengthened in one area, and environmental regulations in the neighboring areas do not change, the neighboring areas will become the receiving site of pollution industries. The enterprises are likely to shift their strategic investment to a neighboring area because of the high local regulatory costs, which will make the industrial green technology development of the neighboring area lag behind. Then industrial green technology progress will not be improved as a whole.

5.2. Analysis of the Heterogeneity of Governance Synergy

Considering the differences in environmental regulation levels between different regions, this study investigated whether there was heterogeneity in the governance synergy. Defining a scope greater than this governance synergy as a high-level governance synergy, the reverse was defined as low-level governance synergy. Each industry was divided into two sub-samples according to the mean of the governance synergy degree, that is, high-level governance synergy (GSH) and low-level governance synergy (GSL). "High-level governance synergy" meant that the governance synergy of all provinces in the industry was high and there was little difference in the intensity of environmental regulations between regions. "Low-level governance synergy" meant that the governance synergy of all provinces in the industry was low and the intensity of environmental regulation varied greatly between regions. After that, a sample-by-sample verification was performed, where the estimated results are shown in models 5 and 6 in Table 3.

On the basis of model 4, models 5 and 6 divided governance synergy into two cases, namely, low-level governance synergy and high-level governance synergy. The results showed that there was a significant difference in the impact of environmental regulation of green technology progress through different governance synergy levels. The result showed that the coefficient of environmental regulation in model 5 was significantly positive at the level of 1%, which indicated that the intensity of environmental regulation promoted the industrial green technology progress under the high-level governance synergy. This may have been due to the fact that there was little variation in the intensity of environmental regulation in the provinces where environmental policies were implemented. Polluters in the industry could hardly reduce regulatory costs by moving to other areas, which helped to encourage industrial firms to innovate green technologies locally to promote the industrial green technology progress as a whole. For a specific industry, if the governance synergy level of all the provinces was high, environmental regulation is beneficial for promoting industrial green technology progress. In model 6, the coefficient of environmental regulation was not significant, which indicated that with the increase in the intensity of environmental regulation, industrial green technology progress was restrained under the low-level governance synergy. The possible explanation for this was that the level of governance coordination was generally low and the intensity of environmental regulation varied greatly between the provinces in a given industry. This easily led to the opportunistic behavior of polluting enterprises avoiding local innovation by relocating, which was detrimental to the industrial green technology progress. For specific industries, environmental regulation could not promote industrial green technology progress if the coordination level of all provinces was low. Environmental regulation impacted the industrial green technology progress via the restriction of the governance synergy degree, which validated hypothesis 2. The specific impact mechanisms are detailed above and will not be repeated here.

5.3. Robustness Test

To verify the robustness of the above results, a robustness test was performed on the replacement governance synergy metric (GS2), as shown in Table 4. After replacing the governance synergy variable, the estimation results of the two proxy variables were basically the same. This did not change the conclusion that environmental regulation promotes industrial green technology progress through governance synergy. Based on the degree of governance synergy, this study divided industries into those with high-level governance synergy (GSH2) and low-level governance synergy (GSL2). This verified that the impact of environmental regulation on the progress of industrial green technology changed substantially with the change in the inter-regional governance synergy level. This showed that the conclusion of this study has strong robustness and authenticity.

Table 4. Robustness test.

	(1)	(2)	(3) (GSH)	(4) (GSL)
Variables	GS	Y	Y	Y
ER	0.0573 **	0.260 **	0.645 ***	−0.0653
	(0.0289)	(0.129)	(0.143)	(0.248)
GS		0.697 ***		
		(0.216)		
innov	−0.00198 **	0.00934 **	−0.00222	0.0375 ***
	(0.000856)	(0.00382)	(0.00361)	(0.00852)
mngt	−0.116 ***	0.625 ***	0.279 **	1.048 ***
	(0.0282)	(0.128)	(0.131)	(0.276)
wage	0.132 ***	0.276 ***	0.371 ***	0.215
	(0.0232)	(0.107)	(0.140)	(0.162)
pat	0.0103	0.0672 *	0.121 ***	0.0634
	(0.00773)	(0.0344)	(0.0405)	(0.0640)
Constant	−0.0702	−3.827 ***	−3.797 ***	−3.082 **
	(0.183)	(0.811)	(1.105)	(1.285)
Observations	462	462	290	172
R-squared	0.246	0.415	0.498	0.378
Number of ids	33	33	27	22

Note: *, **, and *** indicate the significance levels of 10, 5, and 1%, respectively.

6. Conclusions and Policy Recommendations

Environmental pollution is a worldwide problem. Governance synergy is the key to realizing environmental protection and industry development in China. The relationship between environmental regulation, governance synergy, and industrial green technology progress is important for the design of environmental policy and green economy development. Although the central government has been improving the environmental laws and regulations, the strategic behavior of regional mobility of polluting enterprises makes the effect of environmental governance unsatisfactory. This is due to the lack of mechanism design and regulations at the industry level. This study presented the institutional evolution of environmental governance synergy, constructed a new measure of environmental governance synergy, and used the mediating effect model to investigate the transmission mechanism of environmental regulation on industrial green technology progress through governance synergy. The main conclusions are as follows:

(1) There was a mediating effect in the environmental regulation promoting the industrial green technology progress through inter-regional governance synergy. Inter-regional low-level governance synergy hindered the industrial green technology progress. This was because of the lack of a synergetic governance mechanism, which caused some enterprises to migrate to other regions rather than innovate locally. This weakened the effect of environmental policies on encouraging enterprises to engage in green

technology innovation and is not good for the long-term development of industrial green technology.

(2) The impact of environmental regulation on the industrial green technology progress changed substantially with the level of inter-regional governance synergy. At low levels of governance synergy, environmental regulation restrained industrial green technology progress; at high levels of governance synergy, environmental regulation promoted industrial green technology progress.

Based on the above findings, two policy recommendations are made:

(1) The top-level design of governance synergy should be strengthened. Improving the joint prevention system is significant for industrial green technology progress as a whole. The supervision and adaptive incentives to local governments should be strengthened to prevent the enterprises' migration.
(2) The joint governance capacity of different regions in various industries should be improved. The central government should strengthen punishments for violations of regulations and avoid softening the environmental regulation system.

In conclusion, while strengthening environmental regulation, enterprises should be encouraged to innovate based on industrial governance synergy. Environmental regulation promotes the industrial green technology progress only if governance cooperation in the industry is formed.

Author Contributions: J.S.: conceptualization, formal analysis, methodology. Y.Y.: data collection, writing, review and editing, investigation. All authors have read and agreed to the published version of the manuscript.

Funding: Central University Basic Scientific Research Business Expenses (Scientific Research Cultivation) Project "Environmental Regulation, Technological Progress and Energy-Saving Efficiency" (Grant No. 202011023).

Data Availability Statement: The datasets used and/or analyzed during the current study are available from the corresponding author on reasonable request.

Acknowledgments: Central University Basic Scientific Research Business Expenses (Scientific Research Cultivation) Project Environmental Regulation, Technological Progress and Energy-Saving Efficiency (grant no. 202011023).

Conflicts of Interest: The authors declare that they have no known competing financial interests or personal relationships that could have appeared to influence the work reported in this study.

References

1. Gray, W.B.; Shadbegian, R.J. Plant Vintage Technology and Environmental Regulation. *J. Environ. Econ. Manag.* **2003**, *46*, 384–402. [CrossRef]
2. Bartik, T.J. The social value of job loss and its effect on the costs of us environmental regulations. *Rev. Environ. Econ. Policy* **2015**, *9*, 179–197. [CrossRef]
3. Shuai, S.; Fan, Z. Modeling the role of environmental regulations in regional green economy efficiency of China: Empirical evidence from super efficiency DEA-Tobit model. *J. Environ. Manag.* **2020**, *261*, 110227. [CrossRef] [PubMed]
4. Greenstone, M.; List, J.A.; Syverson, C. *The Effect of Environmental Regulation on the Competitiveness of U.S. Manufacturing*; NBER Working Paper; National Bureau of Economic Research: Cambridge, MA, USA, 2012.
5. Millimet, D.L.; Jayjit, R. Empirical Tests of the Pollution Haven Hypothesis When Environmental Regulation is Endogenous. *J. Appl. Econom.* **2016**, *31*, 652–677. [CrossRef]
6. Solarin, S.A.; Al-Mulali, U.; Musah, I.; Ozturk, I. Investigating the Pollution Haven Hypothesis in Ghana: An Empirical Investigation. *Energy* **2017**, *124*, 706–719. [CrossRef]
7. Yuan, B.; Xiang, Q. Environmental regulation, industrial innovation and green development of Chinese manufacturing: Based on an extended CDM model. *J. Clean. Prod.* **2018**, *176*, 895–908. [CrossRef]
8. Xie, R.; Yuan, Y.; Huang, J. Different Types of Environmental Regulations and Heterogeneous Influence on Green Productivity: Evidence from China. *Ecol. Econ.* **2017**, *132*, 104–112. [CrossRef]
9. Shen, C.; Li, S.; Wang, X.; Liao, Z. The effect of environmental policy tools on regional green innovation: Evidence from China. *J. Clean. Prod.* **2020**, *254*, 120122. [CrossRef]

10. Porter, M.E.; Van der Linde, C. Toward a new conception of the environment-competitiveness relationship. *J. Econ. Perspect.* **1995**, *9*, 97–118. [CrossRef]
11. Harrison, A.; Hyman, B.; Martin, L.; Nataraj, S. *When Do Firms Go Green? Comparing Price Incentives with Command and Control Regulations in India*; NBER Working Paper; National Bureau of Economic Research: Cambridge, MA, USA, 2015.
12. Zhai, X.; An, Y. Analyzing influencing factors of green transformation in China's manufacturing industry under environmental regulation: A structural equation model. *J. Clean. Prod.* **2020**, *251*, 119760. [CrossRef]
13. Acemoglu, D.; Aghion, P.; Bursztyn, L.; Hemous, D. The Environment and Directed Technical Change. *Am. Econ. Rev.* **2012**, *102*, 131–166. [CrossRef]
14. Ulucak, R. How do environmental technologies affect green growth? Evidence from BRICS economies. *Sci. Total Environ.* **2020**, *712*, 136504.
15. Borsatto, J.M.L.S.; Amui, L.B.L. Green innovation: Unfolding the relation with environmental regulations and competitiveness. *Resour. Conserv. Recycl.* **2019**, *149*, 445–454. [CrossRef]
16. Wang, Y.; Yu, L. Can the current environmental tax rate promote green technology innovation?—Evidence from China's resource-based industries. *J. Clean. Prod.* **2021**, *278*, 123443. [CrossRef]
17. Song, Y.; Yang, T.; Zhang, M. Research on the impact of environmental regulation on enterprise technology innovation-an empirical analysis based on Chinese provincial panel data. *Environ. Sci. Pollut. Res.* **2019**, *26*, 21835–21848. [CrossRef] [PubMed]
18. Wu, H.T.; Hao, Y.; Ren, S.Y. How do environmental regulation and environmental decentralization affect green total factor energy efficiency: Evidence from China. *Energy Econ.* **2020**, *91*, 33–44. [CrossRef]
19. Bildirici, M.; Gokmenoglu, S.M. The impact of terrorism and FDI on environmental pollution: Evidence from Afghanistan, Iraq, Nigeria, Pakistan, Philippines, Syria, Somalia, Thailand and Yemen. *Environ. Impact Assess. Rev.* **2020**, *81*, 12. [CrossRef]
20. Dong, Z.; Wang, H. The Effect of Environmental Regulation on the Green Technology Progress in 'Local-Adjacent'. *Chinas Ind. Econ.* **2019**, *1*, 100–118. (In Chinese)
21. Li, M.; Du, W.; Tang, S. Assessing the impact of environmental regulation and environmental co-governance on pollution transfer: Micro-evidence from China. *Environ. Impact Assess. Rev.* **2021**, *86*, 106467. [CrossRef]
22. Hu, Z.; Li, G.; Cao, J. Joint Regional Air Pollution Control from the Perspective of Environmental Regulation: Zoning Scheme Design, Collaborative State Assessment and Impact Factor Analysis. *Chinas Ind. Econ.* **2019**, *5*, 24–42. (In Chinese)
23. Rogge, K.S.; Schleich, J. Do Policy Mix Characteristics Matter for Low-carbon Innovation? A Survey-based Exploration of Renewable Power Generation Technologies in Germany. *Res. Policy* **2018**, *47*, 1639–1654. [CrossRef]
24. Fan, Y.; Bai, B.; Qiao, Q.; Kang, P.; Zhang, Y.; Guo, J. Study on eco-efficiency of industrial parks in China based on data envelopment analysis. *J. Environ. Manag.* **2017**, *192*, 107–115. [CrossRef] [PubMed]
25. Huang, J.; Xia, J.; Yu, Y.; Zhang, N. Composite eco-efficiency indicators for China based on data envelopment analysis. Ecological. *Indicators* **2018**, *85*, 674–697. [CrossRef]
26. Xie, H.; Chen, Q.; Lu, F.; Wang, W.; Yao, G.; Yu, J. Spatial-temporal disparities and influencing factors of total factor green use efficiency of industrial land in China. *J. Clean. Prod.* **2019**, *207*, 1047–1058. [CrossRef]
27. Lanoie, P.; Patry, M.; Lajeunesse, R. Environmental Regulation and Producticvity: Testing the Porter Hypothesis. *J. Product. Anal.* **2008**, *30*, 55–66. [CrossRef]
28. Gray, W.B.; Shadbegian, R.J.; Wang, C. Corrigendum to: Do EPA regulations affect labor demand? Evidence from the pulp and paper industry. *J. Environ. Econ. Manag.* **2014**, *68*, 188–202. [CrossRef]
29. Tang, G.; Li, L.H.; Wu, D. Environmental regulation, industry attributes and corporate environmental investment. *J. Account. Res.* **2013**, *6*, 83–89.
30. Brunnermeier, S.B.; Cohen, M.A. Determinants of Environmental Innovation in US Manufacturing Industries. *J. Environ. Econ. Manag.* **2003**, *45*, 278–293. [CrossRef]

processes

Article

Degradation of Landfill Leachate Using UV-TiO_2 Photocatalysis Combination with Aged Waste Reactors

Chunlian Wang [1,2], Xiaojie Sun [1,2,*], Huijun Shan [1,2], Hongxia Zhang [1,2] and Beidou Xi [1,3,4]

1 Guangxi Key Laboratory of Environmental Pollution Control Theory and Technology, Guilin University of Technology, Guilin 541004, China; 1020190273@glut.edu.cn (C.W.); Secret202158@hotmail.com (H.S.); zhx75@glut.edu.cn (H.Z.); xibd@craes.org.cn (B.X.)

2 Guangxi Collaborative Innovation Center for Water Pollution Control and Water Safety in Karst Area, Guilin University of Technology, Guilin 541004, China

3 State Key Laboratory of Environmental Criteria and Risk Assessment, Chinese Research Academy of Environmental Sciences, Beijing 100012, China

4 State Environmental Protection Key Laboratory of Simulation and Control of Groundwater Pollution, Chinese Research Academy of Environmental Sciences, Beijing 100012, China

* Correspondence: sunxiaojie@glut.edu.cn; Tel.: +86-150-7832-9789

Abstract: This study explored the performance of TiO_2 nanoparticles in combination with aged waste reactors to treat landfill leachate. The optimum conditions for synthesis of TiO_2 were determined by a series of characterizations and removal rates of methyl orange. The effect of the ultraviolet irradiation time, amount of the catalyst, and pH on the removal efficiency for the chemical oxygen demand (COD) and color in the leachate was explored to determine the optimal process conditions, which were 500 min, 4 g/L and 8.88, respectively. The removal rates for COD and chroma under three optimal conditions were obtained by the single factor control method: 89% and 70%; 95.56% and 70%; and 85% and 87.5%, respectively. Under optimal process conditions, the overall average removal rates for ammonium nitrogen (NH_4^+–N) and COD in the leachate for the combination of TiO_2 nanoparticles and an aged waste reactor were 98.8% and 32.5%, respectively, and the nitrate (NO_3^-–N) and nitrite nitrogen (NO_2–N) concentrations were maintained at 7–9 and 0.01–0.017 mg/L, respectively. TiO_2 nanoparticles before and after the photocatalytic reaction were characterized by emission scanning electron microscopy, energy dispersive spectroscopy, X-ray diffraction, and Fourier transform infrared spectrometry. In addition, TiO_2 nanoparticles have excellent recyclability, showing the potential of the photocatalytic/biological combined treatment of landfill leachate. This simulation of photocatalysis-landfilling could be a baseline study for the implementation of technology at the pilot scale.

Keywords: TiO_2 nanoparticles; photocatalysis; landfill leachate; aged waste reactors

Citation: Wang, C.; Sun, X.; Shan, H.; Zhang, H.; Xi, B. Degradation of Landfill Leachate Using UV-TiO_2 Photocatalysis Combination with Aged Waste Reactors. *Processes* **2021**, *9*, 946. https://doi.org/10.3390/pr9060946

Academic Editors: Andrea Petrella, Marco Race and Danilo Spasiano

Received: 18 April 2021
Accepted: 24 May 2021
Published: 27 May 2021

1. Introduction

Leachates produced in landfills contain various organic and inorganic compounds, such as refractory organic material, dissolved solid particles, ammonia nitrogen (NH_3–N), and heavy metals, which seriously threaten the environment and local ecosystems [1–3]. Often, combined anaerobic–aerobic biological processes can be utilized to degrade the biodegradable organic pollutants of leachates [4–6]. However, over time, the reduction of the biological oxygen demand/chemical oxygen demand (BOD_5/COD) ratio < 0.3 and microorganisms involved in nitrification–denitrification processes are readily hampered by high concentrations of ammonium nitrogen, making it difficult to be treated by the conventional biological and physicochemical processes [7–10]. Moreover, additional carbon sources are needed to aid the nitrification–denitrification process [11]. Hence, a novel method has been presented to transform landfills into bioreactors via leachate recirculation, which is expected to accelerate the stabilization of landfills and reduce the organic strength of leachates [12]. However, recirculated leachates also lead to the accumulation of higher levels of NH_4^+–N and refractory organics compared with traditional landfills because of the increasing ammonification rate and the lack of carbon sources [13].

Recently, the application of semiconductor-based photocatalysis in the treatment of landfill leachates has attracted attention [14–16]. The technology could effectively mineralize a wide range of organic pollutants into low-toxicity organic small molecules, CO_2, and H_2O without the use of expensive oxidant [17]. Among photocatalysis, TiO_2 has been extensively studied due to its super strong photo-oxidizing ability, easy accessibility and environmental friendliness, and it has been considered to be a green catalyst material with broad development prospects in the field of water treatment [18,19]. TiO_2 absorbs UV light (<380 nm), and electrons are promoted from the valence band (VB) to the conduction band (CB) to generate electron–hole pairs (e^-/h^+). Positive holes typically oxidize organic compounds, inducing their oxidative degradation. In addition, electrons mainly reduce molecular oxygen to superoxide radical anions, thereby leading to a number of reactive oxygen species (OH•, O_2^-•, and HO_2•). Some titanium dioxide-based materials are used to treat refractory pollutants, such as prepared B–TiO_2-graphene oxide ternary nanocomposite in the degradation of bisphenol A with a mineralization rate of 47.66% [20], iron-doped TiO_2 photodegradation for rhodamine B at about 90% [21], and TiO_2-based catalysts obtained by different preparation methods; the simultaneous oxidation of organic matter and the reduction in ammonia were observed during the solar photocatalytic treatment of greywater [22].

Within this context, aiming to increase the effectiveness of the treatment of landfills, researchers have investigated the application of combination techniques, integrating biological processes with advanced oxidation processes [23]. Cai et al. [24] applied cetyltrimethylammonium bromide bentonite–titanium dioxide photocatalytic technology to the pretreatment of aging leachate, which maintained the removal of COD and NH_3–N at 82% and 37% in 60.02 min, respectively. Pellenz et al. [25] used a boron-doped diamond-based photo-electro-Fenton system integrated with biological oxidation to treat landfill leachate, which reduced its toxic potential. Researchers have attempted to implement anaerobic bioreactor technology for effective MSW treatment through leachate decontamination [26]. Simulations of landfilling in anaerobic bioreactors function as anaerobic sludge digesters and facilitate accelerated and economic waste stabilization [27]. The study of integrating biological processes with advanced oxidation processes in simple lab scale reactors is an important stage to enable real-scale integrated applications in the treatment of landfill leachate, which could solve the bottleneck caused by the high toxicity and insufficient carbon source of landfill leachate treatments. Thus, the main objectives of this study were two-fold. First, we intended to explore the best conditions for the synthesis of TiO_2 at different hydrothermal temperatures and the different ratio of titanium to urea based on the efficiency of methyl orange and learn about optimal reaction conditions under different irradiation times, amounts of TiO_2, and initial pH regarding COD and color treatment efficiency in diluted leachate. At this stage, the TiO_2 particles were characterized by energy dispersive spectrum (EDS), X-ray diffraction (XRD), and Fourier transform infrared (FTIR). The second goal was to investigate the possibility of using synthetic TiO_2-guided photocatalysis in combination with an aged waste reactor to treat landfill leachates under the best process conditions by examining the removal rates or concentration of NH_4^+–N, nitrate (NO_3^-–N), nitrite nitrogen (NO_2^-–N), and COD. Currently, there is relatively little published information on dealing with aging leachates by aged waste reactors combined with photocatalytic technology. This study expects to provide some suggestions for accelerating the process of waste stabilization.

2. Materials and Methods

2.1. Sampling

Landfill leachate was obtained from the Chongkou landfill in Guangxi, China. The general characteristics of the raw leachate studied were as follows: COD, 7647 mg/L; pH, 8.88;

NH_4^+–N, 26.73 mg/L. The original water was diluted 30 times prior to treatment given the high contaminant load in the original landfill leachate.

Leachate samples were analyzed by the standard for pollution control on the landfill site of the MSW [28]. COD was determined in accordance with the dichromate titration method. NH_4^+–N, NO_3^-–N, NO_2^-–N, and chromaticity were measured with an ultraviolet-visible spectrophotometer (UV-6100A, Yuanxi Instrument Co., Ltd., Shanghai, China). The pH was measured using the glass electrode method (Yidian Scientific Instrument Co., Ltd., Shanghai, China) and adjusted by adding HCl or NaOH.

2.2. TiO_2 Nanoparticle Synthesis

The TiO_2 nanoparticles were prepared by the green and mild solvothermal method [17,29]. First, 12.6 g of $Ti(SO_4)_2$ and 3.15, 6.3 and 12.6 g of urea were dissolved in 40 mL of ultrapure water, the solution was stirred at room temperature for 20 min, and the mixture was transferred into a Teflon-lined stainless autoclave of 100 mL capacity, sealed, and heated at different temperatures (100 °C, 120 °C, 150 °C and 180 °C) for 12 h. After the system cooled down to room temperature, the products were washed with deionized water three times and separated by centrifugation until the system was neutral. Finally, the products were dried in air at 80 °C for 12 h. Among them, the weight ratio of $Ti(SO_4)_2$ and urea was controlled at about 1:1,1:2 and 1:4.

2.3. Characterization of TiO_2 Nanoparticles

Four methods of characterization were used to describe the characteristics of the synthesized TiO_2 nanoparticles. To determine the structural characteristics and crystalline phases of the nanoparticles, the X-ray diffraction (XRD) method was used. The diffraction angle range of $2\theta = 10°–80°$ was measured by a PANalytical X'Pert3 Powder X-ray diffractometer (the Netherlands). Field emission scanning electron microscopy (FESEM) analysis was used to visualize the morphology of the nanoparticles. This analysis was accomplished with the Electron JSM-7900F FESEM in Japan. Qualitative and quantitative analyses of the sample elements were conducted by energy dispersive spectrum (EDS) analysis and by using the electron JSM-7900F FESEM in Japan at an accelerating voltage of 15 kV. The molecular structure changes of different samples were analyzed by Fourier transform infrared spectrometry (FTIR), using a device produced in the U.S.A. by PerkinElmer, at a scan range of 4000–400 cm^{-1}.

2.4. Photocatalytic Experiments

The visible photocatalytic activities of the obtained samples were investigated by the photodegradation of methyl orange in an aqueous solution. An amount of 20 mg of the sample was dispersed into 50 mL of the methyl orange solution (10 mg/L)/diluted leachate in a Pyrex glass reactor. A 300 W xenon lamp (Naai Precision Instrument Co., Ltd., Shanghai, China) with a UV light source (Figure 1) was used. The solution was allowed to reach an adsorption–desorption equilibrium among the photocatalyst and methyl orange by magnetic stirring in the dark for 30 min before irradiation with UV light. At certain time intervals, the 2 mL suspension was sampled and filtered through the 0.45 m filter membrane to remove the particles. The concentration of methyl orange was determined by a UV-visible spectrophotometer (UV-6100A, Yuanxi, Shanghai, China) according to its absorbance wavelength at 464 nm.

2.5. Cooperative Degradation Experiments

This study was conducted as a follow-up to a previous study, and the same materials were used, including anaerobic reactors formed by domestic waste (numbered C and E) and aged waste reactors (numbered D and F) (Figure 2) [13,30]. These materials were suitable for the microbial domestication of the test stage. The leachate produced by C and E was treated with D and F to form a recharge solution, which circulated back to the corresponding anaerobic landfill device. Groups CD (Figure 2a) and EF (Figure 2b) microbially acclimatized for 14 days. The frequency obtained was 1 time/d when the leachate diluted 30 times was used as the inlet water. From day 15 to day 33, the operation mode of the CD group remained unchanged. However, in groups E and F, the effluent of reactor F was treated by the photocatalytic reactor and then recharged to reactor E. In this study, only the changes in the D and F effluent indexes were investigated.

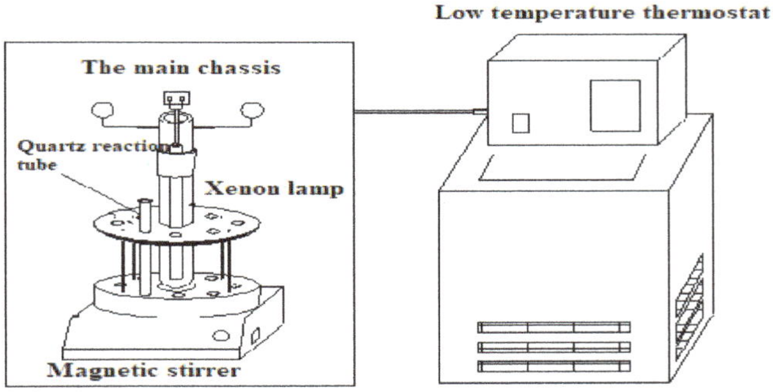

Figure 1. Schematic of the experimental setup.

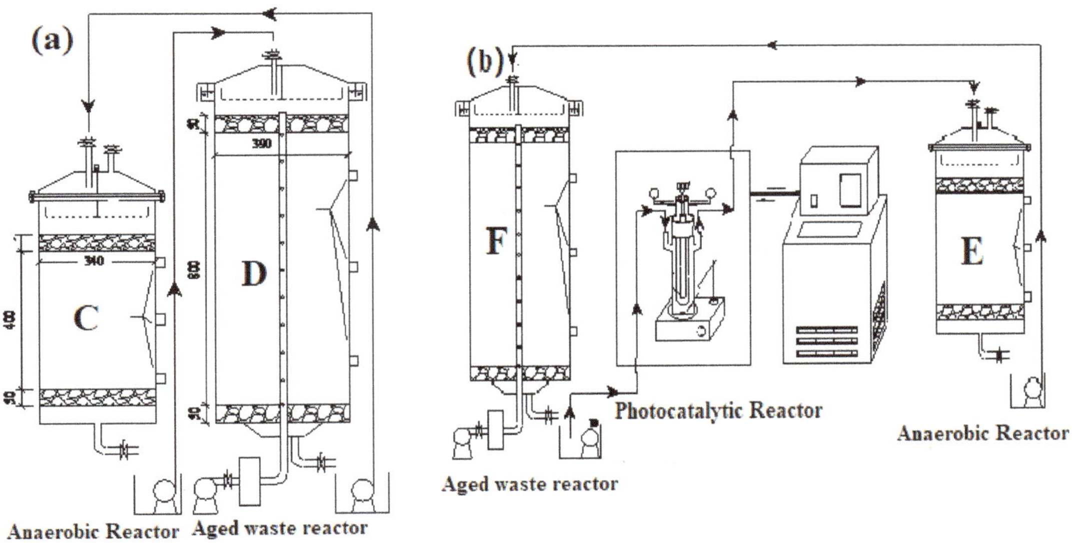

Figure 2. Schematic of bioreactor landfill (**a**), combined bioreactor landfill (**b**).

The photocatalytic reaction phase is as follows: 2 g of the TiO_2 was dispersed into 500 mL (25 °C) of the effluent of reactor F with photocatalytic treatment for 8 h under simulated sunlight (500 W).

3. Results and Discussion

3.1. Characterization of Photocatalysts

With methyl orange as the target pollutant, the best preparation conditions for TiO_2 were obtained by evaluating its degradation efficiency (Figure 3). The removal ratio is calculated by the following equations [31]:

$$\text{removal ratio} = (C_0 - C)/C_0 \times 100\% \quad (1)$$

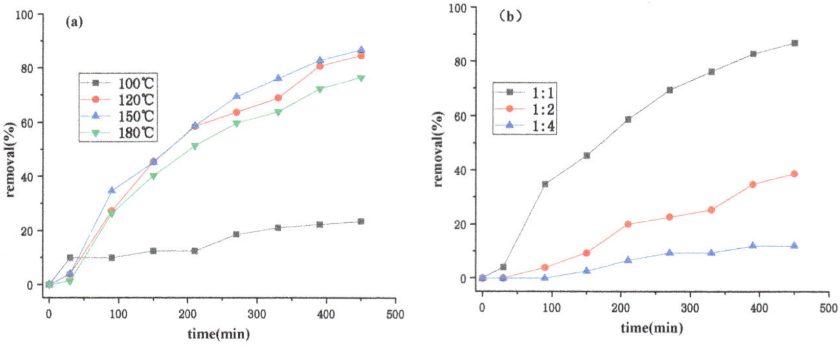

Figure 3. Effect of photocatalysts prepared at different temperature (**a**) and the different ratio of titanium to urea (**b**) on degradation rate of methyl orange concentration.

The surface morphology and surface area were very effective parameters in the photocatalytic activity of TiO_2 [32]. The SEM images revealed that nano-TiO_2 had a smooth surface, and EDS analysis illustrated the content of Ti and O elements, showing that the prepared sample had high purity (Figure 4a,b). The XRD spectrum was used to analyze the crystal structure, as shown in Figure 4c,d. From data analysis, the body-centered tetragonal crystal structure of anatase was identified based on significant diffraction peaks: 20.2°, 37.8°, 47.9°, 53.9°, 55°, and 62.6° corresponding to (101), (004), (200), (105), (211), and (204) planes, respectively (JCPDS 86–1157). As previously reported in the literature, the diffraction peak of the anatase crystal form became stronger as the half-peak width narrowed, and the sample showed a good crystal form at 150 °C. The result of the XRD patterns of products with different ratios of titanium sulfate to urea show that all the samples exhibit high crystallinity, especially when the ratio is 1:2 (Figure 4d). Meanwhile, the particle size of sample was calculated by the Scherrer formula [33]:

$$D = (K\lambda B \cos\theta) \tag{2}$$

where D is the average crystal size of the sample, λ is the X-ray wavelength (1.54056 Å), B is the full width at half maximum of the diffraction peak (radian), K is a coefficient (0.89), and θ is the diffraction angle at the peak maximum. The particle size of TiO_2 at 100 °C, 120 °C, 150 °C, and 180 °C was 5.3, 9.7, 11.8, and 16.4 nm, respectively, and the particle size of TiO_2 at titanium sulfate–urea ratios of 1:1, 1:2, and 1:4 was 11.8, 13.1, and 9.5 nm, respectively. Considering the crystallinity of the synthesized sample and degradation rate for methyl orange, the optimal particle size of titanium dioxide was 11.8 nm in this experiment.

The FTIR spectrum analysis chart of TiO_2 prepared under controlled conditions is shown in Figure 4e,f. The characteristic peak at 150 °C was the strongest, and the crystal structure was the most complete, which was consistent with the XRD analysis. The peak at 460 cm^{-1} of TiO_2 was assigned as the Ti–O stretching vibration and Ti–O–Ti bridging stretching vibration [34]. The two signals at 1640 cm^{-1} and 3420 cm^{-1} can be ascribed to the −OH bending vibration of the water on the surface of TiO_2 (Figure 4e) [35]. The adsorbed water and hydroxyl groups on the surface of the catalyst would interact with the electron-holes generated by the excitation to produce hydroxyl radicals with strong oxidizability (Figure 4e) [36]. Therefore, the best synthesis conditions are a temperature of 150 °C and a titanium sulfate–urea ratio of 1:1.

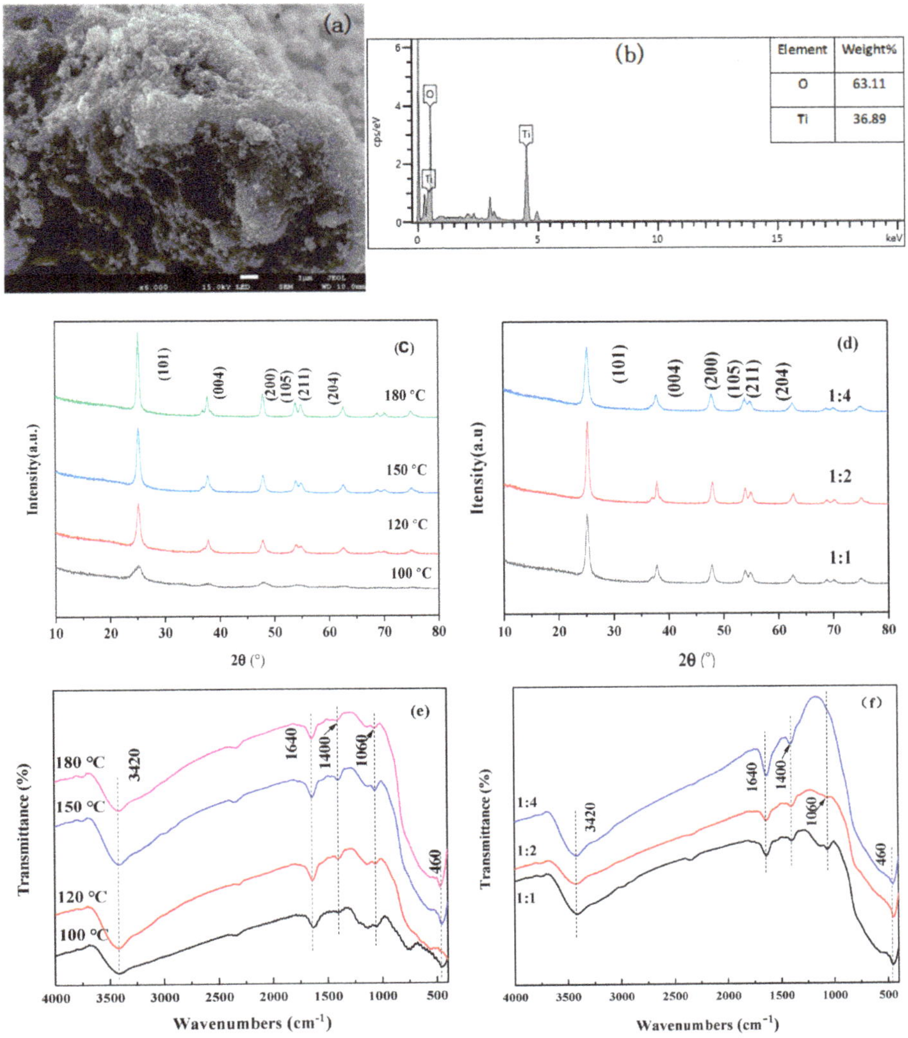

Figure 4. SEM images (**a**) and EDS elemental mapping analysis (**b**), XRD patterns (**c**,**d**), FTIR patterns (**e**,**f**) during TiO_2 nanoparticles.

3.2. Optimization of Photocatalytic Processes

The results of photocatalytic degradation of the landfill leachate by adjusting the irradiation time are shown in Figure 5a. During the dark reaction carried out for 30 min, the removal rates by TiO_2 adsorption of COD and color was 93.94% and 60%, respectively. The maximum COD and color removal rates were 89% and 70% under UV light, respectively. The reason for the increased COD value may be that the photocatalytic reaction oxidizes the complex macromolecular organics into small molecular organics after the light is turned on. Hassan et al. [37] pointed out that an increase in COD concentration at the end of photocatalysis is due to a decrease in catalytic efficiency caused by the deposition of pollutants on the catalyst with time. Sama Azadi et al. [38] also confirmed this conclusion. It is speculated that the reason for a COD removal rate that is higher than the chromaticity removal rate is due to low phenolic substances in the reaction process, as color reduction simply reflects the oxidative opening benzene ring in the phenolic substances to other straight chain compounds [39]. The result is consistent with the removal rate of COD and color treated by AC/TiO_2 for the last 30 min [40]. The reason why the COD decrease was faster than the color removal needs more research. For this batch of the experiment, equal amounts of 200 mg TiO_2 were added into a series of test beakers containing 50 mL of the landfill leachate.

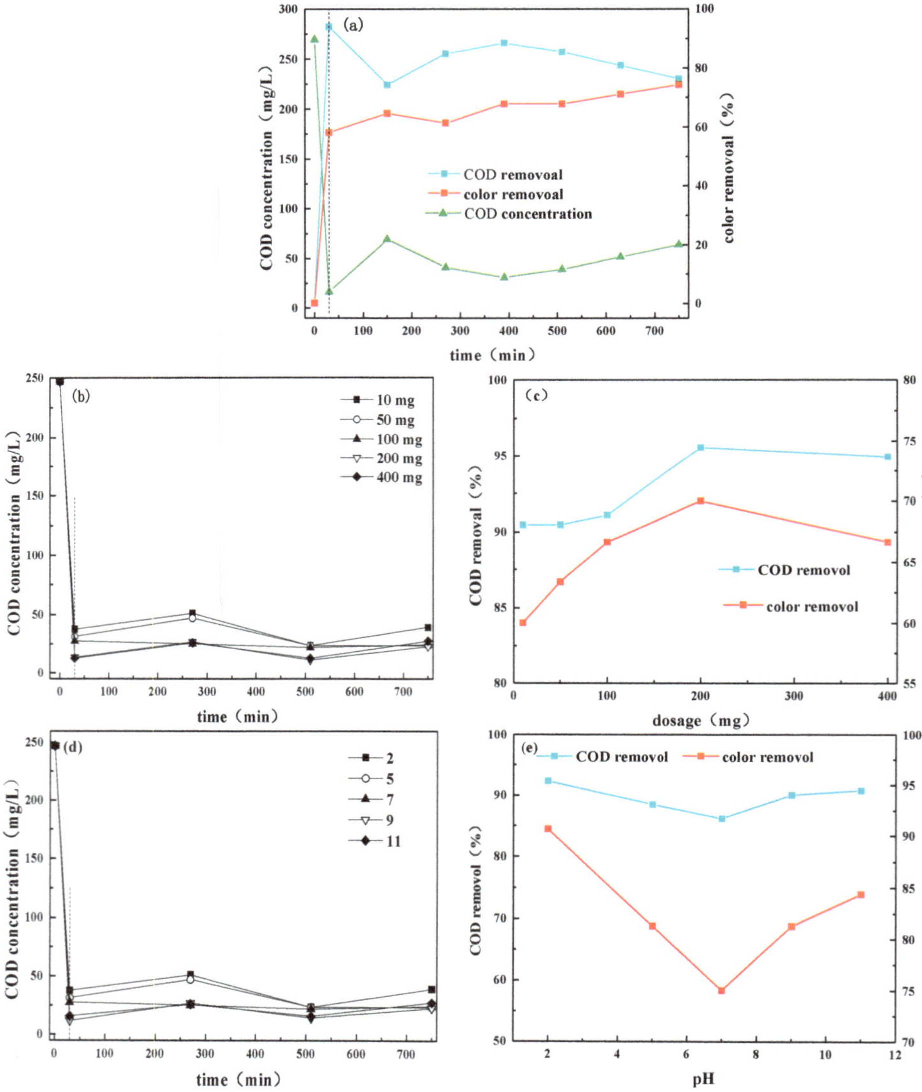

Figure 5. Effects of irradiation time (**a**), TiO_2 dosage (**b**,**c**) and initial pH (**d**,**e**) on COD concentration, removal and color removal.

As shown in Figure 5b,c, when the dosage of TiO_2 was 200 mg (4 g/L), the COD and decolorization removal rate were 94.28% and 70.00%, respectively. The highest photocatalytic efficiency did not appear at the dosage of 400 mg (8 g/L). The study confirmed that with the increase in the amount of TiO_2, the irradiated area and the photocatalytic rate increased [41]. At a TiO_2 dosage of 4 g/L, Jia [42] removed 37.4% COD and 55.5% color, but the degradation ratio decreased, except COD, when the dosage was increased to 16 g/L. Miao et al. [43] noted that when the catalyst dosage reached the saturation level, the light absorption coefficient decreased. Therefore, the optimal dosage TiO_2 was 200 mg (4 g/L) in this study. For this batch of experiment, the duration of experiments performed was 750 min.

The results of photocatalytic degradation of the leachate at different initial pH values (2, 5, 7, 9, and 11) are shown in Figure 5d,e. During the reaction process, different pH values slightly affected the COD removal rate of approximately 85%. The color removal rate was evident under acidic conditions. When the pH value was 2, the color removal increased up to 87.5%. Jia et al. [44] studied the degradation of the landfill leachate by the photocatalytic treatment process and proposed that the optimal pH value was 4. Some researchers suggested that HCO_3^-/CO_3^{2-} anions brought about significant inhibitory effect on photocatalytic oxidation of refractory organics present in the leachate [45]. However, Xu et al. [46] noted that upon decreasing the leachate pH from 5.4 to 3.9, the amount and size of aggregated TiO_2 particles increased. Anpo and Kamat suggested that the highest pollutant adsorption and maximum photocatalytic removal efficiency were observed at a pH close to pH_{zpc} [47]. Azadi et al. [38] indicated that the pH_{zpc} values of C and C–W co-doped TiO_2 C-doped nanoparticles were 6.27 and 6.7, respectively, and the highest COD removal rate was in the range of pH value 6–7. The COD removal efficiency increased with leachate pH as reported in the literature [8]. Therefore, the pH value near pH_{zpc} had the best photocatalytic efficiency. In addition, the reason for the highest color removal rate at pH = 2 may be the conversion of colored substances to colorless substances during the photocatalytic reaction under strong acidity [48]. The original pH of the leachate in this test was alkaline (8.88), and the pH of the leachate was unadjusted from the engineering point of view. For this batch of the experiment, equal amounts of 200 mg TiO_2 were added into a series of test beakers containing 50 mL of the landfill leachate, and the duration of the experiments performed was 750 min.

In addition, combined with the maximum removal rates of COD and color under the three conditions, the optimal reaction time in this study was 500 min.

3.3. *Treatment of Leachate by Aging Reactor Combined with Photocatalysis*

The indicators of each reactor after acclimation are shown in Table 1.

Table 1. The indicators of each reactor after acclimation.

	NH_4^+–N (mg/L)	NO_3–N (mg/L)	NO_2–N (mg/L)	COD (mg/L)
C	5.457	18.209	0.079	245.302
D	1.049	18.226	0.017	185.955
E	18.404	8.5228	0.281	146.390
F	1.049	18.276	0.281	126.607

As shown in Figure 6a, from day 15 to 33, the concentrations of NH_4^+–N in reactors D and F were 0.083–2.523 and 0.083–1.17 mg/L, respectively. The average removal rate was 80.3% for reactor D and 98.8% of reactor F. The NH_4^+–N concentration in the effluent of the photocatalytic reaction was high, which was probably due to the photocatalytic reaction converting the nitrogen-containing organic matter into nitrogen-containing inorganic matter in the leachate. This finding shows that the leachate contains a large amount of heterocyclic nitrogen and amino compounds that readily hampered the nitrification–denitrification processes [9,49].

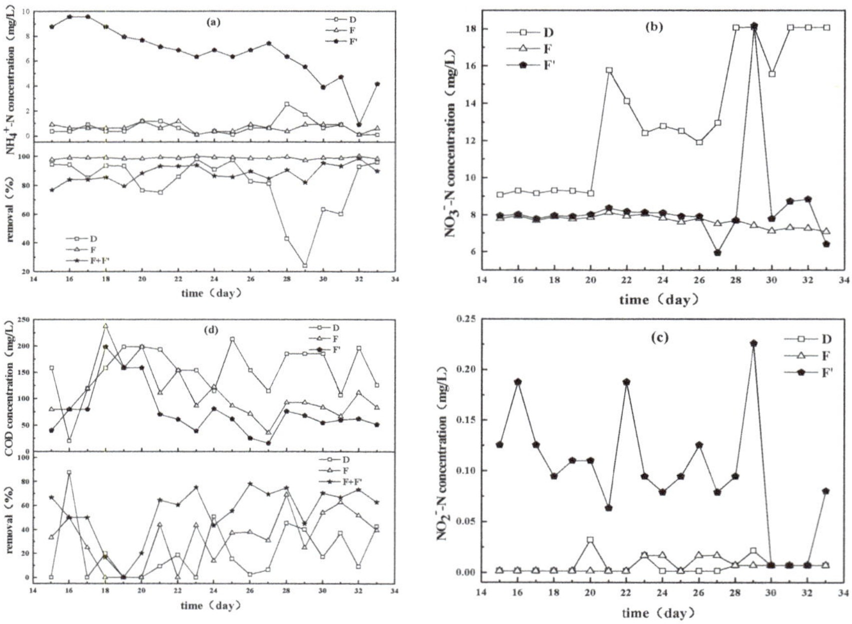

Figure 6. NH_4^+–N concentration and removal (**a**), NO_3^-–N concentration (**b**) NO_2^-–N concentration (**c**), COD concentration and removal (**d**) of leachates.

Figure 6b illustrates that the concentration of NO_3-N in the effluent of reactor D and the NO_3–N concentration of the effluents of reactor F was 7–9 mg/L. Boonnorat et al. found that the leachate produced by a landfill for a long time contained low carbon/nitrogen (C/N) ratios [50]. The photocatalytic reaction of TiO_2 with a circulating leachate transformed the refractory organic matter into biodegradable organic matter, which was looped through the anaerobic landfill reactor E to accelerate the degradation of domestic waste and then recycled at the reactor F to provide a carbon source for denitrifying bacteria. Therefore, the concentration of NO_3^-–N in reactor F was lower than that in reactor D [51,52]. Figure 6c shows that NO_2^-–N concentrations of reactors D and F were 0.01–0.032 and 0.01–0.017 mg/L, respectively. However, the NO_2^-–N concentration of the photocatalytic effluent F′ increased, which confirms the conclusion that the nitrogen-containing organic matter shown in Figure 6a was degraded into NH_4^+–N [53] The N-compound conversion equation is as follows:

$$TiO_2 + hv \rightarrow e^- + h^+ \tag{3}$$

$$h^+ + H_2O \rightarrow OH^\bullet + H^+ \tag{4}$$

$$e^- + O_2 \rightarrow O_2^{-\bullet} \tag{5}$$

$$O_2^{-\bullet} + e^- + 2H^+ \rightarrow H_2O_2 \tag{6}$$

$$H_2O_2 + e^- \rightarrow OH^\bullet + OH^- \tag{7}$$

$$DON + OH^\bullet + O_2^{-\bullet} + H_2O \rightarrow NH_4^+ + OH^- \tag{8}$$

$$NH_4^+ + 2\,O_2^{-\bullet} + 2H_2O \rightarrow NO_2^- + 2\cdot HO_2 + 3H_2 \tag{9}$$

$$2NO_2^- + 2O_2 \rightarrow 2\,NO_3^- \tag{10}$$

$$2NO_2^- + 2\,OH^\bullet \rightarrow NO_3^- + H_2O \tag{11}$$

where $OH^{\bullet}$ and $O_2^{-\bullet}$ are active species produced in the photocatalytic process.

As shown in Figure 6d, at days 15 to 20, the COD concentration difference between the effluent of aged waste reactors D and F was not evident, and the microorganisms were still in a certain adaptation stage. From day 21 to the end of the process, the average COD removal rate (32.5%) of reactor F was also higher than that of reactor D (21.20%). At the end of the research, the COD concentrations of the effluent from reactor D and F were 125.49 and 83.14 mg/L, respectively; thus, reactor D satisfied the standard for pollution control on the landfill site of the MSW [28]. A possible explanation is that in the presence of organic compounds, the generated hydroxyl radicals interact with the aromatic and heterocyclic compounds present in the leachate, which favor the formation of small aliphatic chains of carboxylic acids resulting in compounds that are easily assimilated by microorganisms [54]. This result indicates that the photocatalytic reaction degrades the refractory organics into biodegradable organic matter, which can be degraded by the microorganisms in an aged waste reactor [55,56]. Accordingly, the photocatalytic reaction increases the biodegradable organic matter in the circulating leachate, which is more easily degraded by microorganisms in reactor F. With the combined process, it was possible to treat an effluent with a high organic load, meeting the restrictive standards of release in the recipient water bodies [57].

3.4. Photocatalyst Stability

The experiment of removing COD and color was repeated three times to evaluate the recycling and stability of the nanophotocatalyst. After each cycle, the catalyst was collected and washed by simple centrifugation, washed with deionized water, and dried at 120 °C. As shown in Figure 7, after three cycles, the COD removal rate was 90.00% and the color change rate was 63.33%. These findings indicate that after recovery, the nanophotocatalyst can maintain activity and continue to provide electrons during the photocatalytic reaction to promote substrate degradation. Therefore, it had good industrial application prospects.

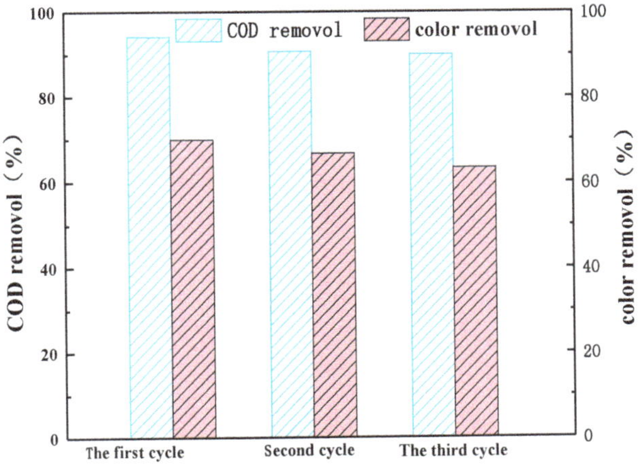

Figure 7. TiO_2 recycling treatment effect.

XRD characterization, FTIR spectroscopy, SEM observation, and EDS analysis were carried out to study the structure, morphology, and chemical composition of the recovered TiO_2 nanoparticles. As shown in Figure 8a, after the reaction, the TiO_2 crystal form, crystal plane, and diffraction peak intensity did not significantly change. According to Scherrer's formula, the grain diameter after the reaction increased to 14.1 nm, but as the grain size increased, the specific surface area and the photocatalytic activity gradually decreased, consistent with the photocatalyst stability test results (Figure 8b). The FTIR diagram shows that, after the photocatalytic reaction, the characteristic peaks at 3420, 1640,

1400, and 1060 cm^{-1} increased. This finding indicates that the surface of TiO_2 adsorbed organic matter containing hydroxyl groups and SO_4^{2-}. In addition, the absorption peaks of CH and CO appeared at 2974 $cm^{-1}$and 2900 cm^{-1}, respectively, which was caused by the organic matter in the leachate that adsorbed on the surface of the TiO_2 particles. However, the characteristic peak intensity of crystalline TiO_2 molecules with a wavelength of 460 cm^{-1} decreased. The reason may be that organic matter occupied the active site of TiO_2, thereby weakening the stretching of Ti–O and Ti–O–Ti vibration (Figure 8c). Accordingly, TiO_2 nanoparticles have the potential to maintain high photocatalytic efficiency during stability tests.

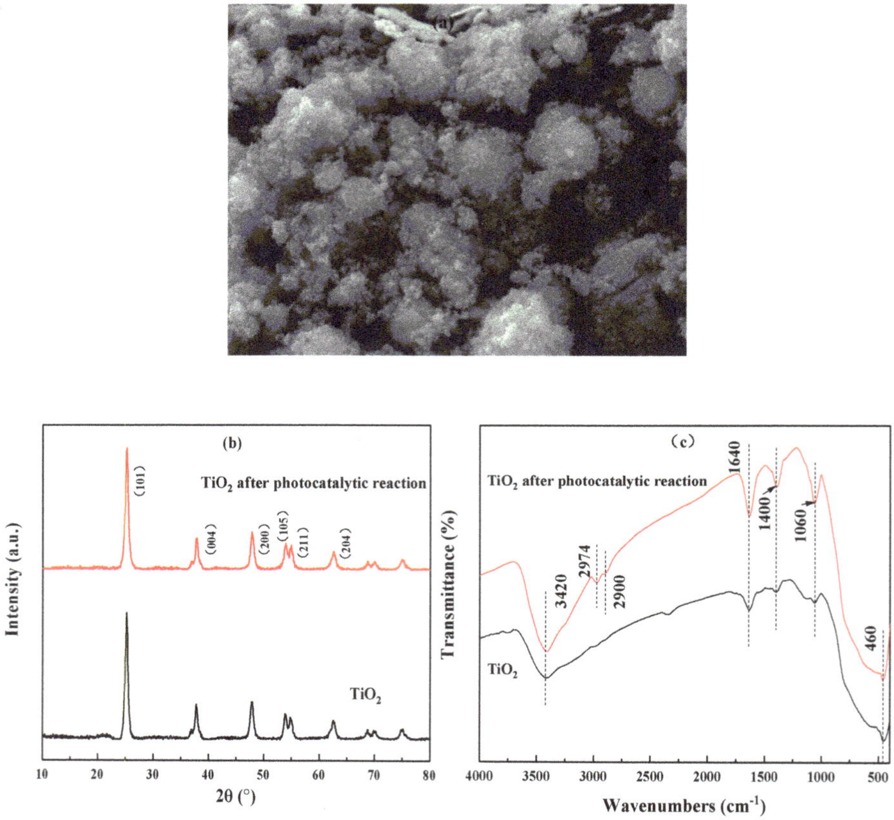

Figure 8. SEM images (**a**), XRD patterns (**b**) and FTIR patterns (**c**) of TiO_2 nanoparticles before and after photocatalysis.

4. Conclusions

The effects of TiO_2 nanoparticles and aged waste reactor combined treatment of NH_4^+–N, NO_3^-–N, NO_2^-–N, and COD in landfill leachate were discussed. SEM, FTIR, and XRD were used to determine the optimal hydrothermal reaction temperature of the synthesized nanoparticles at 150 °C, and the ratio of titanium to urea was 1:1. Then, considering the removal rate for COD and chromaticity as the index, the optimal photocatalytic reaction time was 500 min, the best dosage of TiO_2 nanoparticles was 4 g/L, and the original leachate had pH = 8.88. The removal rates of COD and chroma under three conditions were obtained by the single-factor control method: 89% and 70%; 95.56% and 70%; and 85% and 87.5%, respectively. Under the best process conditions, after the combined treatment with TiO_2 nanoparticles and an aged waste reactor for landfill leachate, the overall average removal rates of NH_4^+–N and COD were 98.8% and 32.5%, respectively. The concentration of NO_3^-–N and NO_2^-–N were maintained at 7–9 and 0.01–0.017 mg/L, respectively. The combined device degraded the refractory biodegradable organic matter into biodegradable organic matter, and the recharge to the anaerobic bioreactor landfill can continue to accelerate the degradation of domestic garbage. This simulation of photocatalysis-landfilling could be a baseline study for the implementation of technology at the pilot scale to accelerate the process of waste stabilization.

Author Contributions: Conceptualization, C.W.; data curation, C.W. and H.S.; funding acquisition, X.S.; investigation, B.X., H.Z.; methodology, H.S.; project administration, B.X.; software, C.W.; supervision, X.S.; writing—original draft, C.W.; writing—review and editing, X.S. and H.Z. All authors have read and agreed to the published version of the manuscript.

Funding: This research was funded by National Natural Science Foundation of China (No. 52070049); Natural Science Foundation of Guangxi (No. 2018GXNSFGA281001); Science and Technology Major Project of Guangxi (GuikeAA18118013); Innovation Project of Guangxi Graduate Education (No. YCSW2021198).

Institutional Review Board Statement: Not applicable.

Informed Consent Statement: Not applicable.

Data Availability Statement: The data presented in this study are available on request from the authors.

Conflicts of Interest: The authors declare no conflict of interest.

References

1. Guo, Y.; Sheehan, J.D.; Chen, Z.; Wang], S. Oxidative degradation of landfill leachate by catalysis of CeMnOx/TiO_2 in super-critical water: Mechanism and kinetic study. *Chem. Eng. J.* **2018**, 578–586.
2. Yang, Y.; Liu, Z.; Demeestere, K.; Van Hulle, S. Ozonation in view of micropollutant removal from biologically treated landfill leachate: Removal efficiency, OH exposure, and surrogate-based monitoring. *Chem. Eng. J.* **2021**, *410*, 128413. [CrossRef]
3. Clarke, B.O.; Anumol, T.; Barlaz, M.; Snyder, S.A. Investigating landfill leachate as a source of trace organic pollutants. *Chemosphere* **2015**, *127*, 269–275. [CrossRef] [PubMed]
4. Gao, J.; Oloibiri, V.; Chys, M.; Audenaert, W.T.M.; Decostere, B.; He, Y.; Van Langenhove, H.; Demeestere, K.; Van Hulle, S.W.H. The present status of landfill leachate treatment and its development trend from a technological point of view. *Rev. Environ. Sci. Biol. Technol.* **2015**, *14*, 93–122. [CrossRef]
5. Duyar, A.; Ciftcioglu, V.; Cirik, K.; Civelekoglu, G.; Uruş, S. Treatment of landfill leachate using single-stage anoxic moving bed biofilm reactor and aerobic membrane reactor. *Sci. Total. Environ.* **2021**, *776*, 145919. [CrossRef] [PubMed]
6. Luo, H.; Zeng, Y.; Cheng, Y.; He, D.; Pan, X. Recent advances in municipal landfill leachate: A review focusing on its characteristics, treatment, and toxicity assessment. *Sci. Total. Environ.* **2020**, *703*, 135468. [CrossRef] [PubMed]
7. Hassan, M.; Xie, B. Use of aged refuse-based bioreactor/biofilter for landfill leachate treatment. *Appl. Microbiol. Biotechnol.* **2014**, *98*, 6543–6553. [CrossRef]
8. Chys, M.; Oloibiri, V.A.; Audenaert, W.T.; Demeestere, K.; Van Hulle, S.W. Ozonation of biologically treated landfill leachate: Efficiency and insights in organic conversions. *Chem. Eng. J.* **2015**, *277*, 104–111. [CrossRef]
9. Oloibiri, V.; Chys, M.; De Wandel, S.; Demeestere, K.; Van Hulle, S.W. Removal of organic matter and ammonium from landfill leachate through different scenarios: Operational cost evaluation in a full-scale case study of a Flemish landfill. *J. Environ. Manag.* **2017**, *203*, 774–781. [CrossRef] [PubMed]
10. Chen, W.; Gu, Z.; Wen, P.; Li, Q. Degradation of refractory organic contaminants in membrane concentrates from landfill leachate by a combined coagulation-ozonation process. *Chemosphere* **2019**, *217*, 411–422. [CrossRef]
11. Chys, M.; Declerck, W.; Audenaert, W.T.; Van Hulle, S. UV/H_2O_2, O_3 and (photo-) Fenton as treatment prior to granular activated carbon filtration of biologically stabilized landfill leachate. *J. Chem. Technol. Biotechnol.* **2014**, *90*, 525–533. [CrossRef]
12. Xu, Q.; Tian, Y.; Wang, S.; Ko, J.H. A comparative study of leachate quality and biogas generation in simulated anaerobic and hybrid bioreactors. *Waste Manag.* **2015**, *41*, 94–100. [CrossRef]
13. Li, J.; Wu, B.; Li, Q.; Zou, Y.; Cheng, Z.; Sun, X.; Xi, B. Ex situ simultaneous nitrification-denitrification and in situ denitri-fication process for the treatment of landfill leachates. *Waste Manag.* **2019**, *88*, 301–308. [CrossRef] [PubMed]
14. Xu, Q.; Siracusa, G.; Di Gregorio, S.; Yuan, Q. COD removal from biologically stabilized landfill leachate using Advanced Oxidation Processes (AOPs). *Process Saf. Environ. Protect.* **2018**, *120*, 278–285. [CrossRef]
15. Fresno, F.; Portela, R.; Suárez, S.; Coronado, J.M. Photocatalytic materials: Recent achievements and near future trends. *J. Mater. Chem. A* **2014**, *2*, 2863–2884. [CrossRef]
16. Yan, M.; Wu, Y.; Liu, X. Photocatalytic nanocomposite membranes for high-efficiency degradation of tetracycline under visible light: An imitated core-shell Au-TiO_2-based design. *J. Alloys Compd.* **2021**, *855*, 157548. [CrossRef]

17. Chen, J.; Xiong, Y.; Duan, M.; Li, X.; Li, J.; Fang, S.W.; Qin, S.; Zhang, R. Insight into the Synergistic Effect of Adsorption-Photocatalysis for the Removal of Organic Dye Pollutants by Cr-Doped ZnO. *Langmuir* **2020**, *36*, 520–533. [CrossRef] [PubMed]
18. Mirzaei, M.; Dabi, B.; Dadvar, M.; Jafarikojour, M. Photocatalysis of Phenol Using a Spinning Disc Photoreactor Immobilized with TiO_2 Nanoparticles: Hydrodynamic Modeling and Reactor Optimization. *Ind. Eng. Chem. Res.* **2017**, *56*, 1739–1749. [CrossRef]
19. Yu, X.; Qiu, H.; Wang, B.; Meng, Q.; Zhao, K. A ternary photocatalyst of all-solid-state Z-scheme TiO_2–Au–BiOBr for efficiently degrading various dyes. *J. Alloys Compd.* **2020**, *839*, 155597. [CrossRef]
20. Altina, I.; Mab, X.; Boffab, V.; Bacaksızc, E.; Magnaccad, G. Hydrothermal preparation of B–TiO_2-graphene oxide ternary nanocomposite, characterization and photocatalytic degradation of bisphenol A under simulated solar irradiation. *Mater. Sci. Semicond. Process.* **2021**, *123*, 105591. [CrossRef]
21. Shao, S.; Yu, J.; Love, J.B.; Fan, X. An economic approach to produce iron doped TiO_2 nanorods from ilmenite for photocatalytic applications. *J. Alloys Compd.* **2021**, *858*, 158388. [CrossRef]
22. Priyanka, K.; Remya, N.; Behera, M. Comparison of titanium dioxide based catalysts preparation methods in the minerali-zation and nutrients removal from greywater by solar photocatalysis. *J. Clean Prod.* **2019**, *235*, 1–10. [CrossRef]
23. Seibert, D.; Borba, F.H.; Bueno, F.; Inticher, J.J.; Modenes, A.N.; Espinoza-Quinones, F.R.; Bergamasco, R. Two-stage integrated system photo-electro-Fenton and biological oxidation process assessment of sanitary landfill leachate treatment: An intermediate products study. *Chem. Eng. J.* **2019**, *372*, 471–482. [CrossRef]
24. Cai, F.F.; Yang, Z.H.; Huang, J.; Zeng, G.M.; Wang, L.K.; Yang, J. Application of cetyltrimethylammonium bromide bentonite–titanium dioxide photocatalysis technology for pretreatment of aging leachate. *J. Hazard. Mater.* **2014**, *275*, 63–71. [CrossRef] [PubMed]
25. Pellenz, L.; Borba, F.H.; Daroit, D.J.; Lassen, M.F.M.; Baroni, S.; Zorzo, C.F.; Guimarães, R.E.; Espinoza-Quiñones, F.R.; Seibert, D. Landfill leachate treatment by a boron-doped diamond-based photo-electro-Fenton system integrated with biological oxidation: A toxicity, genotoxicity and by products assessment. *J. Environ. Manag.* **2020**, *264*, 110473. [CrossRef]
26. Gu, Z.; Chen, W.; Wang, F.; Li, Q. A pilot-scale comparative study of bioreactor landfills for leachate decontamination and municipal solid waste stabilization. *Waste Manag.* **2020**, *103*, 113–121. [CrossRef]
27. Agdag, O.N.; Aktas, D.; Simsek, O. Effects of Different Seed Substances on Anaerobic Degradation of Municipal Solid Waste in Recirculated Bioreactor. *Waste Biomass-Valoriz.* **2021**, *12*, 383–392. [CrossRef]
28. Ministry of Environmental Protection of the People's Republic of China. *Standard for pollution control on the landfill site of the MSW*; Ministry of Environmental Protection of the People's Republic of China: Beijing, China, 2008.
29. Wu, C.; Shen, L.; Zhang, Y.-C.; Huang, Q. Solvothermal synthesis of Cr-doped ZnO nanowires with visible light-driven photocatalytic activity. *Mater. Lett.* **2011**, *65*, 1794–1796. [CrossRef]
30. Sun, X.; Zhang, H.; Cheng, Z. Use of bioreactor landfill for nitrogen removal to enhance methane production through ex situ simultaneous nitrification-denitrification and in situ denitrification. *Waste Manag.* **2017**, *66*, 97–102. [CrossRef]
31. He, J.; Yang, J.; Jiang, F.; Liu, P.; Zhu, M. Photo-assisted peroxymonosulfate activation via 2D/2D heterostructure of $Ti_3C_2/g\text{-}C_3N_4$ for degradation of diclofenac. *Chemosphere* **2020**, *258*, 127339. [CrossRef]
32. Elleuch, L.; Messaoud, M.; Djebali, K.; Attafi, M.; Cherni, Y.; Kasmi, M.; Elaoud, A.; Trabelsi, I.; Chatti, A. A new insight into highly contaminated landfill leachate treatment using Kefir grains pre-treatment combined with Ag-doped TiO_2 photocatalytic process. *J. Hazard. Mater.* **2020**, *382*, 121119. [CrossRef]
33. Ebrahimian, A. Silver Doped TiO_2 Nanoparticles: Preparation, Characterization and Efficient Degradation of 2,4-dichlorophenol Under Visible Light. *J. Water Environ. Nanotechnol.* **2016**, *1*, 135–144.
34. Xu, C.; Yang, F.; Deng, B.; Zhuang, Y.; Li, D.; Liu, B.; Yang, W.; Li, Y. Ti_3C_2/TiO_2 nanowires with excellent photocatalytic performance for selective oxidation of aromatic alcohols to aldehydes. *J. Catal.* **2020**, *383*, 1–12. [CrossRef]
35. Peng, C.; Wang, H.; Yu, H.; Peng, F. (111) TiO_2-x/Ti_3C_2: Synergy of active facets, interfacial charge transfer and Ti^{3+} doping for enhance photocatalytic activity. *Mater. Res. Bull.* **2017**, *89*, 16–25. [CrossRef]
36. Lv, K.; Zuo, H.; Sun, J.; Deng, K.; Liu, S.; Li, X.; Wang, D. (Bi, C and N) codoped TiO_2 nanoparticles. *J. Hazard. Mater.* **2009**, *161*, 396–401. [CrossRef]
37. Hassan, M.; Zhao, Y.; Xie, B. Employing TiO_2 photocatalysis to deal with landfill leachate: Current status and development. *Chem. Eng. J.* **2016**, *285*, 264–275. [CrossRef]
38. Azadi, S.; Karimi-Jashni, A.; Javadpour, S.; Amiri, H. Photocatalytic treatment of landfill leachate: A comparison between N-, P-, and N-P-type TiO_2 nanoparticles. *Environ. Technol. Innov.* **2020**, *19*, 100985. [CrossRef]
39. Qazaq, A.; Hudaya, T.; Lee, I.; Sulidis, A.; Adesina, A. Photoremediation of natural leachate from a municipal solid waste site in a pilot-scale bubble column reactor. *Catal. Commun.* **2007**, *8*, 1917–1922. [CrossRef]
40. Rojviroon, O.; Rojviroon, T.; Sirivithayapakorn, S. Removal of Color and Chemical Oxygen Demand from Landfill Leachate by Photocatalytic Process with AC/TiO_2. *Energy Procedia* **2015**, *79*, 536–541. [CrossRef]
41. Jyothi, K.; Yesodharan, S.; Yesodharan, E. Ultrasound (US), Ultraviolet light (UV) and combination (US + UV) assisted semiconductor catalysed degradation of organic pollutants in water: Oscillation in the concentration of hydrogen peroxide formed in situ. *Ultrason. Sonochem.* **2014**, *21*, 1787–1796. [CrossRef]
42. Youngman, F. Optimization of TiO_2 Photocatalyst in an Advanced Oxidation Process for the Treatment of Landfill Leachate. Ph.D. Thesis, Florida Atlantic University, Boca Raton, FL, USA, 2013.

43. Miao, Z.; Wang, G.; Li, L.; Wang, C.; Zhang, X. Fabrication of black TiO_2/TiO_2 homojunction for enhanced photocatalytic degradation. *J. Mater. Sci.* **2019**, *54*, 14320–14329. [CrossRef]
44. Jia, C.; Wang, Y.; Zhang, C.; Qin, Q. UV-TiO_2 Photocatalytic Degradation of Landfill Leachate. *Water Air Soil Pollut.* **2011**, *217*, 375–385. [CrossRef]
45. Wiszniowski, J.; Robert, D.; Surmacz-Górska, J.; Miksch, K.; Malato, S.; Weber, J.-V. Solar photocatalytic degradation of humic acids as a model of organic compounds of landfill leachate in pilot-plant experiments: Influence of inorganic salts. *Appl. Catal. B Environ.* **2004**, *53*, 127–137. [CrossRef]
46. He, X.; Mitrano, D.M.; Nowack, B.; Bahk, Y.K.; Figi, R.; Schreiner, C.; Bürki, M.; Wang, J. Agglomeration potential of TiO_2 in synthetic leachates made from the fly ash of different incinerated wastes. *Environ. Pollut.* **2017**, *223*, 616–623. [CrossRef] [PubMed]
47. Anpo, M.; Kamat, P.V. *Environmentally Benign Photocatalysts*; Springer: New York, NY, USA, 2010.
48. Chen, C.-Y. Photocatalytic Degradation of Azo Dye Reactive Orange 16 by TiO_2. *Water Air Soil Pollut.* **2009**, *202*, 335–342. [CrossRef]
49. Hassan, M.; Wang, X.; Wang, F.; Wu, D.; Hussain, A.; Xie, B. Coupling ARB-based biological and photochemical (UV/TiO_2 and UV/$S_2O_8^{2-}$) techniques to deal with sanitary landfill leachate. *Waste Manag.* **2017**, *63*, 292–298. [CrossRef]
50. Boonnorat, J.; Honda, R.; Panichnumsin, P.; Boonapatcharoen, N.; Yenjam, N.; Krasaesueb, C.; Wachirawat, M.; Seemuang-On, S.; Jutakanoke, R.; Teeka, J.; et al. Treatment efficiency and greenhouse gas emissions of non-floating and floating bed activated sludge system with acclimatized sludge treating landfill leachate. *Bioresour. Technol.* **2021**, *330*, 124952. [CrossRef]
51. Guo, X.; Li, B.; Zhao, R.; Zhang, J.; Lin, L.; Zhang, G.; Li, R.-H.; Liu, J.; Li, P.; Li, Y.; et al. Performance and bacterial community of moving bed biofilm reactors with various biocarriers treating primary wastewater effluent with a low organic strength and low C/N ratio. *Bioresour. Technol.* **2019**, *287*, 121424. [CrossRef] [PubMed]
52. Gomes, A.I.; Foco, M.L.; Vieira, E.; Cassidy, J.; Silva, T.F.; Fonseca, A.; Saraiva, I.; Boaventura, R.A.; Vilar, V.J. Multistage treatment technology for leachate from mature urban landfill: Full scale operation performance and challenges. *Chem. Eng. J.* **2019**, *376*, 120573. [CrossRef]
53. Gogina, E.; Gulshin, I. Simultaneous Nitrification and Denitrification with Low Dissolved Oxygen Level and C/N ratio. *Procedia Eng.* **2016**, *153*, 189–194. [CrossRef]
54. Liu, T.; He, X.; Jia, G.; Xu, J.; Quan, X.; You, S. Simultaneous nitrification and denitrification process using novel sur-face-modified suspended carriers for the treatment of real domestic wastewater. *Chemosphere* **2020**, *247*, 125831. [CrossRef] [PubMed]
55. Gomes, A.I.; Soares, T.F.; Silva, T.F.; Boaventura, R.A.; Vilar, V.J. Ozone-driven processes for mature urban landfill leachate treatment: Organic matter degradation, biodegradability enhancement and treatment costs for different reactors configuration. *Sci. Total. Environ.* **2020**, *724*, 138083. [CrossRef] [PubMed]
56. Abu Amr, S.S.; Aziz, H.A.; Adlan, M.N.; Aziz, S.Q. Effect of Ozone and Ozone/Fenton in the Advanced Oxidation Process on Biodegradable Characteristics of Semi-aerobic Stabilized Leachate. *Clean-Soil Air Water* **2012**, *41*, 148–152. [CrossRef]
57. Colombo, A.; Módenes, A.N.; Trigueros, D.E.G.; da Costa, S.I.G.; Borba, F.; Espinoza-Quiñones, F.R. Treatment of sanitary landfill leachate by the combination of photo-Fenton and biological processes. *J. Clean. Prod.* **2019**, *214*, 145–153. [CrossRef]

processes

MDPI

Article

Removal of Phenol from Aqueous Solution Using Internal Microelectrolysis with Fe-Cu: Optimization and Application on Real Coking Wastewater

Do Tra Huong [1,*], Nguyen Van Tu [2], Duong Thi Tu Anh [1], Nguyen Anh Tien [3], Tran Thi Kim Ngan [4,5] and Lam Van Tan [5,6,*]

1 Thai Nguyen University of Education, Thai Nguyen University, Thai Nguyen City 250000, Vietnam; duongtuanh@dhsptn.eu.vn
2 Institute of Chemistry and Materials, Ha Noi 100000, Vietnam; nguyenvantu882008@yahoo.com
3 Department of Chemistry, Ho Chi Minh City University of Education, Ho Chi Minh City 700000, Vietnam; tienna@hcmue.edu.vn
4 NTT Hi-Tech Institute, Nguyen Tat Thanh University, Ho Chi Minh City 700000, Vietnam; ttkngan@ntt.edu.vn
5 Institute of Environmental Sciences, Nguyen Tat Thanh University, Ho Chi Minh City 700000, Vietnam
6 Faculty of Environmental and Food Engineering, Nguyen Tat Thanh University, Ho Chi Minh City 700000, Vietnam
* Correspondence: huongdt.chem@tnue.edu.vn (D.T.H.); lvtan@ntt.edu.vn (L.V.T.)

Citation: Huong, D.T.; Van Tu, N.; Anh, D.T.T.; Tien, N.A.; Ngan, T.T.K.; Van Tan, L. Removal of Phenol from Aqueous Solution Using Internal Microelectrolysis with Fe-Cu: Optimization and Application on Real Coking Wastewater. *Processes* **2021**, *9*, 720. https://doi.org/10.3390/pr9040720

Academic Editors: Bipro R. Dhar and Andrea Petrella

Received: 6 March 2021
Accepted: 16 April 2021
Published: 19 April 2021

Abstract: Fe-Cu materials were synthesized using the chemical plating method from Fe powder and $CuSO_4$ 5% solution and then characterized for surface morphology, composition and structure by scanning electron microscopy (SEM), energy dispersive X-ray spectroscopy (EDS) and X-ray diffraction (XRD), respectively. The as-synthesized Fe-Cu material was used for removal of phenol from aqueous solution by internal microelectrolysis. The internal electrolysis-induced phenol decomposition was then studied with respect to various parameters such as pH, time, Fe-Cu material weight, phenol concentration and shaking speed. The optimal phenol decomposition (92.7%) was achieved under the conditions of (1) a pH value of phenol solution of 3, (2) 12 h of shaking at the speed of 200 rpm, (3) Fe-Cu material weight of 10 g/L, (4) initial phenol concentration of 100.98 mg/L and (5) at room temperature (25 ± 0.5 °C). The degradation of phenol using Fe-Cu materials obeyed the second-order apparent kinetics equation with a reaction rate constant of k of 0.009 $h^{-1}L\ mg^{-1}$. The optimal process was then tested against real coking wastewater samples, resulting in treated wastewater with favorable water indicators. Current findings justify the use of Fe-Cu materials in practical internal electrolysis processes.

Keywords: internal microelectrolysis; Fe-Cu material; phenol; wastewater treatment; coking wastewater

1. Introduction

In recent years, the process of industrialization and modernization in Vietnam has taken place rapidly, promoting socio-economic development of the country and accompanying problems of environmental pollution. Phenol is a hazardous pollutant and is listed as one of 129 pollutants that need to be pre-treated, according to US Environmental Protection guidelines. Phenol is often generated in the waste streams of industries such as petrochemicals, oil refining, plastics, steel, textile, paper and pulp, pesticides, pharmaceuticals, synthetic resins, and coking plants [1–3]. Phenol is a less biodegradable chemical that can cause cancer, gene mutation and teratogen. Phenol contaminates water sources, causing tremendous harm to humans and organisms; thus, eradication of phenol pollution in wastewater is being studied in many countries, including Vietnam. In order to treat phenols, physicochemical methods, such as adsorption, flocculation and sedimentation have been used as a traditional treatment. However, they are found to be not very effective.

Internal electrolysis (IE) has been proposed as a remedy for this issue. Usually, based on the electron material used, there are four types of IE, including Fe-Cu, Fe-C, Al-Cu and Al-C. Among them, the Fe-C and Fe-Cu systems are widely used in actual projects. The principle of IE involves two materials with different electrode potentials that generate microelectrode pairs when being in contact. For Fe-C, the Fe-Cu iron system acts as an anode and copper or carbon acts as a cathode, similar to the micro-battery pair in metal corrosion. With a micro-battery with a voltage of about 1.2 V, a small power circuit of μA appears, which acts as a redox agent in the decomposition reaction of adsorbed organic compounds on the electrode surface. Due to this principle, Fe-C and Fe-Cu electrolytic processes are also known as internal microelectrolysis. Therefore, it is possible to dissolve iron without the use of an external current by placing micro-battery pairs as composites of Fe-C and Fe-Cu, providing an advantage in internal microelectrolysis technology in wastewater pretreatment applications [1–26]. The reactions that occur during internal microelectrolysis are as follows [27].

The reaction at the anode (Fe):

$$Fe \rightarrow Fe^{2+} + 2e \; E^0(Fe^{2+}/Fe) = -0.44 \text{ V} \tag{1}$$

The reaction at the cathode (Cu):

$$2H^+ + 2e \rightarrow 2[H] = H_2 \; E^0(H^+/H_2) = 0.00 \text{ V} \tag{2}$$

$$\text{Fe and Cu: } H_2O \rightarrow HO^* + H^+ + e \tag{3}$$

Organic substances that exist in the solution, such as RX (organochlorine compound) and RNO_2 (aromatic ring nitro compound) then receive electrons from the anode surface (metal Fe) and are reduced by the chlorine and amine reaction. The resulting pollutants would become non-toxic or less toxic products, hence being more easily biodegradable.

The internal microelectrolysis process holds two main advantages. First, it could be applied to treat various types of industrial wastewater, including polyester-containing effluent [15], dyes [11,13,16], discharge from coal gasification [12], plant protection products [3,25], nitrate contamination [19], mixed industry (textile, dyeing, paper, plating, mechanic) [6–9], high organic matter [4,17,18], oil contamination [14], TNT and RDX contamination [10]. Second, internal microelectrolysis exhibits high treatment efficiency, fast response time and low operating costs. Fan and Ma [4] used a Fe-Cu electrode system to treat mixed industrial wastewater in Taopu, Shanghai at the capacity of 60,000 m^3/day, achieving a COD removal efficiency of 40%. Yin et al. [12] used this method to connect an external current to treat 4-chlorophenol, reaching a removal efficiency that was higher than 90% after 36 min. Yang et al. [5] also reported that internal electrolysis could be used to treat polyester wastewater, achieving the COD removal efficiency of 58%. The COD of the wastewater decreased from 3353.2 mg/L to 1391.6 mg/L and its BOD_5/COD ratio also increased from 0.27 to 0.42 after treatment. Zhu [17] combined internal microelectrolysis and a bio-membrane to treat mixed industry wastewater, reducing the COD from 150,000 mg/L to 500 mg/L.

Recent trends in enhancing the decomposition of organic pollutants have shifted to the use of bimetallic internal microelectrolytic materials prepared by the deposition of second transition metals on the iron surface. Previous studies showed that transition metals such as Ni, Cu and Co can enhance the catalytic activity of FeO [28]. Two types of catalytic mechanisms of bimetallic internal electrolytic materials have been proposed: (a) indirect reduction by atomic hydrogen ([H] abs) absorbed on the material's surface in bimetallic and transition metal additives form that facilitates the generation of surface-linked hydrogen atoms ([H] abs), and (b) direct reduction on the catalytic active site by receiving electrons during FeO oxidation and surface additives (i.e., transition of metals) to increase FeO oxidation through the formation of a multitude of micro-battery pairs [29,30].

In many studies, Fe-Cu and Fe-C materials are usually made from scrap, iron granules, copper and carbon powder with different sizes or prepared by second transition metal deposition on the surface of Fe, thus offering modest degradation efficiency against pollutants [5,28–31]. However, it has been suggested that the use of chemical plating of Cu on the Fe surface might result in materials with significantly improved degradation capacities for internal microelectrolysis [22]. For example, Xu et al. (2008) fabricated Fe-Cu material using the chemical plating method and used it in treatment of nitrobenzene (100 mg/L) in aqueous solution [31]. The obtained material showed a removal efficiency of approximately 95% at optimal conditions, suggesting a better reactivity of the chemically galvanized Fe-Cu for internal electrolysis. Bo et al. (2014) used a micro-sized Fe-Cu internal electrolyte material prepared by chemical plating to pretreat p-nitrophenol in aqueous solution [22]. The material was prepared with the content of Cu on the surface varying from 30% to 95%. Remarkably, the results pointed out that the Fe-Cu ratio played a key role in degradation of p-nitrophenol.

In this work, we continued this research pathway by fabricating Fe-Cu materials using the chemical plating method and investigated the effect of some factors such as pH, treatment time, mass of Fe-Cu system, shaking rate, and the concentration to efficiency ratio of phenol degradation of Fe-Cu materials in aqueous medium. In addition, the internal electrolysis reaction was applied to treat real coking wastewater from a coal factory in Vietnam.

2. Materials and Methods

2.1. Fabrication of Cu-Fe Material

Fe powder with sizes smaller than 50 µm and 99.9% purity (PA, China) was immersed in 30% NaOH solution for 10 min to remove grease and clean the surface. The surface was activated by treating in HCl 7.4% wt for 3 min. The diluted HCl solution was prepared with an HCl solution concentration of 37% wt. The material was then washed several times with water, followed by drying at 105 °C for 2 h, allowed to cool and stored in a sealed glass jar. Fe-Cu samples were made using the chemical plating method in 5% $CuSO_4$ solution (wt%). To be specific, a total of 100 g of Fe powder was added into 1 L of 5 wt% $CuSO_4$ solution for a period of 2 min. The mixture was then washed several times with water and dried at 105 °C for 3 h under N_2 gas. The material was then stored in a desiccator for further research.

2.2. Characterization of Structure, Composition, Physical Properties, Surface Characteristics of Fe-Cu Materials

The surface characteristics and components of the Fe-Cu material after fabrication were determined by scanning electron microscopy (SEM) and energy-dispersive X-ray spectroscopy (EDS) (on an SEM-EDS machine, JSM 6610 LA—JEOL, Tokyo, Japan), respectively. Measurements were made at the Institute of Materials Chemistry, Institute of Military Science and Technology, Vietnam. The structure of the material was determined by the method of X-ray diffraction (XRD) (on a Bruker D5000, Siemens, München, Germany). The measurement was conducted at the department of Chemistry—Hanoi University of Natural Sciences.

2.3. Study on Decomposition of Phenol

Factors affecting degradation of phenol were investigated, including pH, time, dosage of Fe-Cu materials, shaking rate and initial concentration of phenol. The experiments were carried out at room temperatures (25 ± 0.5 °C). Parameters for the experiments are shown in Table 1.

Table 1. Parameters for single factor investigations.

Investigation	pH	Time (h)	Fe-Cu Dosage (g)	Shaking Speed (rpm)	Initial Phenol Concentration (mg/L)
Effect of pH	2, 3, 4, 5, 6, 7 and 8	12	1	200	100
Effect of time	3	2, 4, 6, 8, 12, 20 and 24	1	200	100
Effect of material dosage	3	12	0.25, 0.5, 0.75, 1.0, 1.25 and 1.5	200	100
Effect of shaking speed	3	12	1	100, 120, 150, 180 and 200	100
Effect of initial phenol concentration	3	12	1	200	50, 100, 150, 200, 250 and 300

All experiments were conducted three times to check the repeatability. The result of each experiment is the average result of the three times, with P values less than 0.05 indicating a significant difference between the means.

The phenol degradation efficiency was calculated by the formula:

$$H\% = \frac{(C_0 - C_{cb})}{C_0} \times 100\% \quad (4)$$

In which: C_0 is the concentration of the phenol solution before decomposition (mg/L), C_{cb} is the concentration of the phenol solution after decomposition (mg/L) and H is the degradation efficiency (%).

The initial and post-treatment phenolic concentrations were determined on the HPLC Waters Acquity Arc instrument at the University of Education, Thai Nguyen University, Thai Nguyen Province, Vietnam. The instrument was equipped with chromatographic column C18 Inertsil ODS (5 μm, 250 × 3 mm, GL Sciences Inc., Tokyo, Japan). The optimal conditions for the determination of phenol content are as follows: wavelength of 272 nm, ratio of phosphate buffer solution mixture (pH = 4) to acetonitrile solution (pH = 3) of 30:70 (v/v), flow rate of 1.0 mL/ min, column temperature of 30 °C. TSS, BOD_5, COD, total N, total P and NH_4^+-N indicators were determined at the Thai Nguyen Center for Natural Resources and Environment Monitoring.

3. Results and Discussion

3.1. Survey Results on Surface Characteristics and Physical Properties of Fe-Cu Materials

From the analysis results of the SEM-EDS images of the materials and synthetic materials shown in Figures 1–4, it was found that the surface composition of the synthetic materials was different from the original materials. The Fe powder particles were arranged overlapping each other in blocks, whereas the Fe and Cu powder particles in Fe-Cu material were distributed relatively evenly on the surface with sizes of less than 50 μm. This shows that there was an even distribution of Cu plating alternating Fe powder particles to form Fe-Cu micro-cell pairs. The results of analyzing the EDS spectrum of Fe and Fe-Cu materials shown in Figures 3 and 4 and Tables 2 and 3 show specifically the presence of the elements and their content in each sample.

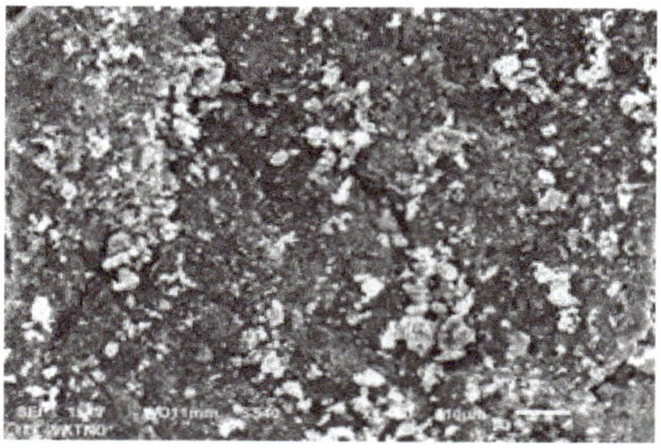

Figure 1. SEM image of Fe material.

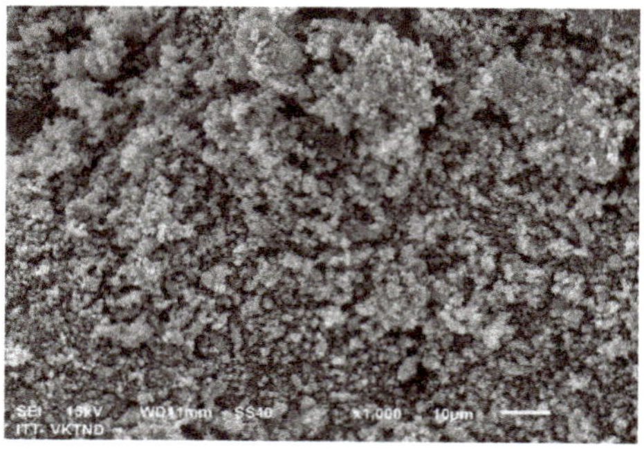

Figure 2. SEM image of Fe-Cu material.

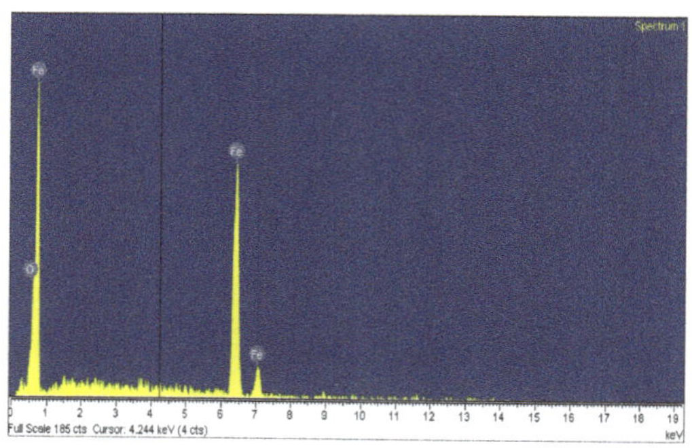

Figure 3. The EDS spectrum of material Fe.

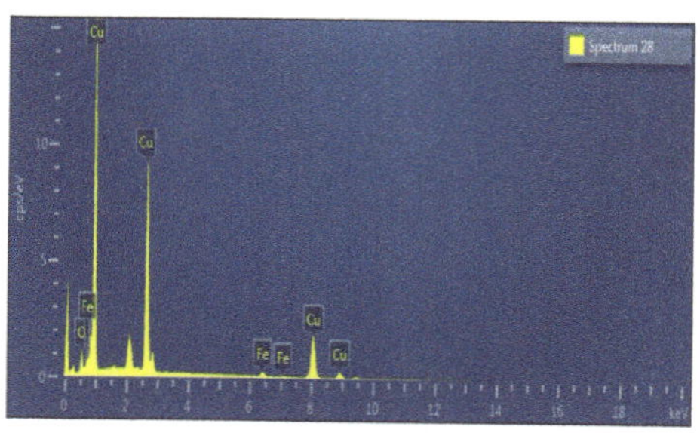

Figure 4. EDS spectrum of Fe-Cu material.

Table 2. Results of analyzing Fe sample elements.

Elements	% Mass	% Atom
O	8.95	25.55
Fe	91.05	74.45
Total	100.00	100.00

Table 3. Results of analyzing Fe-Cu sample elements.

Elements	% Mass	% Atom
O	12.11	24.97
Fe	18.59	21.83
Cu	69.30	53.20
Total	100.00	100.00

Analysis of the EDS of the Fe-Cu samples has shown Cu appearance, which proves successful copper plating. The reaction in the process of dissolving and chemically plating copper follows the processes below:

$$Fe \rightleftharpoons Fe^{2+} + 2e$$

$$Fe^{2+} \rightleftharpoons Fe^{3+} + e \text{ (There may be a part of } Fe^{2+} \text{ oxidized to } Fe^{3+}\text{)}$$

$$Cu^{2+} + 2e \rightleftharpoons Cu$$

On the other hand, the analysis of the structure of the synthesized Fe and Fe-Cu materials shown in Figure 5 shows that the components of the two spectra were distinctly different. The Fe accounted for 18.59% of the mass according to EDX. On the XRD pattern, the peaks assigned to Fe were quite weak, which suggests that Fe may be converted into other compounds such as Fe•Cu_2O. However, this is not always the case because Fe might be present in the Fe-Cu materials in amorphous forms, which does not show up in XRD results. Therefore, further studies are needed to confirm the exact material composition.

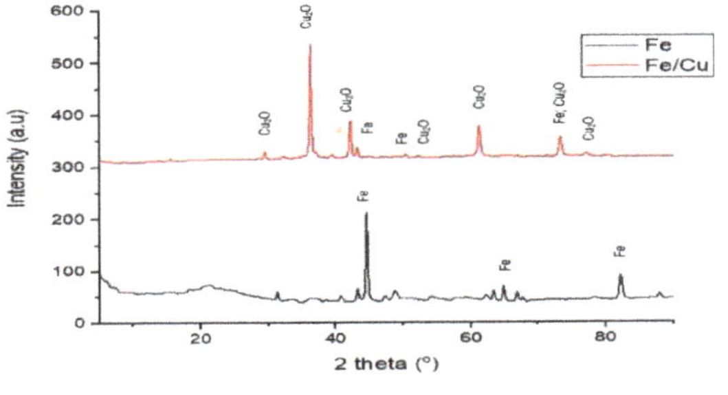

Figure 5. XRD spectrums of Fe and Fe-Cu materials.

This proves that the Fe-Cu bimetal material has been successfully fabricated and that Cu has coated the Fe surface to form Fe-Cu micro-battery pairs.

3.2. Decomposition of Phenol

3.2.1. Effect of pH

According to Equations (1) and (2), the pH value has a great influence on the reaction rate and the redox ability to create [H]. When the pH is lower, the amount of H^+ provided for the reaction becomes adequate or excessive, therefore accelerating the internal electrolysis or the corrosion of the electrode system. A lower initial pH value is associated with a higher concentration of [H]. Furthermore, in the presence of O_2, the cathode reduction of the internal electrolytic reaction can also occur in the following reaction:

$$O_2 + 4H^+ + 4e \rightarrow 2O^* + 4[H] \rightarrow 2H_2O;\ E^0\ (O_2/H_2O) = 1.23\ V \quad (5)$$

Thus, more H^+ would produce more [H] and O*, enhancing the ability to redox and reduce phenol and leading to a better phenol treatment efficiency. The initial pH value also affects the rate of corrosion reactions of Fe/Cu materials to form Fe^{2+}, Fe^{3+}, Fe $(OH)_2$ and $Fe(OH)_3$. In a more acidic environment than Fe^{2+}, Fe^{3+} is easy to form, yet it is difficult to precipitate $Fe(OH)_2$ and $Fe(OH)_3$. Conversely, when the pH is high and in the presence of dissolved oxygen, $Fe(OH)_2$ and $Fe(OH)_3$ concentration could be increased gradually in response to reaction time. Iron hydroxides are also factors that indirectly remove the phenol part as well as the intermediate compounds of the treatment by adsorption, flocculation and precipitation.

The results shown in Figure 6 show that when the pH value increases from 4 to 9, the phenol decomposition efficiency decreases. This can be explained by three main phenol decomposition processes, including decomposition due to the impact of internal electrolyte materials, adsorption and coagulation with iron hydroxide. As the concentration of Fe (II) and Fe (III) ions exceeds 10^{-5} mol/L in the material, precipitates of $Fe(OH)_2$ and $Fe(OH)_3$ will appear at pH values higher than 3, which is favorable for the flocculation of Fe (II) and Fe (III). At high pH (pH > 3), the phenol decomposition process was hindered, and coagulation was accelerated. As a result, phenol decomposition efficiency was decreased. Therefore, a pH value of 3 gave the maximum phenol decomposition efficiency and was selected for further studies.

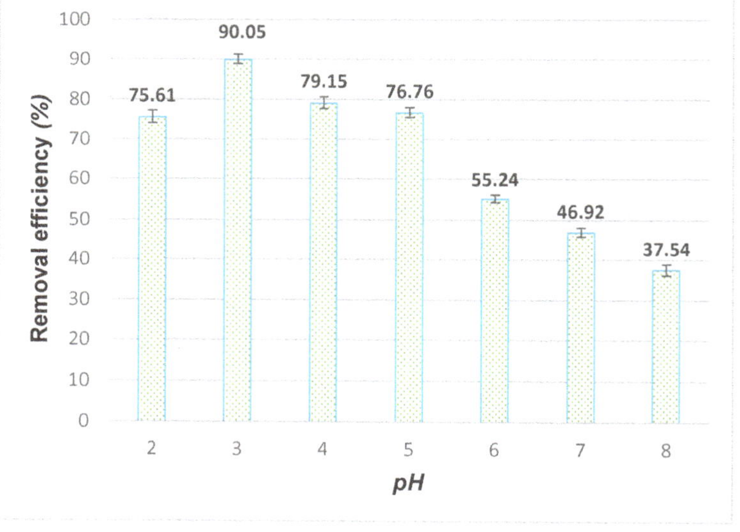

Figure 6. Effect of pH on the phenol removal performance of Fe-Cu material.

3.2.2. Effect of Time

Variations of phenol removal with respect to reaction time are illustrated in Figure 7. The results show that when increasing the time from the 2 to 12 h, the phenol decomposition efficiency increased rapidly to a maximum value of 92.39%. Thereafter, over a period from 12 to 24 h, decomposition efficiency decreased slowly and then became almost stable.

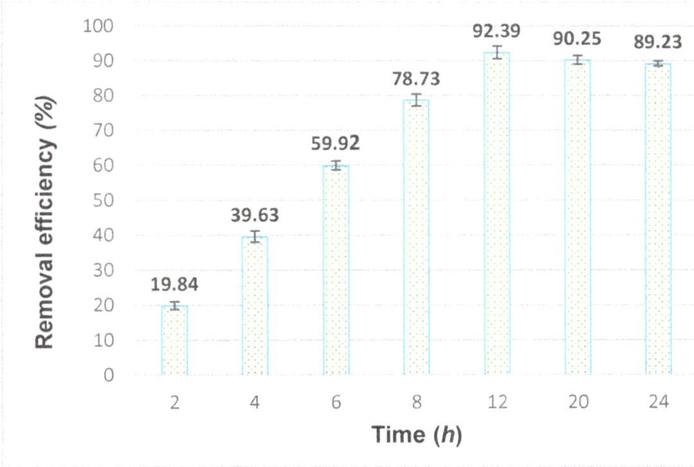

Figure 7. The effect of time on the phenol removal performance of Fe-Cu material.

The current result is in line with that of a previous study [19], which used Fe-Cu to treat polyester wastewater. This can be explained as follows: when the time increases from 2 to 12 h, the ongoing occurrence of the interior microelectrolysis reactions accumulates ferrous and ferric hydroxides, thus promoting phenol degradation efficiency. However, increasing the time from 12 to 24 h also leads to increased precipitation of hydroxides on the Fe-Cu surface and impedes the electron transmission between Fe-Cu and wastewater, thereby neutralizing the Fe-Cu surface and terminating the internal electrolysis, so the phenol degradation efficiency is reduced [20]. Therefore, we chose 12 h as the optimal time for phenol decomposition of Fe-Cu materials.

3.2.3. Effect of Dosage of Material

Removal efficiencies achieved at different dosages are shown in Figure 8. As the dosage of Fe-Cu material increased from 0.25 to 1.0 g, the phenol decomposition efficiency increased gradually. Increasing the dosage from 1.0 to 6.0 g seemed to impair the phenol degradation efficiency. In general, increasing the dosage of Fe-Cu resulted in a higher generation of microscopic galvanic cells, possibly leading to improved phenol removal efficiency. However, excessive amount of Fe-Cu in the solution might cause particle agglomeration, thus reducing the contacting area among Fe-Cu and wastewater. Moreover, the excess iron would react with H^+ present in the solution, leading to weakened reaction from Fe-Cu [21]. The Fe-Cu utilization efficiency would decrease remarkably if its dosage were too high. Therefore, the material weight of 1.0 g or 10 g/L was selected as the optimal material weight to decompose the phenol of Fe-Cu material.

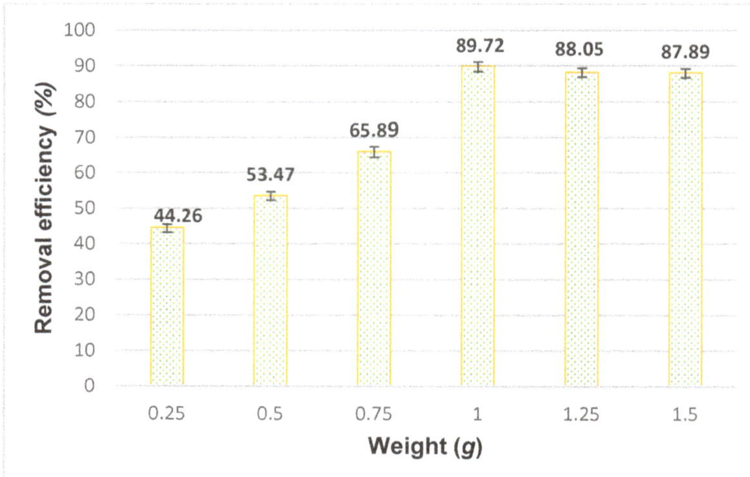

Figure 8. The effect of material dosage on phenol removal performance of Fe-Cu material.

3.2.4. Effect of Shaking Speed

The effect of shaking speed on phenol decomposition efficiency is shown as in Figure 9. As the speed was accelerated, the phenol decomposition efficiency also improved. This can be explained as follows. The shaking speed increases the dissolved oxygen content into the solution and enhances the ability to diffuse pollutants to the surface in contact with the Fe-Cu electrode, as well as the rapid dispersion of the products treated at the electrode in the solution. However, in acidic environments with low pH, the dissolved oxygen content is lower than in alkaline media. The effect of dissolved oxygen content on phenol degradation efficiency can be explained by the following reasons:

(1) When the shaking rate increases, the dissolved oxygen concentration in the electrolyte solution will also increase, in turn accelerating the subsequent cathode process when the pH changes to a neutral medium. This contributes to the corrosion rate as well as the rate of reaction with electrolytic internal materials [22].

(2) Oxygen could combine with H^+ to forms H_2O_2 hyperoxides, which then react with newly generated Fe^{2+} ions to form $Fe(OH)_2$ and $Fe(OH)_3$ ions. These are good phenol flocculation agents and intermediate products of phenol degradation.

(3) One previous study [22] suggested that the increased shaking speed caused the decomposition of substance molecules and dispersion of the intermediate decomposition products in the solution. At that time, the possibility of contact between the decomposed substances and the intermediate products with the surface of the Fe-Cu electrode system are increased, causing oxidation in the solution, improving electrochemical reduction on the cathode surface, and improving processing speed and efficiency.

When shaking speed increased from 100 to 150 rpm, the phenol decomposition speed increased rapidly from 180 to 200 rpm. This could be the reason why at this time the dissolved oxygen concentration in the solution was almost saturated. Therefore, we chose the shaking speed of 200 rpm to proceed to subsequent experiments.

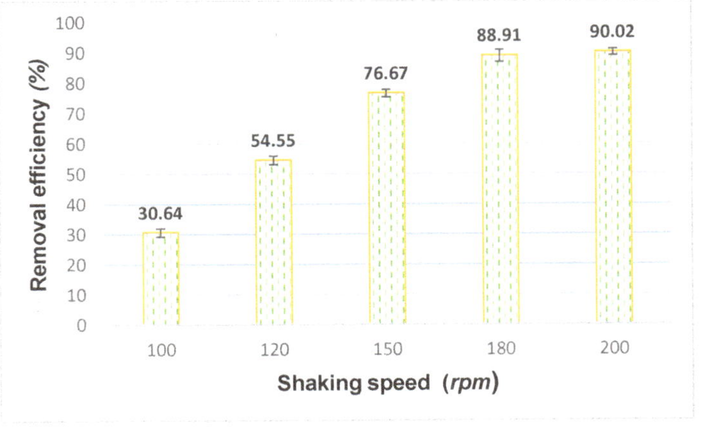

Figure 9. The effect of shaking speed on the phenol removal efficiency of Fe-Cu material.

3.2.5. Effect of Initial Phenol Concentration

Phenol removal efficiencies at different initial phenol concentrations are shown as in Figure 10. The results from Figure 10 show that the phenol decomposition efficiency increased proportionally when the concentration increased from 53.38 to 100.98 mg/L. Afterwards, in the concentration range from 146.69 to 250.76 mg/L, the phenol decomposition performance decreased. At a phenol concentration value of 100.98 mg/L, the degradation efficiency reached the maximum value of 92.7%, which indicates almost complete phenol decomposition. There was a sharp decrease in performance at higher concentrations of phenols (higher than 100.98 mg/L), possibly due to higher required amount of internal electrolytic material. At low phenol concentrations, a low concentration gradient would obstruct the mass transportation. Simultaneously, the short lifetime of HO* is also a contributing factor to reduce the number of reactions with phenol. At higher phenol concentrations, it is more likely for phenol and HO* to mutually react, which results in improved phenol removal efficiency. However, at very high phenol concentrations (146.69 mg/L), the phenol removal efficiency decreased to 72.71% due to limited formation of HO* in the interior micro-electrolysis system Therefore, there should be further studies and surveys to handle phenol at high concentrations.

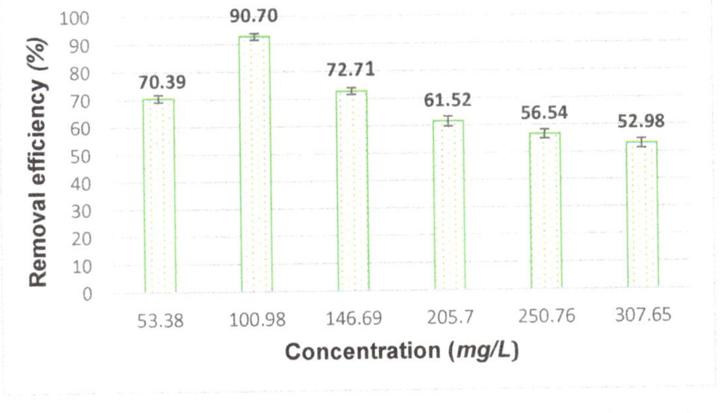

Figure 10. The effect of initial phenol concentration on the phenol removal capability of Fe-Cu material.

3.2.6. Decomposition Analysis Phenol Concentrations by HPLC

Figure 11 illustrates HPLC results of different phenol solutions (initial concentration of 100.98 mg/L) treated with different masses of Fe-Cu internal electrolysis material. It was indicated that phenol was completely decomposed when being treated with Fe-Cu material with the weight of 10 g/L, under 12 h of shaking at 200 rpm and at a pH value of 3.

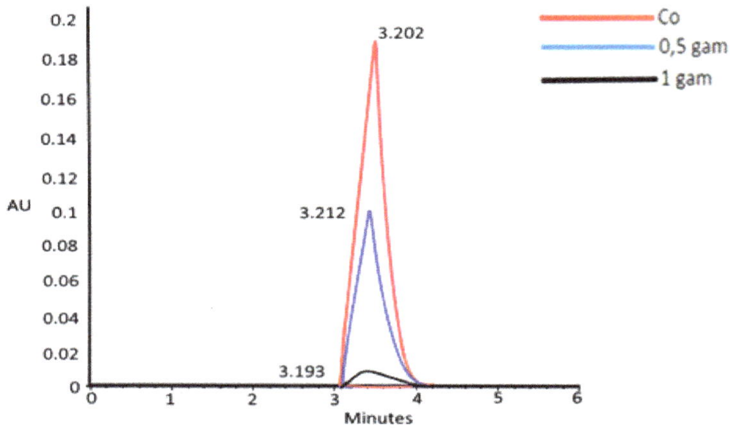

Figure 11. Chromatographic lines of a phenol solution sample depend on the amount of Fe-Cu material. Red curve: chromatogram of an untreated phenol sample (100.98 mg/L). Blue curve: chromatogram of a phenol sample treated with the following conditions: initial phenol concentration of 100.98 mg/L, Fe-Cu material weight of 5 g/L, shaking time of 12 h, shaking speed of 200 rpm, at pH = 3. Black curve: chromatogram of a phenol sample treated with the following conditions: initial phenol concentration of 100.98 mg/L, Fe-Cu material weight of 10 g/L, shaking time of 12 h, shaking speed of 200 rpm, at pH = 3.

3.2.7. Degradation Kinetics of Phenol Using Fe-Cu Material

The classical kinetics is that of the first-order and second-order chemical reaction kinetics. The equations are shown as follows:

$$\text{First-order kinetic model: } \ln C_t = -k_1 \cdot t + A_1 \quad (6)$$

$$\text{Second-order kinetic model: } 1/C_t = k_2 \cdot t + A_2 \quad (7)$$

$$\text{Third-order kinetic model: } 1/C_t^2 = 2k_3 \cdot t + A_3 \quad (8)$$

where: k_1 and k_2 are the first-order and second-order reaction rate constants, respectively; A_1, A_2 and A_3 are constants. C_0 is the initial concentration of the phenol solution before decomposition (mg/L), which is 100 mg/L.

Based on the investigations on the efficiency of phenol degradation over time, we surveyed the kinetics of phenol degradation according to the first-, second-, and third-order kinetic equations, as shown in Figures 12–14.

The results show that the degradation of phenol by internal microelectrolysis material of Fe-Cu seemed to follow the second-order apparent kinetics due to a higher linear regression coefficient ($R^2 = 0.9507$) than those of other kinetics. The calculated reaction rate constant of k of the second-order model was 0.009 $h^{-1}Lmg^{-1}$.

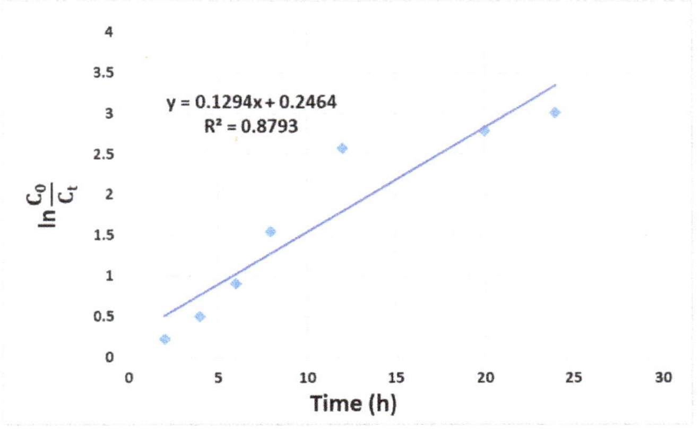

Figure 12. Model of the first-order apparent kinetics.

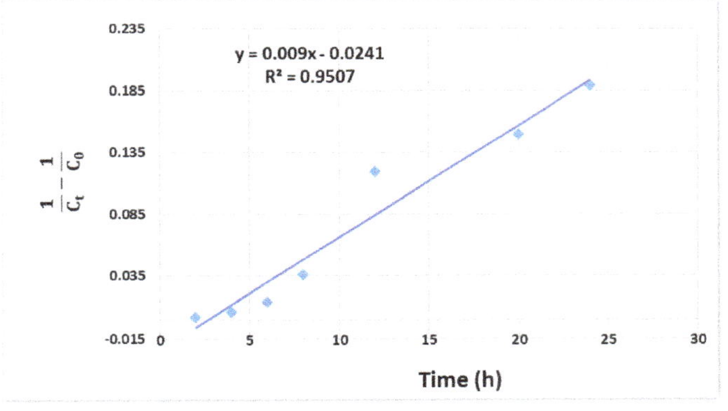

Figure 13. Model of the second-order apparent kinetics.

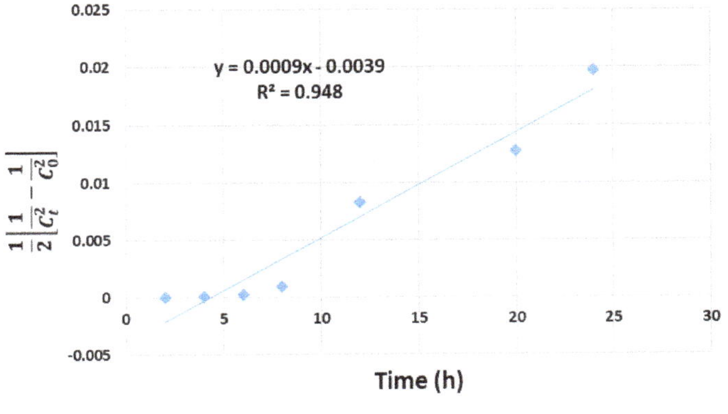

Figure 14. Model of the third-order apparent kinetics.

3.2.8. Real Sample Analysis

In this study, the optimal experimental parameters, including a Fe-Cu mass of 10 g/L, a shaking time of 12 h, a pH of and a shaking rate of 200 rpm, were adopted for pretreating real wastewater samples collected from a coal factory (Thai Nguyen Iron and Steel Joint Stock Company, Thai Nguyen Province, Vietnam). The results are shown in Figures 15 and 16 and Table 4.

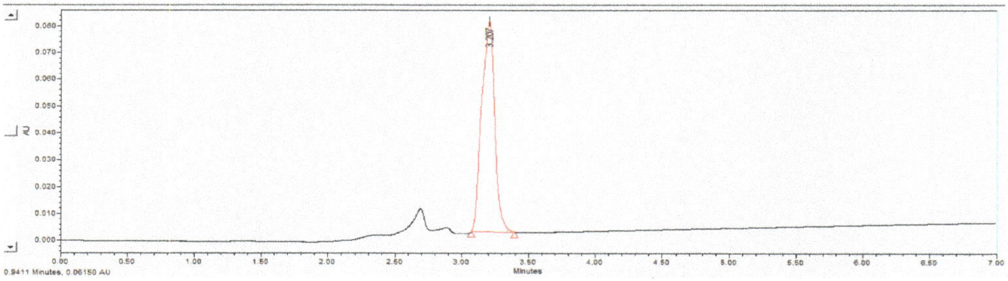

Figure 15. Chromatograms of wastewater samples containing phenols before treatment using Fe-Cu internal microelectrolysis material.

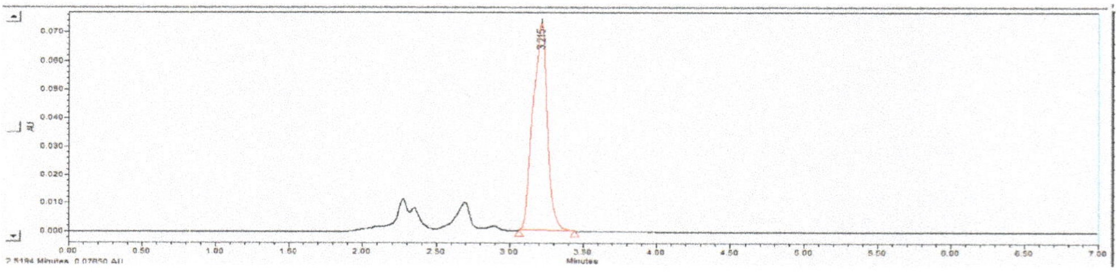

Figure 16. Chromatograms of wastewater samples containing phenols after treatment using Fe-Cu internal microelectrolysis material.

Table 4. Parameters of coking wastewater before and after treatment using Fe-Cu materials.

Parameters	Unit	Method of Analysis	Result (mg/L)		Efficiency H (%)
			Before	After	
DO	mg/L	TCVN 7325:2004	0.6	2.8	-
TSS	mg/L	SMEWW 2540 D:2012	124	63.4	48.87
BOD_5(20 °C)	mg/L	TCVN 6001-1:2008	1215	540.6	55.50
COD	mgO_2/L	SMEWW 5220C:2012	2379	1189	50.02
Phenol	mg/L	TCVN 6216:1996	173.70	50.86	70.07
CN^-	mg/L	SMEWW4500 CN^-B:2012	0.05	<0.01	-
Total N	mg/L	TCVN 6638:2000	876	644	26.48
NH_4^+-N	mg/L	TCVN 6179-1:1996	473	165.2	65.07
Total P	mg/L	TCVN 6202:2008	15.6	9.3	40.38

The mechanism of phenol degradation in the interior micro-electrolysis has been presented previously [25], and it involved the conversion of decomposed phenol into a less toxic intermediate compound. To be specific, during the micro-electrolysis process, radicals and oxidants are produced and oxidize organic compounds. This causes the destruction of structures of benzene ring and chemical bonds on its side chain, transforming toxic compounds into biodegradable intermediates [20]. Simultaneously, microelectron currents in the galvanic cell reaction also cause electron transfer, which promotes the growth and biodegradation capacity of microorganisms and stimulates active metabolic enzymes [32]. Furthermore, the justification of interior microelectrolysis technology for biological treatment is corroborated by its high degradation efficiency of refractory compounds and the ability to improve wastewater biodegradability [33].

The results in Table 3 show that the highest treatment efficiency was observed in the phenol parameter (70.07%), followed by the NH_4^+-N, BOD_5 and COD. The remaining parameters exhibited lower removal efficiencies (<50%). Thus, further biological treatment methods are recommended in order to achieve steel industry wastewater standards [34].

4. Conclusions

A sample of Fe-Cu material for internal electrolysis was synthesized from Fe powder material by using the chemical plating method. The Cu content at the material surface reached 69.30% (by weight). The surface, structure and composition of the as-synthesized materials were characterized by scanning electron microscopy (SEM), energy-dispersive X-ray spectroscopy (EDS) and X-ray diffraction diagram (XRD), respectively.

The internal electrolysis-induced phenol decomposition was then studied with respect to various parameters, including pH, time, Fe-Cu material weight, phenol concentration and shaking speed. The optimal phenol decomposition (92.7%) was achieved at the pH value of 3, the shaking time of 12 h, the shaking speed of 200 rpm, the weight of Fe-Cu material of 10 g/L, the initial phenol concentration of 100.98 mg/L and at room temperature (25 ± 0.5 °C). The degradation of phenol using Fe-Cu materials obeyed the second-order apparent kinetics equation with a reaction rate constant of k of 0.009 $L \times mg^{-1}h^{-1}$. Further evaluation using real coking wastewater resulted in treated effluents with favorable water indicators, suggesting the suitability of Fe-Cu materials in practical processes to treat coking wastewater before biological treatment. Further studies should contemplate the evaluation of material stability through cyclic reactions.

Author Contributions: Investigation, D.T.H., N.V.T., D.T.T.A., N.A.T. and T.T.K.N.; supervision, L.V.T.; writing—original draft, D.T.H.; writing—review and editing, L.V.T. All authors have read and agreed to the published version of the manuscript.

Funding: This work was funded by the Ministry of Education and Training of Vietnam, under the project B2019-TNA-10.

Institutional Review Board Statement: Not applicable.

Informed Consent Statement: Not applicable.

Data Availability Statement: The data presented in this study are available on request from the corresponding author.

Acknowledgments: This work was completed with financial support from the Ministry of Education and Training of Vietnam, under the project B2019-TNA-10.

Conflicts of Interest: The authors declare no conflict of interest.

References

1. Ma, L.; Zhang, W. Enhanced Biological Treatment of Industrial Wastewater with Bimetallic Zero-Valent Iron. *Environ. Sci. Technol.* **2008**, *42*, 5384–5389. [CrossRef]
2. El-Gohary, F.A.; Kamel, G. Characterization and biological treatment of pre-treated landfill leachate. *Ecol. Eng.* **2016**, *94*, 268–274. [CrossRef]
3. Visvanathan, C.; Abeynayaka, A. Developments and future potentials of anaerobic membrane bioreactors (AnMBRs). *Membr. Water Treat.* **2012**, *3*, 1–23. [CrossRef]
4. Fan, J.-H.; Ma, L.-M. The pretreatment by the Fe–Cu process for enhancing biological degradability of the mixed wastewater. *J. Hazard. Mater.* **2009**, *164*, 1392–1397. [CrossRef]
5. Yang, X.; Xue, Y.; Wang, W. Mechanism, kinetics and application studies on enhanced activated sludge by interior microelectrolysis. *Bioresour. Technol.* **2009**, *100*, 649–653. [CrossRef]
6. Jin, Y.-Z.; Zhang, Y.-F.; Li, W. Micro-electrolysis technology for industrial wastewater treatment. *J. Environ. Sci.* **2003**, *15*, 334–338.

7. Fan, L.; Ni, J.; Wu, Y.; Zhang, Y. Treatment of bromoamine acid wastewater using combined process of micro-electrolysis and biological aerobic filter. *J. Hazard. Mater.* **2009**, *162*, 1204–1210. [CrossRef]
8. Zhu, Y.; Fang, Z.; Xia, Z. Study on the reaction materials for micro-electrolysis treatment of wastewater. *Membr. Sci. Technol. Lanzhou* **2001**, *21*, 56–60.
9. Pan, L.; Wu, J.; Wang, J. Treatment of high mass concentration coking wastewater using enhancement catalytic iron carbon internal-electrolysis. *J. Jiangsu Univ. Nat. Sci. Ed.* **2010**, *31*, 348–352.
10. Oh, S.-Y.; Chiu, P.C.; Kim, B.J.; Cha, D.K. Enhancing Fenton oxidation of TNT and RDX through pretreatment with zero-valent iron. *Water Res.* **2003**, *37*, 4275–4283. [CrossRef]
11. Huang, L.; Sun, G.; Yang, T.; Zhang, B.; He, Y.; Wang, X. A preliminary study of anaerobic treatment coupled with micro-electrolysis for anthraquinone dye wastewater. *Desalination* **2013**, *309*, 91–96. [CrossRef]
12. Yin, X.; Bian, W.; Shi, J. 4-chlorophenol degradation by pulsed high voltage discharge coupling internal electrolysis. *J. Hazard. Mater.* **2009**, *166*, 1474–1479. [CrossRef]
13. Cui, D.; Guo, Y.-Q.; Lee, H.-S.; Wu, W.-M.; Liang, B.; Wang, A.-J.; Cheng, H.-Y. Enhanced decolorization of azo dye in a small pilot-scale anaerobic baffled reactor coupled with biocatalyzed electrolysis system (ABR–BES): A design suitable for scaling-up. *Bioresour. Technol.* **2014**, *163*, 254–261. [CrossRef] [PubMed]
14. Li, G.; Guo, S.; Li, F. Treatment of oilfield produced water by anaerobic process coupled with micro-electrolysis. *J. Environ. Sci.* **2010**, *22*, 1875–1882. [CrossRef]
15. Chen, R.; Chai, L.; Wang, Y.; Liu, H.; Shu, Y.; Zhao, J. Degradation of organic wastewater containing Cu–EDTA by Fe–C micro-electrolysis. *Trans. Nonferrous Met. Soc. China* **2012**, *22*, 983–990. [CrossRef]
16. Qin, L.; Zhang, G.; Meng, Q.; Xu, L.; Lv, B. Enhanced MBR by internal micro-electrolysis for degradation of anthraquinone dye wastewater. *Chem. Eng. J.* **2012**, *210*, 575–584. [CrossRef]
17. Zhu, Q.; Guo, S.; Guo, C.; Dai, D.; Jiao, X.; Ma, T.; Chen, J. Stability of Fe–C micro-electrolysis and biological process in treating ultra-high concentration organic wastewater. *Chem. Eng. J.* **2014**, *255*, 535–540. [CrossRef]
18. Guan, X.; Xu, X.; Lu, M.; Li, H. Pretreatment of Oil Shale Retort Wastewater by Acidification and Ferric-Carbon Micro-Electrolysis. *Energy Procedia* **2012**, *17*, 1655–1661. [CrossRef]
19. Yang, X. Interior microelectrolysis oxidation of polyester wastewater and its treatment technology. *J. Hazard. Mater.* **2009**, *169*, 480–485. [CrossRef] [PubMed]
20. Cheng, H.; Xu, W.; Liu, J.; Wang, H.; He, Y.; Chen, G. Pretreatment of wastewater from triazine manufacturing by coagulation, electrolysis, and internal microelectrolysis. *J. Hazard. Mater.* **2007**, *146*, 385–392. [CrossRef]
21. Ruan, X.C.; Liu, M.Y.; Zeng, Q.F.; Ding, Y.H. Degradation and decolorization of reactive red X-3B aqueous solution by ozone integrated with internal micro-electrolysis. *Sep. Purif. Technol.* **2010**, *74*, 195–201. [CrossRef]
22. Bo, L.; Zhang, Y.; Chen, Z.Y.; Yang, P.; Zhou, Y.X.; Wang, J.L. Removal of p-nitrophenol (PNP) in aqueous solution by the micron-scale iron-copper (Fe/Cu) bimetallic particles. *Appl. Catal. B Environ.* **2014**, *144*, 816–830.
23. Gang, Q.; Dan, G. Pretreatment of petroleum refinery wastewater by microwaveenhanced Fe^0/GAC micro-electrolysis. *Desalination Water Treat.* **2014**, *52*, 2512–2518. [CrossRef]
24. Xiaoying, Z.; Mengqi, J.; Xiang, Z.; Wei, C.; Dan, L.; Yuan, Z.; Xiaoyao, S. Enhanced removal mechanism of iron carbon micro-electrolysis constructed wetland on C, N, and P in salty permitted effluent of wastewater treatment plant. *Sci. Total Environ.* **2019**, *1*, 21–30.
25. Weiwei, M.; Yuxing, H.; Chunyan, X.; Hongjun, H.; Wencheng, M.; Hao, Z.; Kun, L.; Dexin, W. Enhanced degradation of phenolic compounds in coal gasification wastewater by a novel integration of micro-electrolysis with biological reactor (MEBR) under the micro-oxygen condition. *Bioresour. Technol.* **2018**, *251*, 303–310. [CrossRef]
26. Peng, L.; Zhipeng, L.; Xuegang, W.; Yadan, G.; Lizhang, W. Enhanced decolorization of methyl orange in aqueous solution using ironecarbon micro-electrolysis activation of sodium persulfate. *Chemosphere* **2017**, *180*, 100–107.
27. Tianguo, L.; Zhengyang, D.; Ronggao, Q.; Xiaojun, X.; Bo, L.; Yue, L.; Ming, J.; Fangdong, Z.; Yongmei, H. Enhanced characteristics and mechanism of Cu(II) removal from aqueous solutions in electrocatalytic internal micro-electrolysis fluidized-bed. *Chemosphere* **2020**, 126225. [CrossRef]
28. Cwiertny, D.M.; Bransfield, S.J.; Livi, K.J.T.; Fairbrother, D.H.; Roberts, A.L. Exploring the Influence of Granular Iron Additives on 1,1,1- Trichloroethane Reduction. *Environ. Sci. Technol.* **2006**, *40*, 6837–6843. [CrossRef]
29. Xu, W.Y.; Gao, T.Y. Dechlorination of carbon tetrachloride by the catalyzed Fe-Cu process. *J. Environ. Sci.* **2007**, *19*, 792–799. [CrossRef]
30. Ma, L.M.; Ding, Z.G.; Gao, T.Y.; Zhou, R.F.; Xu, W.Y.; Liu, J. Discoloration of methylene blue and wastewater from a plant by a Fe/Cu bimetallic system. *Chemosphere* **2004**, *55*, 1207–1212. [CrossRef]
31. Xu, W.Y.; Gao, T.Y.; Fan, J.H. Reduction of nitrobenzene by the catalyzed Fe/Cu process. *J. Environ. Sci.* **2008**, *20*, 915–992. [CrossRef]
32. Yang, Z.; Ma, Y.; Liu, Y.; Li, Q.; Zhou, Z.; Ren, Z. Degradation of organic pollutants in near-neutral pH solution by Fe-C micro-electrolysis system. *Chem. Eng. J.* **2017**, *315*, 403–414. [CrossRef]

33. Wu, S.; Qi, Y.; Fan, C.; Dai, B.; Huang, J.; Zhou, W.; He, S.; Gao, L. Improvement of anaerobic biological treatment effect by catalytic micro-electrolysis for monensin production wastewater. *Chem. Eng. J.* **2016**, *296*, 260–267. [CrossRef]
34. Huong, D.T.; Nguyen, V.T.; Ha, X.L.; Nguyen Thi, H.L.; Duong, T.T.; Nguyen, D.C.; Nguyen Thi, H.-T. Enhanced Degradation of Phenolic Compounds in Coal Gasification Wastewater by Methods of Microelectrolysis Fe-C and Anaerobic-Anoxic—Oxic Moving Bed Biofilm Reactor (A2O-MBBR). *Processes* **2020**, *8*, 1258. [CrossRef]

processes

Article

Liquid-Phase Removal of Methylene Blue as Organic Pollutant by Mesoporous Activated Carbon Prepared from Water Caltrop Husk Using Carbon Dioxide Activation

Yu-Quan Lin and Wen-Tien Tsai *

Graduate Institute of Bioresources, National Pingtung University of Science and Technology, Pingtung 912, Taiwan; wsx55222525@gmail.com

* Correspondence: wttsai@mail.npust.edu.tw; Tel.: +886-877-032-02

Abstract: In this work, a mesoporous activated carbon (AC) was prepared from a unique lignocellulosic biomass (water caltrop husk) in triplicate using a single-step physical activation process at lower temperature (i.e., 750 °C) and longer holding time (i.e., 90 min). Based on the pore properties and adsorption properties for removal of methylene blue (MB) as organic pollutant, the results proved that the resulting AC possesses a mesoporous feature with the Brunauer–Emmett–Teller (BET) surface area of 810.5 m^2/g and mesopore volume of about 0.13 cm^3/g. Due to its fast adsorption rate and maximal adsorption capacity fitted (126.6 mg/g), the mesoporous carbon material could be used as an excellent adsorbent for liquid-phase removal of MB. In addition, the pseudo-second-order model is well suited for describing the adsorption system between the cationic adsorbate and the resulting AC with oxygen surface groups.

Keywords: water caltrop husk; CO_2 activation; mesoporous activated carbon; methylene blue; adsorptive removal; kinetic modeling

Citation: Lin, Y.-Q.; Tsai, W.-T. Liquid-Phase Removal of Methylene Blue as Organic Pollutant by Mesoporous Activated Carbon Prepared from Water Caltrop Husk Using Carbon Dioxide Activation. *Processes* **2021**, *9*, 238. https://doi.org/10.3390/pr9020238

Academic Editor: Andrea Petrella
Received: 31 December 2020
Accepted: 25 January 2021
Published: 27 January 2021

1. Introduction

Organic chemicals (e.g., dyes, pesticides) released from anthropogenic activities have caused serious environmental issues because these pollutants have deteriorated the environmental quality in water, groundwater and soil phases. Furthermore, exposure to them will increase the health risks in both the food chain and drinking water. Therefore, many advanced treatment methods for the removal of organic pollutants from aqueous solutions have been developed in recent years [1]. Among these processes, activated carbon (AC) adsorption may be the most used method for the rapid removal of environmental pollutants from their existing phase (gas or liquid). An adsorption results in the removal of adsorbate molecules from the solution by diffusing them into porous adsorbent due to the concentration difference between the liquid and solid phases. Eventually, the concentration of the adsorbate remained in the solution will be in a dynamic equilibrium with that on the solid phase. In general, this recuperative process has significant features, including easy operation, low energy consumption, simple design, and high efficiency [2,3]. However, it should be noted that the adsorption process may be an expensive method due to the relatively high costs involved in commercial AC and its disposal, particularly when exhausted and not regenerated by steam for hydrolyzable adsorbates [4]. In this regard, there are many review papers focusing on the preparation of biomass-based AC and its applications for the removal of toxic pollutants from water in recent years [5–11]. The motivations for reusing agricultural residues as AC precursors also include sustainable waste management and climate change mitigation. On one hand, AC is usually a microporous carbonaceous material with high adsorption capacity [12], which is related to its specific surface area, pore volume, pore size distribution, and internal porosity [13]. However, the microporous feature in the AC materials could limit or restrict the transport diffusion of large molecule adsorbates into the micropores with pore widths (or diameters) less than 2.0 nm [14], thus reducing the adsorption capacity and retarding the adsorption rate for reaching adsorption equilibrium [15,16]. To mitigate from the so-called wall effect, many studies have focused on the production of biomass-based AC materials with a pore width (or diameter) distribution in the range of 2–50 nm for the removal of large dyes from aqueous phases [17–27]. It should be noted that these mesoporous AC materials were mostly produced by chemical activation processes.

Water caltrop (*Trapa natans*) is an aquatic floating plant commonly found in tropical and sub-tropical Asian countries. The pulp within the water caltrop fruit is a popular food because of its richness in starch. However, its outer layer (husk, shell, peel, or pericarps) is stripped off, thus generating water caltrop husk (WCH). This agricultural residue is often discarded in farmlands, or sometimes reused as an organic fertilizer. Due to its lignocellulosic compositions, there are some studies on the reuse of WCH as a biosorbent for dye removal [28–30], and a precursor for producing carbon materials in recent years [31–35]. For example, Kumar et al. [34] investigated the preparation of H_3PO_4-activated carbon from WCH and its removal performance of hexavalent chromium, showing a maximum adsorption capacity of 87.31 mg/g according to the Thomas model.

The adsorbate methylene blue (MB) is a cationic dye, commonly adopted as a probe molecule for determining the specific surface area or adsorption capacity of various materials quickly, including AC, charcoal, graphite, and clay [12,36–38]. It is also used as a chemical indicator, industrial dye, drug (e.g., treatment of malaria), and biological strain [39,40]. However, MB is also a toxic dye, which can result in harmful effects on humans (e.g., diarrhea, vomiting, and cyanosis) and environmental problems [23]. In this regard, the removal of MB as organic pollutant from aqueous solution or effluent is important for environmental protection and human health. Santoso et al. [11] reviewed research on the removal of MB using carbon-based adsorbents (focusing on activated carbon and biochar), and also discussed their structural properties influencing MB adsorption performance.

In a previous study [41], the pore and chemical properties of mesoporous AC produced from coconut shell using a single-step CO_2 activation process were studied, showing that a significant increase in the pore properties of resulting AC was found between 700 and 750 °C. As mentioned above, there is a scarcity of research that has investigated the production of mesoporous AC from WCH using physical activation by gasification gas CO_2. Therefore, this work aims to produce mesoporous AC from WCH by CO_2 activation at lower temperatures (i.e., 750 °C) and longer holding time (i.e., 90 min), and also characterize its pore properties and chemical compositions on the surface. Subsequently, the resulting AC was used to evaluate its removal performance of MB from the aqueous solution under various adsorption conditions.

2. Materials and Methods

2.1. Materials

The starting material (i.e., WCH) for producing mesoporous AC was obtained from a local farmers' association at Guantian District. The sun-dried WCH was first milled and then sieved to obtain particle sizes in the range of 0.420–0.841 mm. The pretreatment of WCH and its thermochemical properties (including proximate analysis, elemental analysis, calorific value, and thermogravimetric analysis) have been described in the previous study [42]. The dried WCH sample with average particle size of 0.63 mm (i.e., passed through mesh No. 20 and retained on mesh No. 40) was used in the physical activation experiments. In this work, the adsorbate MB, which was purchased from Merck Co., was modeled as a toxic pollutant for determining the adsorption capacity of the resulting AC in aqueous solution.

2.2. Physical Activation Experiments

The procedures for the preparation of AC from biomass precursor have been described in the previous study [41]. A vertical electric heating tube reactor (length 80 cm, inner diameter 10 cm) with a mesh-made sample holder was used to perform the carbonization-activation experiments. In order to produce mesoporous AC, the single-step physical activation experiments were performed in triplicate as follows: the temperature of WCH precursor (about 5 g) was increased to 500 °C at a heating rate of about 10 °C/min under the flow of 500 cm^3/min (N_2). The prepared charcoal was subsequently activated by a gasification gas (CO_2) of 50 cm^3/min under the desired conditions (i.e., activation temperature of 750 °C for holding 90 min). The yield of resulting AC (WCH-AC) was found to be about 10%. Before determining the pore properties of resulting AC (WCH-AC), the product sample was dried at about 105 °C overnight.

2.3. Characterization of Resulting Activated Carbon

The determinations of pore properties for the resulting AC (WCH-AC) were measured at −196 °C by using the ASAP-2020 porosimetry system (Micromeritics Co., Norcross, GA, USA). Using the N_2 adsorption–desorption isotherms obtained, the pore properties, including specific surface area, pore volume, pore size distribution, and average pore size (or width), can be calculated by the corresponding methods or models [13]. For example, the Brunauer-Emmett-Teller (BET) method was used to obtain the so-called BET surface area, which was determined in the relative pressure (P/P_0) range of 0.05–0.30. The micropore surface area and micropore volume were estimated from the *t*-method [13]. Accordingly, the mesopore volume was obtained from total pore volume minus micropore volume. In addition, the modified Kelvin equation using the Harrett–Joyner–Halenda method served as the basis for the analysis of pore size distribution [13]. Based on the helium displacement method, the true densities of resulting AC were determined by the AccuPyc-1340 pycnometer (Micromeritics Co., Norcross, GA, USA). The calculations of other pore properties like particle density and average pore width are referred to in the previous study [43]. The textural microstructures and elemental compositions of resulting AC on the surface were observed by the S-3000N scanning electron microscopy – energy dispersive X-ray spectroscopy (SEM-EDS) (Hitachi Co., Tokyo, Japan). Prior to the SEM-EDS analysis, the sample was plated by the E1010 ion sputter (Hitachi Co., Tokyo, Japan) to form a thin film with conductive gold.

2.4. Experiments of Adsorption Performance

The adsorption experiments for the MB removal from the water solution were similar to the previous study [41]. In this work, the data on the adsorption capacity of MB in the solution (2 L) were determined at the fixed conditions of 25 °C and 200 rpm. The determining process parameters included initial concentrations (i.e., 5, 10, 15, and 20 mg/L), WCH-AC adsorbent dosages (i.e., 0.1, 0.3, and 0.5 g/2 L) and initial pH values (i.e., 3.0, 7.0, and 11.0). For each adsorption experiment, an aliquot solution (about 15 cm^3) was drawn out at specified intervals (i.e., 1, 5, 10, 20, 30, 40, 50, and 60 min). The U-3900 UV/Visible spectrophotometer (Hitachi Co., Tokyo, Japan) was used to analyze the residual MB concentration (i.e., C_t), which was measured at the wavelength of 664 nm.

3. Results and Discussion

3.1. Pore Properties of Resulting Activated Carbon

As listed in Table 1, the pore properties of resulting AC (WCH-AC), including surface area, pore volume, average pore diameter, densities, and porosity were determined in the present study. By comparison, the BET surface area (i.e., 810.5 m^2/g) of WCH-AC was slightly larger than that of commercial AC (i.e., 660 m^2/g) [42]. In addition, this surface area was comparable to other studies on WCH-AC. In the work by Hsu et al. [33], the authors activated the microporous WCH biochar with ZnO and KOH at 900 °C, showing a BET surface area in the range of 1175–1537 m^2/g. In the study by Kumar et al. [34], the BET surface area and *t*-plot micropore volume of WCH-AC by H_3PO_4-activation were 782.89 m^2/g and 0.134 cm^3/g, respectively. Based on the data in Table 1, the ratio of micropore surface area to BET surface area was close to 0.76, which is also consistent with the ratio of micropore volume to total pore volume (i.e., 0.71). Therefore, the mesopore volume with pore diameter ranging from 2.0 nm to 50.0 nm could be estimated by subtracting the micropore volume from the total pore volume. Although the resulting AC (WCH-AC) is mainly a microporous carbon material, it still featured a mesoporosity of at least 20% or more. Figure 1 depicts its N_2 adsorption and desorption isotherms at −196 °C. From the very high potential for adsorption, the type I isotherms appeared in WCH-AC due to its microporous structure (i.e., pore diameter < 2 nm) [44], which is indicative of micropore filling at P/P_0 less than 0.05. With increases in P/P_0, the so-called hysteresis loop takes place in the WCH-AC, exhibiting adsorbate (i.e., nitrogen molecule) filling by capillary condensation in mesoporous solids (adsorption isotherm), but differing from that of mesopore emptying (desorption isotherm). It can be seen that the data were in accordance with mesopore volume and average pore diameter (0.130 cm^3/g and 2.17 nm, respectively, as listed in Table 1). More consistently, the pore size distribution of WCH-AC had two peaks as shown in Figure 2. One appeared as a narrow curve with the pores with less than 2 nm (micropores), but another was observed at ca. 4 nm (mesopores). Herein, the average pore size of WCH-AC was estimated by its BET surface area and total pore volume assuming cylindrical geometry in all pores [45].

Table 1. Pore properties of resulting activated carbon (WCH-AC).

Property	Value [a]
Single point surface area (m^2/g) [b]	790.8 ± 42.5
BET surface area (m^2/g) [c]	810.5 ± 2 5.7
Langmuir surface area (m^2/g)	1198.5 ± 38.6
Micropore surface area (m^2/g) [d]	618.9 ± 16.6
External surface area (m^2/g) [e]	191.6 ± 19.6
Total pore volume (cm^3/g) [f]	0.441 ± 0.024
Micropore volume (cm^3/g) [d]	0.311 ± 0.013
Pore diameter (Å) [g]	21.7 ± 0.8
True density (g/cm^3) [h]	1.787
Particle density (g/cm^3) [i]	0.999
Porosity (-) [j]	0.441

[a] Average ± standard deviation (n = 3). [b] Measured at 0.30 of relative pressure (P/P_0). [c] Measured in the relative pressure (P/P_0) range of 0.05-0.30. [d] Determined by the *t*-plot method. [e] Equal to BET surface area minus micropore surface area. [f] Measured at 0.995 of P/P_0. [g] Estimated from the total pore volume and BET surface area. [h] Determined by the helium-displacement measurement. [i] Obtained from the values of total pore volume and true density [45]. [j] Obtained from the values of particle density and true density [45].

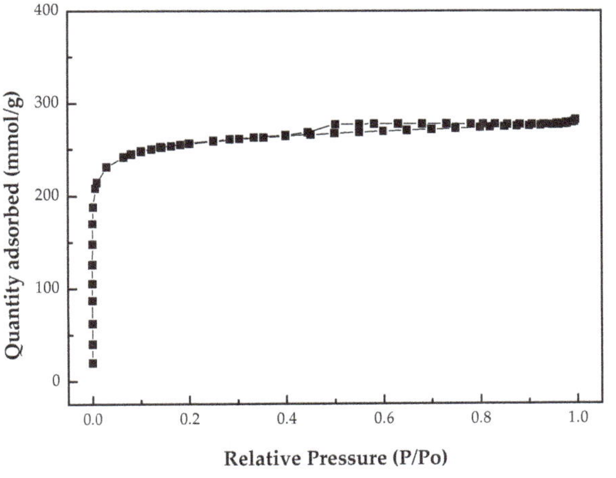

Figure 1. N_2 adsorption–desorption isotherms of resulting activated carbon (WCH-AC).

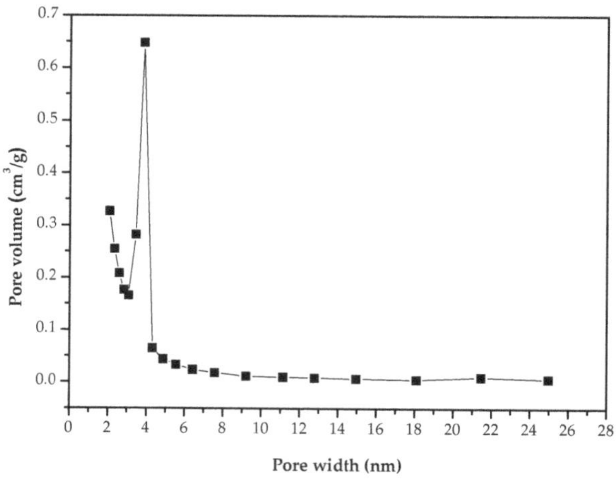

Figure 2. Pore size distribution of resulting activated carbon (WCH-AC).

3.2. SEM-EDS Observations of Resulting Activated Carbon

For confirming the porous microstructures of resulting AC (WCH-AC), the SEM was used to observe its morphological texture with two different magnifications (i.e., 1000 and 3000). As shown in Figure 3, many small pores appeared on the surface of the resulting AC. In addition, the AC also exhibited a rigid frame on the hard surface due to the physical activation of lignocellulosic constituents at a lower temperature (i.e., 750 °C) and longer holding time (i.e., 90 min). Therefore, the porous structure observed by the SEM was highly related to the pore properties, as listed in Table 1. Furthermore, the elemental distributions on the surface of resulting AC were also analyzed by the EDS while observing by the SEM. As illustrated in Figure 4, the main elements, including carbon (85.63 wt%) and oxygen (9.54 wt%), were present in the WCH-AC. It should be noted that the high content of oxygen in the resulting AC was indicative of its richness in functional groups containing oxygen (e.g., carbonyl and hydroxyl) on the surface. The presence of oxygen and other organic elements in surface groups had a profound effect on the adsorption properties of the AC, which is in connection with its slightly polar nature (i.e., hydrophilicity) [12]. Furthermore, some inorganic elements, including potassium (3.69 wt%), sodium (0.39 wt%), and chlorine (0.74 wt%), were found in the resulting AC. They were formed from metal oxides (e.g., K_2O and Na_2O) and metal chlorides (e.g., KCl and NaCl).

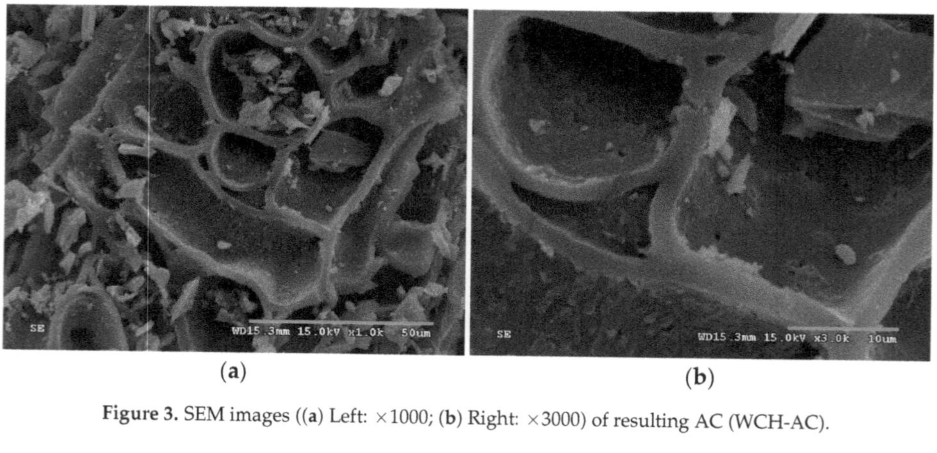

Figure 3. SEM images ((**a**) Left: ×1000; (**b**) Right: ×3000) of resulting AC (WCH-AC).

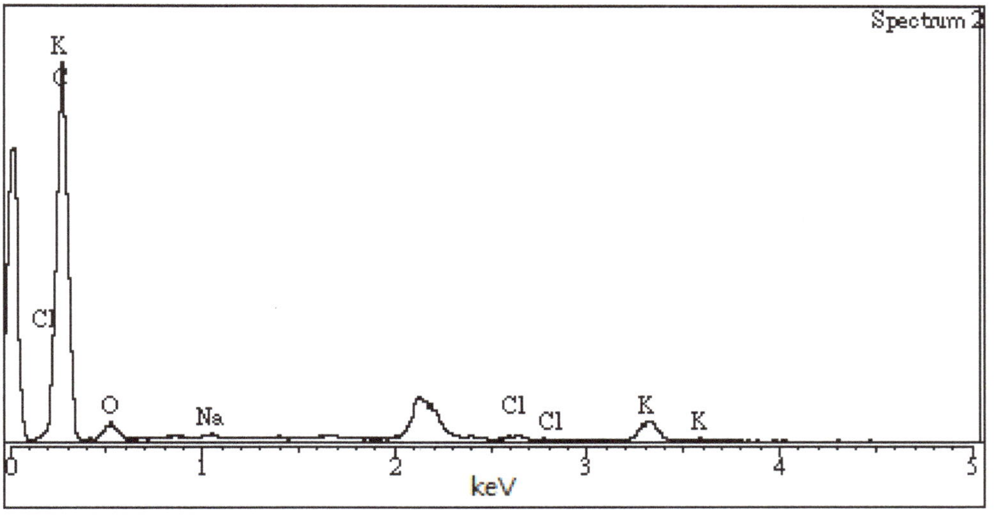

Figure 4. EDS spectra of resulting AC (WCH-AC) and its relative compositions of elements detected.

3.3. Adsorption Performances of Resulting Activated Carbon

Figures 5 and 6 show the variations on residual MB concentration at the specific adsorption conditions under various WCH-AC adsorbent dosages and initial MB concentrations, respectively. It can be seen that the dimensionless residual MB concentration (C_t/C_0) indicated a fast decrease during the limited adsorption time. This indicates a strong interaction between the MB adsorbate and the WCH-AC adsorbent. As mentioned above, the resulting AC should be negatively-polar on its surface. Therefore, a pseudo-second-order model was used to fit the adsorption system with a linear form [46]:

$$t/q_t = 1/(k \times q_e{}^2) + (1/q_e) \times t \quad (1)$$

where q_t is the amount of MB adsorbed at time t (mg/g), q_e is the amount of MB adsorbed at equilibrium (mg/g), and k is the rate constant of this model (g/(mg.min)). Therefore, the adsorption time ($t_{1/2}$) necessary to adsorb half of the adsorption amount of MB at equilibrium ($q_e/2$) by the WCH-AC adsorbent was obtained as follows:

$$t_{1/2} = 1/(k \times q_e) \quad (2)$$

Table 2. Pseudo-second-order model parameters for MB adsorption onto WCH-AC at various WCH-AC dosages. [a]

Adsorbent Dosage (g/2 L)	k (g/(mg.min))	q_e (mg/g)	Correlation Coefficient	$t_{1/2}$ (min)	h (mg/(g.min))
0.1	0.0012	126.58	0.986	6.58	19.23
0.3	0.0147	68.03	1.000	1.00	68.03
0.5	0.1442	40.16	1.000	0.17	232.57

[a] Process conditions: initial MB concentration = 10 mg/L, initial pH = 7.0.

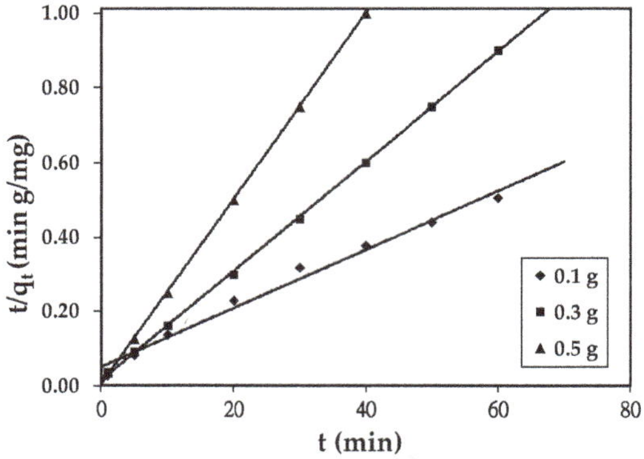

Figure 5. Variations of dimensionless concentration (C_t/C_0) vs. time under various WCH-AC dosages (initial MB concentration = 10 mg/L, initial pH = 7.0); symbols: experimental data, full lines: obtained from the pseudo-second-order model parameters (Table 2).

Table 3. Pseudo-second-order model parameters for MB adsorption onto WCH-AC at various initial MB concentrations. [a]

Initial MB Concentration (mg/L or ppm)	k (g/(mg.min))	q_e (mg/g)	Correlation Coefficient	$t_{1/2}$ (min)	h (mg/(g.min))
5	0.1625	33.44	1.000	0.18	181.71
10	0.0147	68.03	1.000	1.00	68.03
15	0.0024	104.17	0.998	4.00	26.04
20	0.0016	125.00	0.994	5.00	25.00

[a] Process conditions: WCH-AC dosage = 0.3 g/2L, initial pH = 7.0.

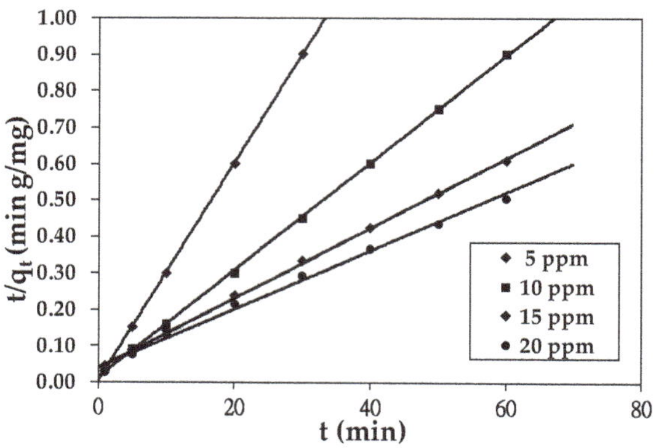

Figure 6. Variations of dimensionless concentration (C_t/C_0) vs. time under various initial MB concentrations (WCH-AC dosage = 0.3 g/2 L, initial pH = 7.0); symbols: experimental data, full lines: obtained from the pseudo-second-order model parameters (Table 3).

In addition, the initial adsorption rate (h) was obtained by the equation [47,48]:

$$h = k \times {q_e}^2 \tag{3}$$

By using the model fitting, the values of the adsorption parameters for the AC-MB system were summarized in Table 2, Table 3, and Table 4, which correspond to the processes' parameters of adsorbent dosage, initial MB concentration, and pH, respectively. With high correlation coefficients (>0.98), the adsorption of MB into WCH-AC followed this kinetic model well. Regarding the discussion on the relationships between the values of the fitted model parameters (i.e., q_e, k, and h) and process parameters (i.e., adsorbent dosage and C_0), this has been elucidated in previous studies [43]. The data in Table 4 indicates an important significance. At the acidic solution, the surface of resulting AC was protonized by excessive protons. As a consequence, the cationic dye (i.e., MB) was repelled more from the positively charged surface because of the repulsive force [49]. In contrast, the surface of resulting AC was prone to lose the protons at a higher basicity, thus resulting in the negatively charged surface for adsorbing the adsorbate MB preferably due to the electrostatic attraction.

Table 4. Pseudo-second-order model parameters for MB adsorption onto WCH-AC at three different pH values. [a]

Initial pH	k (g/(mg.min))	q_e (mg/g)	Correlation Coefficient	$t_{1/2}$ (min)	h (mg/(g.min))
3	0.0101	68.97	1.000	1.44	48.04
7	0.0147	68.03	1.000	1.00	68.03
11	0.0207	69.93	1.000	0.69	101.23

[a] Process conditions: WCH-AC dosage = 0.3 g/2L, initial concentration = 10 mg/L.

4. Conclusions

In this study, the pore properties and adsorption performances of the mesoporous AC from water caltrop husk prepared at lower temperature (i.e., 750 °C) and longer holding time (i.e., 90 min) are summarized as follows:

- The resulting AC possessed a mesoporous feature with the BET specific surface area of 810.5 m^2/g and mesopore volume of 0.13 cm^3/g, which are superior to commercial AC products.
- Due to its fast adsorption rate and maximal adsorption capacity fitted by the model (126.6 mg/g), the mesoporous carbon material could be used as an excellent adsorbent for liquid-phase removal of MB.
- The pseudo-second-order model is well suited for describing the adsorption system, which includes the cationic adsorbate and the resulting AC with hydrophilicity of oxygen surface groups.

Author Contributions: Conceptualization, W.-T.T.; methodology, Y.-Q.L.; validation, Y.-Q.L.; data curation, Y.-Q.L.; formal analysis, Y.-Q.L.; writing—original draft preparation, W.-T.T.; writing—review and editing, W.-T.T. All authors have read and agreed to the published version of the manuscript.

Funding: This research received no external funding.

Institutional Review Board Statement: Not applicable.

Informed Consent Statement: Not applicable.

Acknowledgments: The SEM-EDS observations were assisted by the Instrument Center at National Pingtung University of Science and Technology.

Conflicts of Interest: The authors declare no conflict of interest.

References

1. Teodosiu, C.; Gilca, A.F.; Barjoveanu, G.; Fiore, S. Emerging pollutants removal through advanced drinking water treatment: A review on processes and environmental performances assessment. *J. Clean. Prod.* **2018**, *197*, 1210–1221. [CrossRef]
2. Jeirani, Z.; Niu, C.H.; Soltan, J. Adsorption of emerging pollutants on activated carbon. *Rev. Chem. Eng.* **2017**, *33*, 491–522. [CrossRef]
3. Al-Ghouti, M.A.; Al-Kaabi, M.A.; Ashfag, M.Y.; Da'na, D.A. Produced water characteristics, treatment and reuse: A review. *J. Water Process Eng.* **2019**, *28*, 222–239. [CrossRef]

4. Tsai, W.T.; Chang, C.Y.; Ho, C.Y.; Chen, L.Y. Adsorption properties and breakthrough model of 1,1-dichloro-1-fluoroethane on activated carbons. *J. Hazard. Mater.* **1999**, *69*, 53–66. [CrossRef]
5. Loannidou, O.; Zabaniotou, A. Agricultural residues as precursors for activated carbon production—A review. *Renew. Sust. Energy Rev.* **2007**, *11*, 1966–2005. [CrossRef]
6. Paraskeva, P.; Kalderis, D.; Diamadopoulos, E. Production of activated carbon from agricultural by-products. *J. Chem. Technol. Biotechnol.* **2008**, *83*, 581–592. [CrossRef]
7. Alslaibi, T.M.; Abustan, I.; Ahmad, M.A.; Foul, A.A. A review: Production of activated carbon from agricultural byproducts via conventional and microwave heating. *J. Chem. Technol. Biotechnol.* **2013**, *88*, 1183–1190. [CrossRef]
8. Yahya, M.A.; Al-Qodah, Z.; Ngah, C.W.Z. Agricultural bio-waste materials as potential sustainable precursors used for activated carbon production: A review. *Renew. Sust. Energy Rev.* **2015**, *46*, 218–235. [CrossRef]
9. Gonzalez-Garcia, P. Activated carbon from lignocellulosics precursors: A review of the synthesis methods, characterization techniques and applications. *Renew. Sust. Energy Rev.* **2018**, *82*, 1393–1414. [CrossRef]
10. Qureshi, U.A.; Hameed, B.H.; Ahmed, M.J. Adsorption of endocrine disrupting compounds and other emerging contaminants using lignocellulosic biomass-derived porous carbons: A review. *J. Water Process Eng.* **2020**, *38*, 101380. [CrossRef]
11. Santoso, E.; Ediati, R.; Kusumawati, Y.; Bahruji, H.; Sulistiono, D.O.; Prasetyoko, D. Review on recent advances of carbon based adsorbent for methylene blue removal from waste water. *Mater. Today Chem.* **2020**, *16*, 100233. [CrossRef]
12. Marsh, H.; Rodriguez-Reinoso, F. *Activated Carbon*; Elsevier: Amsterdam, The Netherlands, 2006.
13. Lowell, S.; Shields, J.E.; Thomas, M.A.; Thommes, M. *Characterization of Porous Solids and Powders: Surface Area, Pore Size and Density*; Springer: Dordrecht, The Netherlands, 2006.
14. Ruthven, D.M. *Principles of Adsorption and Adsorption Processes*; John Wiley & Sons: New York, NY, USA, 1984.
15. Hsieh, C.T.; Teng, H. Influence of mesopore volume and adsorbate size on adsorption capacities of activated carbons in aqueous solutions. *Carbon* **2000**, *38*, 863–869. [CrossRef]
16. Piai, L.; Dykstra, J.E.; Adishakti, M.G.; Blockland, M.; Langenhoff, A.A.M.; van der Wal, A. Diffusion of hydrophilic organic micropollutants in granular activated carbon with different pore sizes. *Water Res.* **2019**, *162*, 518–527. [CrossRef] [PubMed]
17. Hu, Z.; Srinivasan, M.P.; Ni, Y. Preparation of mesoporous high-surface-area activated carbon. *Adv. Mater.* **2000**, *12*, 62–65. [CrossRef]
18. Macedo, J.S.; Junior, N.B.C.; Almeida, L.E.; Vieira, E.F.S.; Cestari, A.R.; Gimenez, I.F.; Carreno, N.L.V.; Barreto, L.S. Kinetic and calorimetric study of the adsorption of dyes on mesoporous activated carbon prepared coir dust. *J. Colloid Interface Sci.* **2006**, *298*, 515–522. [CrossRef]
19. Gao, J.; Kong, D.; Wang, Y.; Wu, J.; Sun, S.; Xu, P. Production of mesoporous activated carbon from tea fruit peel residues and its evaluation of methylene blue removal from aqueous solutions. *BioResources* **2013**, *8*, 2145–2160. [CrossRef]
20. Salman, J.M. Preparation of mesoporous-activated carbon from branches of pomegranate trees: Optimization on removal of methylene blue using response surface methodology. *J. Chem.* **2013**, 489670. [CrossRef]
21. Yu, L.; Luo, Y.M. The adsorption mechanism of anionic and cationic dyes by Jerusalem artichoke stalk-based mesoporous activated carbon. *J. Environ. Chem. Eng.* **2014**, *2*, 220–229. [CrossRef]
22. Islam, A.A.; Ahmed, M.J.; Khanday, W.A.; Asif, M.; Hameed, B.H. Mesoporous activated carbon prepared from NaOH activation of rattan (*Lacosperma secundiflorum*) hydrochar for methylene blue removal. *Ecotoxicol. Environ. Saf.* **2017**, *138*, 279–285. [CrossRef]
23. Jawad, A.H.; Rashid, R.A.; Ismail, K.; Sabar, S. High surface area mesoporous activated carbon developed from coconut leaf by chemical activation with H_3PO_4 for adsorption of methylene blue. *Desalin. Water Treat.* **2017**, *74*, 326–335. [CrossRef]
24. Marrakchi, F.; Ahmed, M.J.; Khanday, W.A.; Asif, M.; Hameed, B.H. Mesoporous-activated carbon prepared from chitosan flakes via single-step sodium hydroxide activation for the adsorption of methylene blue. *Int. J. Biol. Macromol.* **2017**, *98*, 233–239. [CrossRef]
25. Wongcharee, S.; Aravinthan, V.; Erdei, L.; Sanongraj, W. Mesoporous activated carbon prepared from macadamia nut shell waste by carbon dioxide activation: Comparative characterization and study of methylene blue removal from aqueous solution. *Asian-Pac. J. Chem. Eng.* **2018**, *13*, e2179. [CrossRef]
26. Khasri, A.; Bello, O.S.; Ahmad, M.A. Mesoporous activated carbon from *Pentace* species sawdust via microwave-induced KOH activation: Optimization and methylene blue adsorption. *Res. Chem. Intermed.* **2018**, *44*, 5737–5757. [CrossRef]
27. Boudia, R.; Mimanne, G.; Benhabib, K.; Pirault-Roy, L. Preparation of mesoporous activated carbon from date stones for the adsorption of Bemacid Red. *Water Sci Technol.* **2019**, *79*, 1357–1366. [CrossRef] [PubMed]
28. Khan, T.A.; Nazir, M.; Khan, E.A. Adsorptive removal of rhodamine B from textile wastewater using water chestnut (*Trapa natans L.*) peel: Adsorption dynamics and kinetic studies. *Toxicol. Environ. Chem.* **2013**, *95*, 919–931. [CrossRef]
29. Rehman, R.; Salariya, B.; Mitu, L. Batch scale adsorptive removal of brilliant green dye using *Trapa natans* peels in cost effective manner. *Rev. Chim.* **2016**, *67*, 1333–1337.
30. Kumar, S.; Naryanasamy, S.; Venkatesh, R.P. Removal of Cr (VI) from synthetic solutions using water caltrop shell as a low-cost biosorbent. *Sep. Sci. Technol.* **2019**, *54*, 2783–2799. [CrossRef]
31. Rao, L.L.; Liu, S.F.; Wang, L.L.; Ma, C.D.; Wu, J.Y.; An, L.Y.; Hu, X. N-doped porous carbons from low-temperature and single-step sodium amide activation of carbonized water chestnut shell with excellent CO_2 capture performance. *Chem. Eng. J.* **2019**, *359*, 428–435. [CrossRef]

32. Wang, P.; Fan, L.; Yan, L.; Shi, Z. Low-cost water caltrop shell-derived hard carbons with high initial coulombic efficiency for sodium-ion battery anodes. *J. Alloys Compd.* **2019**, *775*, 1028–1035. [CrossRef]
33. Hsu, C.H.; Pan, Z.B.; Chen, C.R.; Wei, M.X.; Chen, C.A.; Lin, H.P.; Hsu, C.H. Synthesis of multiporous carbons from the water caltrop shell for high-performance supercapacitors. *ACS Omega* **2020**, *5*, 10626–10632. [CrossRef]
34. Kumar, S.; Patra, C.; Naryanasamy, S.; Rajaraman, P.V. Performance of acid-activated water caltrop (*Trapa natans*) shell in fixed bed column for hexavalent chromium removal from simulated wastewater. *Environ. Sci. Pollut. Res.* **2020**, *27*, 28042–28052. [CrossRef] [PubMed]
35. Yin, W.J.; Zhang, Z.H.; Liu, T.C.; Xu, J.; Xiao, S.Z.; Xu, Y. N-doped animal keratin waste porous biochar derived from *Trapa natans* husks. *Materials* **2020**, *13*, 987. [CrossRef] [PubMed]
36. Kaewprasit, C.; Hequet, E.; Abidi, N.; Gourlot, J.P. Application of methylene blue adsorption to cotton fiber specific surface area measurement: Part I. methodology. *J. Cotton Sci.* **1998**, *2*, 164–173.
37. Chen, G.; Pan, J.; Han, B.; Yan, H. Adsorption of methylene blue on montmorillonite. *J. Dispers. Sci. Technol.* **1999**, *20*, 1179–1187. [CrossRef]
38. Hegyesi, N.; Vad, R.T.; Pukanszky, B. Determination of the specific surface area of layered silicates by methylene blue adsorption: The role of structure, pH and layer charge. *Appl. Clay Sci.* **2017**, *146*, 50–55. [CrossRef]
39. Jia, P.; Tan, H.; Liu, K.; Gao, W. Removal of methylene blue from aqueous solution by bone char. *Appl. Sci.* **2018**, *8*, 1903. [CrossRef]
40. Lu, G.; Nagbanshi, M.; Goldau, N.; Mendes Jorge, M.; Meissner, P.; Jahn, A.; Mockenhaupt, F.P.; Muller, O. Efficacy and safety of methylene blue in the treatment of malaria-a systematic review. *BMC Med.* **2018**, *16*, 59. [CrossRef]
41. Tsai, W.T.; Jiang, T.J. Mesoporous activated carbon produced from coconut shell using a single-step physical activation process. *Biomass Conver. Biorefin.* **2018**, *8*, 711–718. [CrossRef]
42. Tsai, W.T.; Lin, Y.Q.; Tsai, C.H.; Chung, M.H.; Chu, M.H.; Huang, H.J.; Jao, Y.H.; Yeh, S.I. Conversion of water caltrop husk into biochar by torrefaction. *Energy* **2020**, *195*, 116967. [CrossRef]
43. Tsai, W.T.; Huang, P.C.; Lin, Y.Q. Reusing cow manure for the production of activated carbon using potassium hydroxide (KOH) activation process and its liquid-phase adsorption performance. *Processes* **2019**, *7*, 737. [CrossRef]
44. Gregg, S.J.; Sing, K.S.W. *Adsorption, Surface Area, and Porosity*; Academic Press: London, UK, 1982.
45. Smith, J.M. *Chemical Engineering Kinetics*, 3rd ed.; McGraw-Hill: New York, NY, USA, 1981.
46. Ho, Y.S.; Chiang, C.C.; Hsu, Y.C. Sorption kinetics for dye removal from aqueous solution using activated clay. *Sep. Sci. Technol.* **2001**, *36*, 2473–2488. [CrossRef]
47. Hamadi, N.K.; Swaminathan, S.; Chen, X.D. Adsorption of paraquat dichloride from aqueous solution by activated carbon derived from used tires. *J. Hazard. Mater.* **2004**, *B112*, 133–141. [CrossRef] [PubMed]
48. Paska, O.M.; Pacurariu, C.; Muntean, S.G. Kinetic and thermodynamic studies on methylene blue biosorption using corn-husk. *RSC Adv.* **2014**, *4*, 62621–62630. [CrossRef]
49. Zaini, M.A.A.; Okayama, R.; Machida, M. Adsorption of aqueous metal ions on cattle-manure-compost based activated carbons. *J. Hazard. Mater.* **2009**, *170*, 1119–1124. [CrossRef] [PubMed]

processes

MDPI

Article

Methyl Orange Photo-Degradation by TiO_2 in a Pilot Unit under Different Chemical, Physical, and Hydraulic Conditions

Andrea Petrella [1,*], Danilo Spasiano [1], Pinalysa Cosma [2], Vito Rizzi [2], Marco Race [3], Maria Cristina Mascolo [3] and Ezio Ranieri [4]

[1] Dipartimento di Ingegneria Civile, Ambientale, Edile, Del Territorio e di Chimica, Politecnico di Bari, Via E. Orabona 4, 70125 Bari, Italy; danilo.spasiano@poliba.it

[2] Dipartimento di Chimica, Università degli Studi di Bari "Aldo Moro", Via E. Orabona 4, 70125 Bari, Italy; pinalysa.cosma@uniba.it (P.C.); vito.rizzi@uniba.it (V.R.)

[3] Dipartimento di Ingegneria Civile e Meccanica, Università degli Studi di Cassino e del Lazio Meridionale, Via di Biasio 43, 03043 Cassino, Italy; marco.race@unicas.it (M.R.); mc.mascolo@unicas.it (M.C.M.)

[4] Dipartimento di Biologia, Università degli Studi di Bari "Aldo Moro", Via E. Orabona 4, 70125 Bari, Italy; ezio.ranieri@uniba.it

* Correspondence: andrea.petrella@poliba.it; Tel.: +39-(0)8-0596-3275; Fax: +39-(0)8-0596-3635

Abstract: The photo-catalytic degradation of a textile azo-dye as Methyl Orange was studied in an innovative unit constituted by a channel over which a layer of titanium dioxide (TiO_2) catalyst in anatase form was deposited and activated by UVB irradiation. The degradation kinetics were followed after variation of the chemical, physical, and hydraulic/hydrodynamic parameters of the system. For this purpose, the influence of the TiO_2 dosage (g/cm^3), dye concentration (mg/L), pH of the solution, flow-rate (L/s), hydraulic load (cm), and irradiation power (W) were evaluated on the degradation rates. It was observed that the maximum dosage of TiO_2 was 0.79 g/cm^3 while for higher dosage a reduction of homogeneity of the cement conglomerate occurred. The Langmuir–Hinshelwood (LH) kinetic model was followed up to a dye concentration around 1 mg/L. It was observed that with the increase of the flow rate, an increase of the degradation kinetics was obtained, while the further increase of the flow-rate associated with the modification of the hydraulic load determined a decrease of the kinetic rates. The results also evidenced an increase of the kinetic rates with the increase of the UVB intensity. A final comparison with other dyes such as Methyl Red and Methylene Blue was carried out in consideration of the pH of the solution, which sensibly affected the removal efficiencies.

Keywords: photo-catalysis; TiO_2; azo dye; kinetic study; hydraulic and hydrodynamic parameters

Citation: Petrella, A.; Spasiano, D.; Cosma, P.; Rizzi, V.; Race, M.; Mascolo, M.C.; Ranieri, E. Methyl Orange Photo-Degradation by TiO_2 in a Pilot Unit under Different Chemical, Physical, and Hydraulic Conditions. *Processes* **2021**, *9*, 205. https://doi.org/10.3390/pr9020205

Academic Editor: María V. López-Ramón

Received: 5 January 2021

Accepted: 19 January 2021

Published: 21 January 2021

1. Introduction

Emerging contaminants in water and wastewater are chemical compounds produced by industrial practices and anthropogenic activities [1–6] that must be removed due to their potential toxicological effects on human health and the environment [7–13].

For this purpose, Advanced Oxidation Processes (AOPs) are efficient methods to remove from water and wastewater contaminants of organic nature that are not degradable by biological processes [14–16]. Specifically, these treatments result effective for the removal of micro-pollutants as pesticides, personal care products, pharmaceuticals, flame retardants, antifoulants, stabilizers, and plasticizers, which have harmful effects on the reproductive system [8,17–20].

AOPs are processes involving the production of very reactive radical species able to degrade a wide range of biopersistent organic substrates [21–23]. Among these, photo-catalysis is an efficient treatment carried out with a catalyst that generates radicals when irradiated with light of a suitable wavelength [24–28]. TiO_2 has been extensively studied as a catalyst due to its wide band gap (3.2 eV), strong oxidizing power, high resistance to chemicals, nontoxic nature, and low cost [24,29].

The TiO_2 photocatalysis can be applied in the removal of xenobiotic azo dyes from the textile industry such as Methylene Blue, Methyl Red, and Methyl Orange, which tend to bioaccumulate in the environment and have allergenic, carcinogenic, mutagenic, and teratogenic properties. Methylene Blue is used in dye manufacturing industries, plastics, cosmetics, and printing [30,31]. It is a toxic compound since it can cause eye burns, vomiting, jaundice, and diarrhea [32,33]. Methyl Red is used in textile dyeing and paper printing and it is hazardous in case of skin contact (irritant), inhalation, and ingestion [34,35]. Methyl Orange is widely used in dyeing, printing textiles, and paper industries. Methyl Orange is the common name of a water-soluble aromatic synthetic compound ($C_{14}H_{14}N_3NaO_3S$) containing an azo group (–N=N–). This molecule is a toxic compound and can cause hypersensitivity, allergies and may be fatal if inhaled [31,36]. The main environmental damage caused by the textile industry is the release of untreated effluents into the water bodies, corresponding to ~80% of the total emissions [37].

The aim of this paper was to study the UVB photo-catalytic degradation of Methyl Orange in water and wastewater by the use of an innovative unit [38–40]. The photocatalytic system was a recirculating plant formed of two tanks connected by a channel over which anatase TiO_2 was deposited and UVB irradiated. The influence of the TiO_2 dosage (g/cm^3), dye concentration c_0 (mg/L), pH of the solution, flow-rate Q (L/s), hydraulic load h_w (cm), and irradiation power (W) were evaluated on the degradation kinetics of this hazardous dye. Moreover, the removal efficiencies of Methyl Orange were compared with those of other textile dyes such as Methyl Red and Methylene Blue in order to have more detailed information about the treatment of these harmful compounds through this unit.

2. Experimental Section

Anatase titanium dioxide (TiO_2) was provided by Adriatica Legnami s.r.l., Italy, and was characterized by 0.15 µm average grain size and 3.85 g/cm^3 specific gravity. Methyl Orange (MO), pure chemical from Sigma Aldrich, was used to prepare the synthetic solutions in tap water (pH = 7.5) and distilled water (pH = 6). Moreover, Methyl Red (MR) and Methylene Blue (MB), pure chemicals from Sigma Aldrich, were also used to prepare tap and distilled water solutions in order to make a comparison.

The determination of the residual dye concentrations was obtained by a UV–Vis spectrophotometer Mod. UVIKON 942 from Kontron Instruments, Augsburg, Germany.

The unit which was used for the experimental tests is depicted in Figure 1. Specifically, Figure 1A,B report the overview and the top view schemes of the system, respectively. Figure 1C represents the real overview of the laboratory scale pilot plant. It was characterized by a channel (c) (15 cm width, 185 cm length) over which layers of cement mortars (0.5 mm thickness) with different TiO_2 concentrations (0.16 g/cm^3, 0.39 g/cm^3, 0.55 g/cm^3, 0.79 g/cm^3) were deposited. The dye influent solution was kept under UVB irradiation in contact with the catalyst. For this purpose, three low-pressure UVB lamps (40 W each, λem = 312 nm), Vilber-Lourmat (Collégien, France), were used. The system was characterized by an upper reservoir with manifolds (a) which allowed for the change of the treated volume solution associated with the modification of the hydraulic load (hw) with consequent variation of the flow-rates (Q). The introduction of layers of different thicknesses (1.5, 1.0 and 0.5 mm; 11 × 70 cm) into an opening of this tank allowed for the modification of the flow-rate (Q), keeping constant the hydraulic load (hw). The system was also characterized by a bottom reservoir (b) with similar capacity to the former and equipped with a piezometric tube. A pump (d) (Mod. CPm 130, Pedrollo, Milan, Italy; 0.37 kW, 230 V, 50 Hz) was used for the recirculation of the influent solution through the unit. Table 1 represents a summary of the tests carried out in the present investigation.

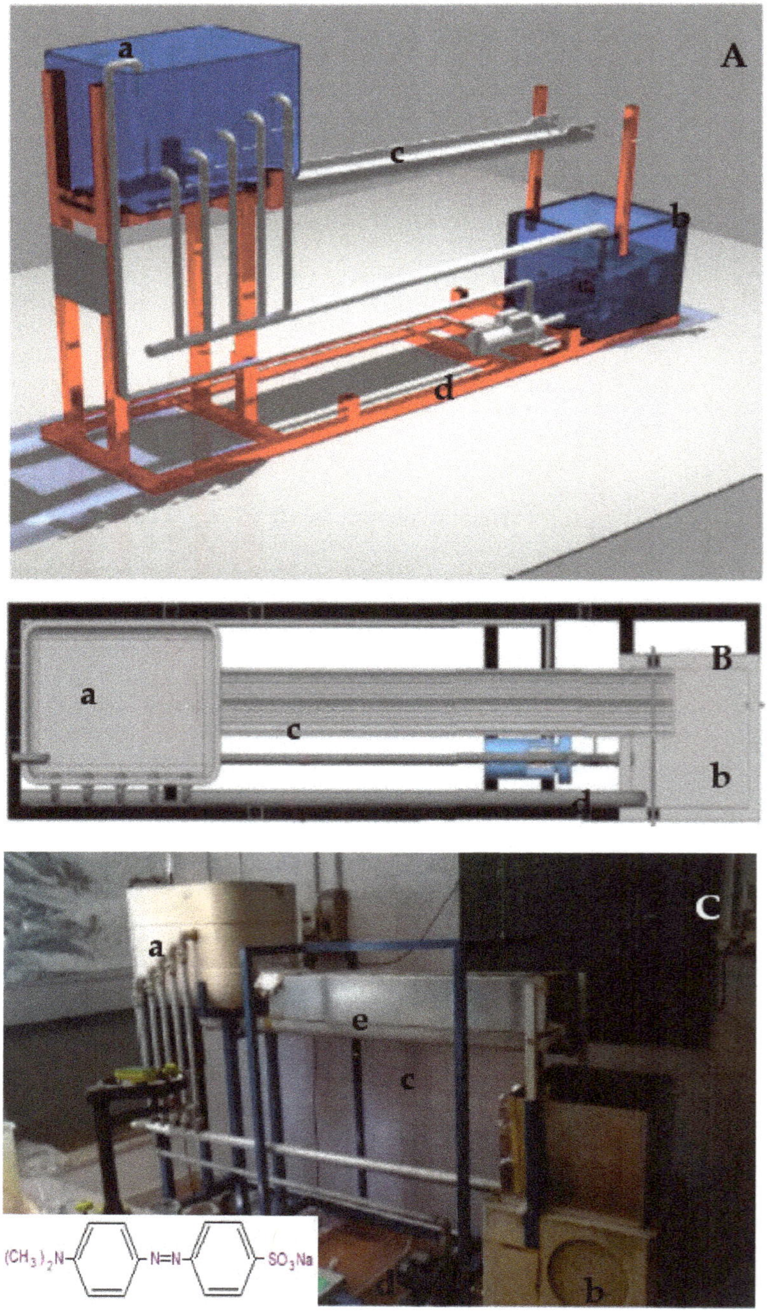

Figure 1. (**A**) Overview scheme and (**B**) top view scheme of the laboratory scale pilot plant. (**C**) Picture of the unit. (a) Upper reservoir, (b) bottom reservoir, (c) channel, (d) pump, (e) UVB lamp.

Table 1. Summary of the tests carried out in the present investigation. c_0 = initial dye concentration, V_{sol} = volume of the influent solution, Q = flow-rate, h_w = hydraulic load, l_w = irradiated liquid width in the channel, l_c = irradiated liquid length in the channel, l_d = irradiated liquid depth in the channel, V_{irr} = irradiated volume ($l_w \times l_c \times l_d$) of the liquid in the channel, I_{rt} = irradiated retention time (V_{irr}/Q), irradiation power (W), *pH*.

Test No.	c_0 (mg/L)	TiO_2 (g/cm^3)	V_{sol} (L)	Q (L/s)	h_w (cm)	l_w (cm)	l_c (cm)	l_d (cm)	V_{irr} (L)	I_{rt} (s)	*Light*	*Power* (W)	*pH*
1	0.7	0.16	60	0.066	13.5	14	140	0.65	1.27	17.9	yes	120	7.5
2	0.7	0.39	60	0.066	13.5	14	140	0.65	1.27	17.9	yes	120	7.5
3	0.7	0.55	60	0.066	13.5	14	140	0.65	1.27	17.9	yes	120	7.5
4	0.7	0.79	60	0.066	13.5	14	140	0.65	1.27	17.9	yes	120	7.5
5	0.7	0.85	60	0.066	13.5	14	140	0.65	1.27	17.9	yes	120	7.5
6	0.7	0.95	60	0.066	13.5	14	140	0.65	1.27	17.9	yes	120	7.5
7	0.3, 0.7, 1.2, 2.5, 5	0.79	60	0.066	13.5	14	140	0.65	1.27	17.9	yes	120	7.5
8	0.3, 0.7, 1.2, 2.5, 5	0.79	60	0.147	13.5	15	140	0.79	1.66	10.5	yes	120	7.5
9	0.3, 0.7, 1.2, 2.5, 5	0.79	60	0.210	13.5	15	140	0.85	1.78	7.9	yes	120	7.5
10	0.3, 0.7, 1.2, 2.5, 5	0.79	60	0.305	13.5	15	140	1.05	2.20	6.7	yes	120	7.5
11	0.7	0.79	72.5	0.355	18	15	140	1.08	2.30	5.8	yes	120	7.5
12	0.7	0.79	90	0.408	22.5	15	140	1.12	2.35	5.2	yes	120	7.5
13	0.7	0.79	105	0.441	27	15	140	1.16	2.45	4.9	yes	120	7.5
14	0.7	0	60	0.305	13.5	15	140	1.05	2.20	6.7	yes	120	7.5
15	0.7	0.79	60	0.305	13.5	15	140	1.05	2.20	6.7	no	no	7.5
16	0.7	0.79	60	0.305	13.5	15	140	1.05	2.20	6.7	yes	40	7.5
17	0.7	0.79	60	0.305	13.5	15	140	1.05	2.20	6.7	yes	80	7.5
18	0.7	0.79	60	0.305	13.5	15	140	1.05	2.20	6.7	yes	120	6.0

3. Results and Discussion

The first set of photocatalytic experiments was carried out at different TiO_2 dosage, tests no. 1–6 (Figure 2), constant flow-rate (Q = 0.066 L/h), the hydraulic load (h_w = 13.5 cm), and the influent substrate concentration (c_0 = 0.7 mg/L) as reported in Table 1. The processes described in this paper, at low initial dye concentration, may be approximated by the "pseudo-first-order" equation,

$$ln(c/c_0) = -k \times \theta \times t \quad (1)$$

i.e., the Langmuir–Hinshelwood (LH) kinetic model. In this equation, c_0 (mg/L) is the starting concentration of the substrate and c (mg/L) is the concentration of the dye at specific time intervals, while k is the apparent rate constant (min^{-1}) and θ represents the number of the surface active sites present on the catalyst surface [41–43]. The trend reported in the inset of Figure 2 is in agreement with this model, accordingly after linear correlation the apparent rate constant is represented by the slope. Figure 2 shows that the degradation rate increased with the increase of the catalyst concentration due to the growing number of the active sites necessary for the photocatalytic oxidation. Theoretically, an improvement of the results could be obtained with a further increase of oxide concentration but a reduction of homogeneity was observed which was detrimental for the process since leaching of the catalyst from the cement conglomerate occurred. In this respect, the maximum dosage of TiO_2 was 0.79 g/cm^3.

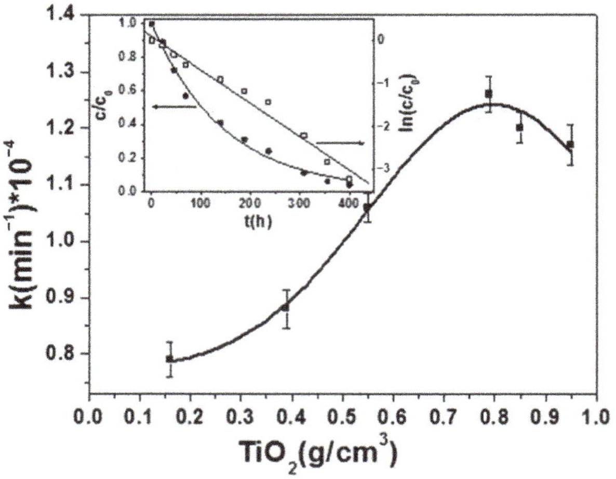

Figure 2. Influence of the TiO_2 dosage on the degradation rate (Methyl Orange (MO) = 0.7 mg/L; Q = 0.066 L/s, h_w = 13.5 cm, tests no. 1–6). In the inset: c/c_0 vs. t and $ln(c/c_0)$ vs. t correlations relative to test no. 4 (MO = 0.7 mg/L; TiO_2 = 0.79 g/cm^3; Q = 0.066 L/s, h_w = 13.5 cm).

The second set of experiments (Table 1, test no. 7) was carried out at different substrate concentration (0.3–5.0 mg/L range, Figure 3), constant flow-rate (0.066 L/h), the hydraulic load (h_w = 13.5 cm), and with 0.79 g/cm^3 catalyst dosage because, as formerly reported, it is the titania concentration corresponding to the highest degradation rate. It can be observed that the best performance was obtained with a substrate concentration corresponding to 0.7 mg/L (Figure 3A), while lower values were obtained with the [MO] increase. Basically, the Langmuir–Hinshelwood (LH) kinetic model was not followed at higher concentrations due to the absorption of the UV radiation operated by the dye molecules which limited the photocatalytic process thus becoming bare photolysis.

In fact, the photocatalysis of the adsorbed substrate is based on two simultaneous reactions, oxidation from photogenerated holes (h^+) and reduction from photogenerated electrons (e^-) after excitation of titania with UVB light. Specifically, the oxidation of the adsorbed water by the holes generates hydroxyl radicals –OH while the reduction of the oxygen by the electrons generates superoxide radicals $-O_2^-$, both reacting with the organic molecule [44,45].

The increased concentration of the dye solution in the 1.2–5 mg/L range decreased the interaction of light with the catalyst surface; accordingly, the combined effect of UV radiation and titania started to be less effective.

Figure 3B shows the temporal evolution of the MO UV−vis absorption spectrum for a 5 mg/L influent solution. The spectrum shows a maximum corresponding to the $\pi \rightarrow \pi^*$ transitions of the dimethylamino electron donors at 470 nm and a 270 nm peak associated to $\pi \rightarrow \pi^*$ transitions of the aromatic rings. A UV−vis quenching and blue-shift of the MO absorption peak during the kinetic experiments was observed, which confirmed the degradation of the substrate associated with the removal of the N-methyl groups [46].

The third set of experiments (Table 1, tests no. 7, 8, 9, 10) was carried out at different flow rates (0.066 L/s, 0.147 L/s, 0.210 L/s, 0.305 L/s) with a substrate concentration in the range of 0.3–5.0 mg/L, constant catalyst dosage 0.79 g/cm^3. The variation of the flow rate, constant hydraulic load (h_w = 13.5 cm) was obtained by the introduction of layers of different thicknesses (1.5, 1.0, and 0.5 mm; 11 × 70 cm) into an opening of this tank. It was observed that with the increase of the flow rate, an increase of the degradation kinetics was obtained. This result can be ascribed to an increasingly higher dissolution of the dissolved oxygen into the water solution, which affected the kinetic rates because of the increase of the photogenerated radicals reacting with the substrate. Moreover, a faster re-circulation of the solution and an increase of the irradiated volume (V_{irr}, Table 1) can also explain these trends. Furthermore, in this case, it can be observed that the best performance was obtained with a substrate concentration corresponding to 0.7 mg/L, while lower values were obtained with a further increase up to 5 mg/L. Figure 4B shows the kinetic trend and the temporal evolution of the MO UV–vis absorption spectrum for a 0.7 mg/L influent solution characterized by quenching and blue-shift of the maximum absorption peak.

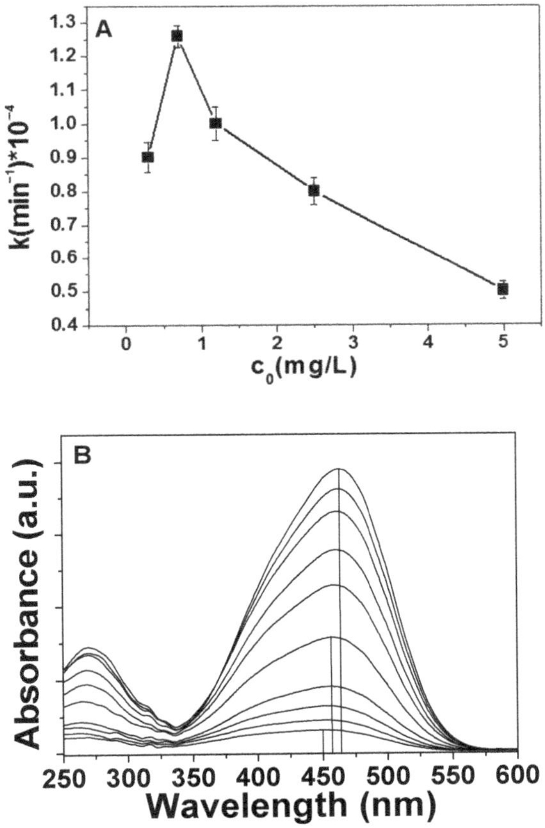

Figure 3. (**A**) Influence of the dye concentration c_0 on the degradation rate (TiO_2 = 0.79 g/cm^3; Q = 0.066 L/s, h_w = 13.5 cm, test no. 7. (**B**) MO UV–Vis absorption spectrum decay during the photo-degradation. MO = 5 mg/L; TiO_2= 0.79 g/cm^3; Q = 0.066 L/s, h_w = 13.5 cm.

The fourth set of experiments (Table 1, tests no. 10, 11, 12, 13) was carried out at different hydraulic loads (h_w) corresponding to different solution volumes V_{sol} (60 L, 72.5 L, 90 L and 105 L) and flow-rates Q (0.305 L/s, 0.355 L/s, 0.408 L/s, 0.441 L/s), constant substrate concentration (c_0 = 0.7 mg/L), and the catalyst dosage (0.79 g/cm^3). These results were combined with the former results regarding the variation of the flow rates at the same hydraulic load (0.066 L/s, 0.147 L/s, 0.210 L/s, 0.305 L/s) and reported in Figure 4C. On the contrary of what expected, the further increase of the flow rates associated with the modification of the hydraulic load determined a decrease of the kinetics because of the increasingly large volume (V_{sol}) of the dye solution to treat (60 L for 0.305 L/s, 72.5 L for

0.355 L/s, 90 L for 0.408 L/s, 105 L for 0.441 L/s) and of the ever shorter catalysts/substrate contact times (I_{rt}) (6.7 s for 0.305 L/s, 5.8 s for 0.355 L/s, 5.2 s for 0.408 L/s, 5.2 s for 0.441 L/s, Table 1).

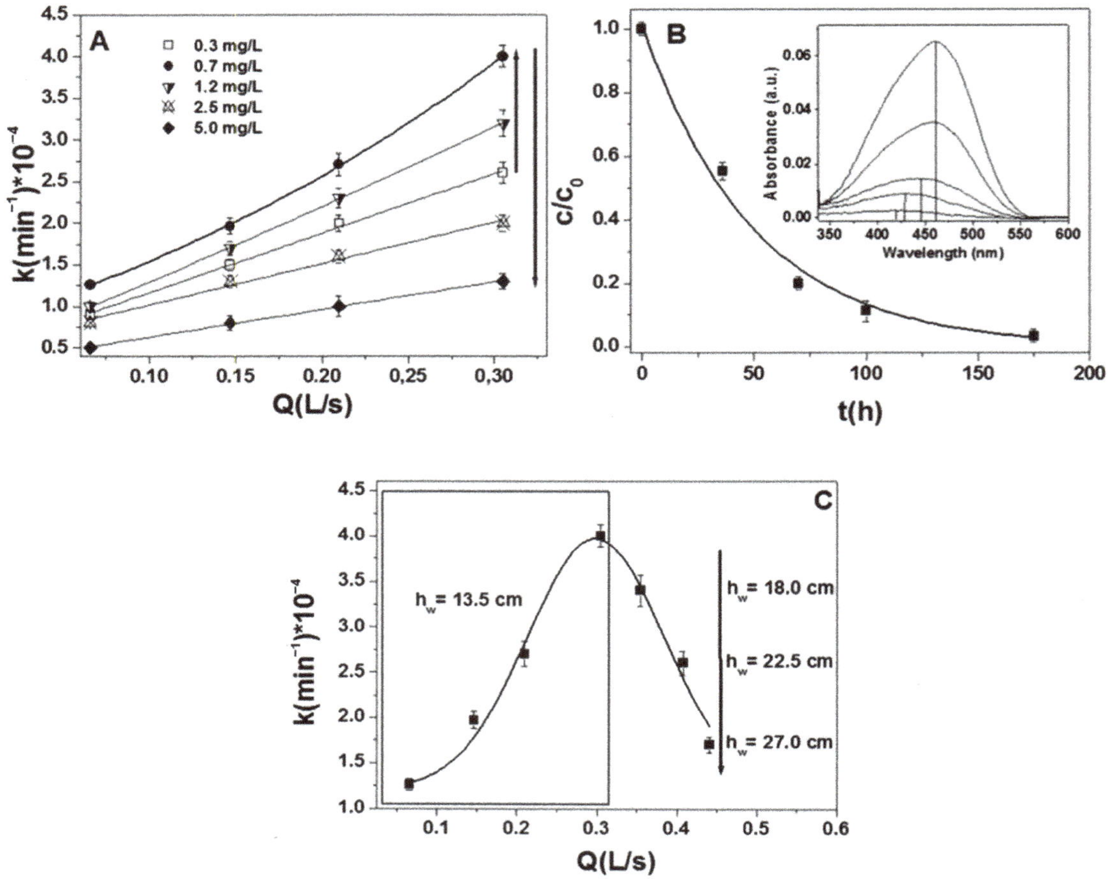

Figure 4. (**A**) Influence of the flow rate at different dye concentrations (TiO_2 = 0.79 g/cm^3, h_w = 13.5 cm, tests no. 7, 8, 9, 10). (**B**) kinetic curve relative to the experiment carried out with MO = 0.7 mg/L; TiO_2 = 0.79 g/cm^3, Q = 0.305 L/h, h_w = 13.5 cm and in the inset the relative UV–vis absorption spectrum decay. (**C**) kinetics obtained at different hydraulic parameters (flow rate Q, hydraulic load h_w, MO = 0.7 mg/L; TiO_2 = 0.79 g/cm^3, tests no. 7–13).

From the kinetic study carried out at variable hydraulic and hydrodynamic conditions, it was observed that the best results were detected at Q = 0.305 L/s and h_W = 13.5 cm hydraulic load (corresponding to the maximum of the curve, Figure 4C) with a dye concentration and TiO_2 dosage, respectively, ranging 0.7 mg/L and 0.79 g/cm^3.

Figure 5 reports how the combined effect of UVB light and catalyst (test no. 10) affects the degradation of the substrate and influences the kinetic rates (fifth set of tests). For this purpose, the photolytic and the bare adsorption tests (tests no. 14 and 15) resulted very slow, thus showing limited effects on the dye removal.

The synergistic combination of irradiation and catalyst can be influenced by the intensity of UVB light. It was studied and represented in Figure 6 reporting the measurements carried out at Q = 0.305 L/s, h_W = 13.5 cm, c_0 = 0.7 mg/L, and TiO_2 = 0.79 g/cm^3. In the present case, the tests were performed with 40 W, 80 W, and 120 W (tests no. 10, 16, and

17) and the results demonstrated the increase of the kinetic rates with the increase of the UVB intensity, although it was observed that passing from 40 W to 120 W the apparent rate constant only doubled.

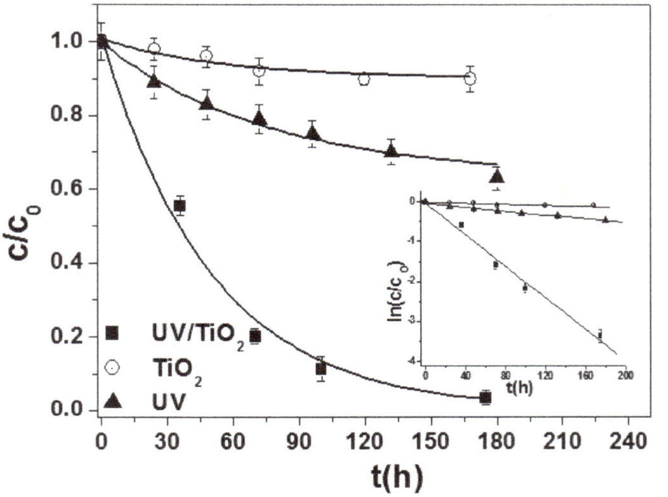

Figure 5. Kinetic trends for the UV/TiO_2 photo-catalysis (test no. 10), UVB photolysis (test no. 14), adsorption (only TiO_2, test no. 15), MO = 0.7 mg/L; TiO_2 = 0.79 g/cm^3; Q = 0.305 L/s; h_w = 13.5. In the inset: lnc/c_0 vs. t correlations.

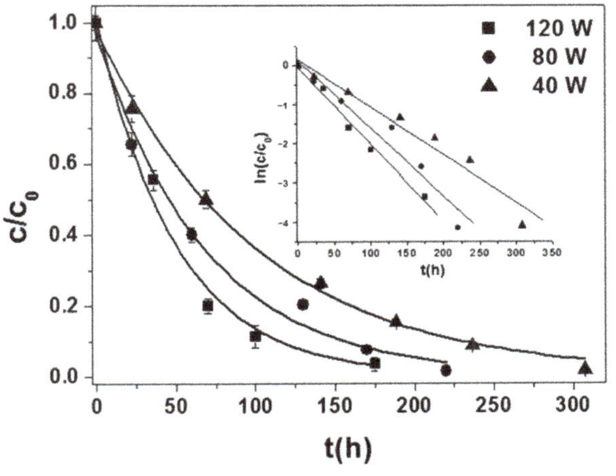

Figure 6. Kinetic trends for the UV/TiO_2 photo-catalysis at different UVB intensity (tests no. 10, 16, 17), MO = 0.7 mg/L; TiO_2 = 0.79 g/cm^3; Q = 0.305 L/s; h_w = 13.5. In the inset: lnc/c_0 vs. t correlations.

Similar results were also observed in the case of other dye substrates as Methyl Red while in the case of Methylene Blue, the kinetic rates dramatically increased from 40 W to 120 W, all the other chemical and hydraulic/hydrodynamic parameters were constant (Table 2).

Basically, the pH of the solution and the pH at zero point charge of TiO_2 (pH_{ZPC}) can influence these results because the surface state of the catalyst and the charge of the substrate functional groups are affected by pH variations [47].

Specifically, the TiO_2 surface is negative (Ti-O^-) at pH higher than the catalyst pH_{zpc} (6.8), while the TiO_2 surface is positive (Ti-OH_2^+) at pH lower than the catalyst pH_{zpc} [41,47]. In the present case, the solutions were prepared in tap water with a pH around 7.5.

The non-polar Methyl Red is not affected by these operative conditions, while Methylene Blue is a cationic dye at all pH [41,47,48]. Methyl Orange is instead characterized by a negative charge; accordingly, the lowest kinetics of MO can be ascribed to the Coulombic repulsion between the negative sulfonate groups of the dye and the negative charged surface of TiO_2 [41,47]. Methyl Red is sorbed by secondary Van der Waals bonds between the hydroxyl and amino functional groups of this substrate and the catalyst, while the best performances with Methylene Blue can be explained by the adsorption of the positive charged dye onto the negative charged catalyst surface (Table 2) [48]. The better interaction between the MB functionalities and the TiO_2 surface can also explain the large increase of the kinetic rate with the increase of the UVB intensity as regard to the other dyes.

Table 2. (A) Apparent rate constants for the UV/TiO_2 photo-catalysis at different UVB intensity (tests no. 10, 16, 17) for Methyl Orange, Methyl Red, and Methylene Blue, [dye] = 0.7 mg/L; TiO_2 = 0.79 g/cm^3; tap water (pH = 7.5); Q = 0.305 L/s; hw = 13.5.

Dye	Molecular Structure	Power (W)	k (min^{-1}) $\times 10^{-4}$
Methyl Orange	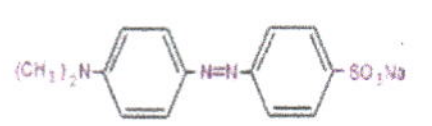	40 80 120	2 ± 0.2 2.9 ± 0.3 4 ± 0.1
Methyl Red	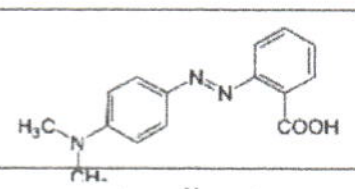	40 80 120	2.6 ± 0.1 3.7 ± 0.1 5.7 ± 0.3
Methylene Blue	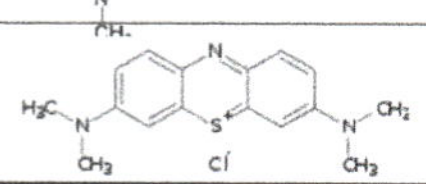	40 80 120	8.8 ± 0.4 19 ± 0.5 35 ± 0.5

It can be also observed that if the measurements were carried out in distilled water (pH = 6), with all the other chemical, physical, and hydraulic/hydrodynamic parameters constant, the kinetics resulted different (tests no. 18). Under these conditions, the interactions of the Methyl Orange functional groups with the catalyst surface were improved with an increase of the value of the apparent rate constant in the range of 40%, quite similar to that obtained with Methyl Red which, as reported before, is not affected by the pH. The kinetic rate of the Methylene Blue showed a decrease in the range of the 20% associated with a decrease of the interactions between the positive charged dye and the more positive catalyst surface (Table 3, Figure 7).

Table 3. (A) Apparent rate constants for the UV/TiO_2 photo-catalysis at 120 W (tests no. 18) for Methyl Orange, Methyl Red, and Methylene Blue, [dye] = 0.7 mg/L; TiO_2 = 0.79 g/cm^3; distilled water (pH = 6); Q = 0.305 L/s; hw = 13.5.

Dye	Molecular Structure	Power (W)	k (min^{-1}) $\times 10^{-4}$
Methyl Orange	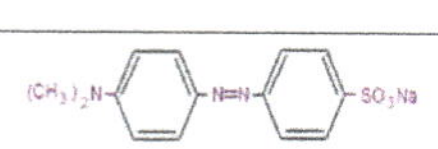	120	5.6 ± 0.3
Methyl Red	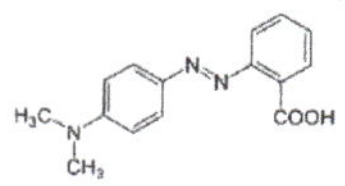	120	5.2 ± 0.3
Methylene Blue	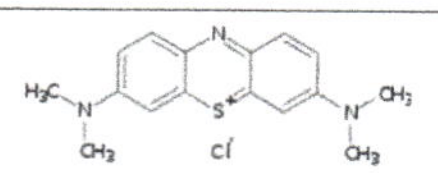	120	29 ± 0.5

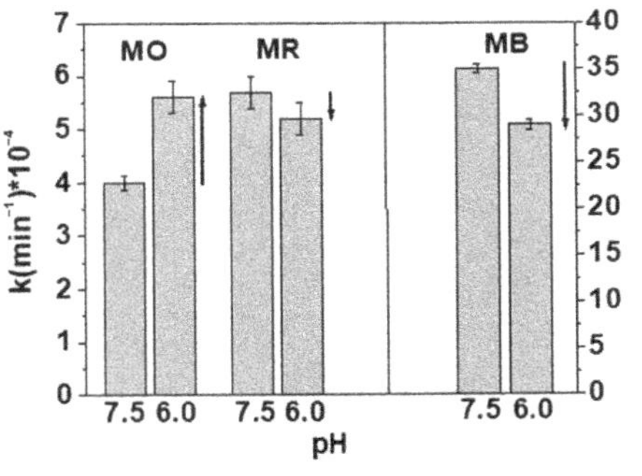

Figure 7. Apparent rate constants in tap water (pH = 7.5) and distilled water (pH = 6) at 120 W (tests no. 10 and 18) for Methyl Orange (MO), Methyl Red (MR), and Methylene Blue (MB), [dye] = 0.7 mg/L; TiO_2 = 0.79 g/cm^3; Q = 0.305 L/s; hw = 13.5.

4. Conclusions

A laboratory-scale unit was employed to study the UVB photo-catalytic degradation of Methyl Orange by anatase TiO_2 embedded in a cement matrix and deposited onto a channel of a recirculating system. The influence of the TiO_2 dosage (g/cm^3), dye concentration (mg/L), flow-rate (L/s), hydraulic load (cm) and irradiation power (W), and pH of the solution were evaluated on the degradation rates.

The degradation rate increased with the increase of the catalyst concentration due to the growing number of the active sites necessary for the photocatalytic oxidation. The maximum dosage of TiO_2 was 0.79 g/cm^3 and with a further increase of oxide concentration, a reduction of homogeneity was observed, which was detrimental for the process since leaching of the catalyst from the cement conglomerate occurred.

The best performance was obtained with a substrate concentration corresponding to 0.7 mg/L, while lower values were obtained with a further increase up to 5 mg/L. The Langmuir–Hinshelwood (LH) kinetic model was followed up to ~1 mg/L concentration; at higher concentrations, a bare photolysis process occurred.

It was observed that with the increase of the flow rate an increase of the degradation kinetics was obtained due to the increasingly higher dissolution of the oxygen into the water solution, the faster re-circulation of the solution, and the increase of the irradiated volume.

The further increase of the flow rates associated with the modification of the hydraulic load determined a decrease of the kinetic rates because of the increasingly large volume of the dye solution to treat and of the ever shorter catalysts/substrate contact times.

The photolytic and the bare adsorption tests showed very slow rates thus demonstrating the effective synergistic action of the UVB light/catalyst system on the dye removal.

An increase of the kinetic rates with the increase of the UVB intensity was observed, although the values only doubled from 40 W to 120 W.

A comparison with other dyes was carried out. Similar results were observed in the case of Methyl Red, while in the case of Methylene Blue, the kinetic rates dramatically increased from 40 W to 120 W.

The pH of the solution influenced these results because the charge of the catalyst surface and the charge of the substrate functional groups were affected by pH variations. For this reason, different results were observed with the different dyes at the pH of tap water and at the pH of distilled water.

As a final remark, the kinetic trends reported in this paper are not easily comparable with literature results due to the different operative conditions of the systems. In particular, in this paper, the measurements were carried out in a re-circulating unit where TiO_2 was immobilized onto a channel. Moreover, in this case, the TiO_2/dye molar ratio was 2 or 3 orders of magnitude lower and the catalyst particle size on the order of micron (lower specific surface area) was deposited and not suspended.

The operations were also carried out with low-pressure UVB lamps and no thermal activation of the film or additions of other oxidants as O_2, H_2O_2, $S_2O_8^{-2}$ were carried out to improve sorption/degradation of Methyl Orange.

Author Contributions: Conceptualization, D.S.; methodology, M.R.; validation, M.R.; formal analysis, V.R.; investigation, A.P.; resources, A.P.; data curation, M.C.M.; writing—original draft preparation, A.P.; writing—review and editing, P.C.; visualization, E.R.; supervision, D.S. All authors have read and agreed to the published version of the manuscript.

Funding: This research received no external funding.

Data Availability Statement: Data sharing not applicable.

Conflicts of Interest: The authors declare no conflict of interest.

References

1. Joseph, L.; Jun, B.M.; Jang, M.; Park, C.M.; Muñoz-Senmache, J.C.; Hernández-Maldonado, A.J.; Heyden, A.; Yu, M.; Yoon, Y. Removal of contaminants of emerging concern by metal-organic framework nanoadsorbents: A review. *Chem. Eng. J.* **2019**, *369*, 928–946. [CrossRef]
2. García-Córcoles, M.T.; Rodríguez-Gómez, R.; de Alarcón-Gómez, B.; Çipa, M.; Martín-Pozo, L.; Kauffmann, J.M.; Zafra-Gómez, A. Chromatographic methods for the determination of emerging contaminants in natural water and wastewater samples: A review. *Crit. Rev. Anal. Chem.* **2019**, *49*, 160–186. [CrossRef] [PubMed]
3. Alimi, O.S.; Budarz, J.F.; Hernandez, L.M.; Tufenkji, N. Microplastics and nanoplastics in aquatic environments: Aggregation, deposition, and enhanced contaminant transport. *Environ. Sci. Technol.* **2018**, *52*, 1704–1724. [CrossRef] [PubMed]
4. Petrella, A.; Petruzzelli, V.; Basile, T.; Petrella, M.; Boghetich, G.; Petruzzelli, D. Recycled porous glass from municipal/industrial solid wastes sorting operations as a lead ion sorbent from wastewaters. *React. Funct. Polym.* **2010**, *70*, 203–209. [CrossRef]
5. Petrella, A.; Petrella, M.; Boghetich, G.; Basile, T.; Petruzzelli, V.; Petruzzelli, D. Heavy metals retention on recycled waste glass from solid wastes sorting operations: A comparative study among different metal species. *Ind. Eng. Chem. Res.* **2012**, *51*, 119–125. [CrossRef]
6. Tammaro, M.; Fiandra, V.; Mascolo, M.C.; Salluzzo, A.; Riccio, C.; Lancia, A. Photocatalytic degradation of atenolol in aqueous suspension of new recyclable catalysts based on titanium dioxide. *J. Environ. Chem. Eng.* **2017**, *5*, 3224–3234. [CrossRef]
7. Quesada, H.B.; Baptista, A.T.A.; Cusioli, L.F.; Seibert, D.; de Oliveira Bezerra, C.; Bergamasco, R. Surface water pollution by pharmaceuticals and an alternative of removal by low-cost adsorbents: A review. *Chemosphere* **2019**, *222*, 766–780. [CrossRef]
8. Sillanpää, M.; Ncibi, M.C.; Matilainen, A.; Vepsäläinen, M. Removal of natural organic matter in drinking water treatment by coagulation: A comprehensive review. *Chemosphere* **2018**, *190*, 54–71. [CrossRef]
9. Petrella, A.; Spasiano, D.; Rizzi, V.; Cosma, P.; Race, M.; De Vietro, N. Thermodynamic and kinetic investigation of heavy metals sorption in packed bed columns by recycled lignocellulosic materials from olive oil production. *Chem. Eng. Comm.* **2019**, 1–16. [CrossRef]
10. Bai, X.; Acharya, K. Removal of seven endocrine disrupting chemicals (EDCs) from municipal wastewater effluents by a freshwater green alga. *Environ. Pollut.* **2019**, *247*, 534–540. [CrossRef] [PubMed]
11. Spasiano, D.; Luongo, V.; Petrella, A.; Alfè, M.; Pirozzi, F.; Fratino, U.; Piccinni, A.F. Preliminary study on the adoption of dark fermentation as pretreatment for a sustainable hydrothermal denaturation of cement-asbestos composites. *J. Clean. Prod.* **2017**, *166*, 172–180. [CrossRef]
12. Petrella, A.; Petruzzelli, V.; Ranieri, E.; Catalucci, V.; Petruzzelli, D. Sorption of Pb(II), Cd(II) and Ni(II) from single- and multimetal solutions by recycled waste porous glass. *Chem. Eng. Commun.* **2016**, *203*, 940–947. [CrossRef]
13. Rizzi, V.; D'Agostino, F.; Gubitosa, J.; Fini, P.; Petrella, A.; Agostiano, A.; Semeraro, P.; Cosma, P. An alternative use of olive pomace as a wide-ranging bioremediation strategy to adsorb and recover disperse orange and disperse red industrial dyes from wastewater. *Separations* **2017**, *4*, 29. [CrossRef]
14. Zazou, H.; Afanga, H.; Akhouairi, S.; Ouchtak, H.; Addi, A.A.; Akbour, R.A.; Assabane, A.; Douch, J.; Elmchaour, A.; Duplay, J.; et al. Treatment of textile industry wastewater by electrocoagulation coupled with electrochemical advanced oxidation process. *J. Water Process Eng.* **2019**, *28*, 214–221. [CrossRef]
15. Brillas, E. A review on the photoelectro-Fenton process as efficient electrochemical advanced oxidation for wastewater remediation. Treatment with UV light, sunlight, and coupling with conventional and other photo-assisted advanced technologies. *Chemosphere* **2020**, 126198. [CrossRef] [PubMed]
16. Amor, C.; Marchão, L.; Lucas, M.S.; Peres, J.A. Application of advanced oxidation processes for the treatment of recalcitrant agro-industrial wastewater: A review. *Water* **2019**, *11*, 205. [CrossRef]
17. Vandenberg, L.N.; Luthi, D.; Quinerly, D.A. Plastic bodies in a plastic world: Multi-disciplinary approaches to study endocrine disrupting chemicals. *J. Clean. Prod.* **2017**, *140*, 373–385. [CrossRef]
18. Plahuta, M.; Tišler, T.; Toman, M.J.; Pintar, A. Toxic and endocrine disrupting effects of wastewater treatment plant influents and effluents on a freshwater isopod Asellus aquaticus (Isopoda, Crustacea). *Chemosphere* **2017**, *174*, 342–353. [CrossRef]

19. Gubitosa, J.; Rizzi, V.; Lopedota, A.; Fini, P.; Laurenzana, A.; Fibbi, G.; Fanelli, F.; Petrella, A.; Laquintana, V.; Denora, N.; et al. One pot environmental friendly synthesis of gold nanoparticles using Punica Granatum Juice: A novel antioxidant agent for future dermatological and cosmetic applications. *J. Colloid Interface Sci.* **2018**, *521*, 50–61. [CrossRef]
20. Petrella, A.; Spasiano, D.; Rizzi, V.; Cosma, P.; Race, M.; De Vietro, N. Lead ion sorption by perlite and reuse of the exhausted material in the construction field. *Appl. Sci.* **2018**, *8*, 1882. [CrossRef]
21. Tayo, L.L.; Caparanga, A.R.; Doma, B.T.; Liao, C.H. A Review on the removal of pharmaceutical and personal care products (PPCPs) using advanced oxidation processes. *J. Adv. Oxid. Technol.* **2018**, *21*, 196–214. [CrossRef]
22. Wang, C.; Kim, J.; Malgras, V.; Na, J.; Lin, J.; You, J.; Zhang, M.; Li, J.; Yamauchi, Y. Metal–organic frameworks and their derived materials: Emerging catalysts for a sulfate radicals-based advanced oxidation process in water purification. *Small* **2019**, *15*, 1900744. [CrossRef]
23. Oppenlander, T. *Advanced oxidation processes (AOPs): Principles, Reaction Mechanisms, Reactor Concepts*; Wiley VCH: Weinheim, Germany, 2007.
24. Al-Mamun, M.R.; Kader, S.; Islam, M.S.; Khan, M.Z.H. Photocatalytic activity improvement and application of UV-TiO_2 photocatalysis in textile wastewater treatment: A review. *J. Environ. Chem. Eng.* **2019**, *7*, 103248. [CrossRef]
25. Mascolo, M.C.; Ring, T.A. Recyclable aggregates of mesoporous titania synthesized by thermal treatment of amorphous or peptized precursors. *Materials* **2018**, *11*, 381. [CrossRef] [PubMed]
26. Moreira, N.F.; Narciso-da-Rocha, C.; Polo-López, M.I.; Pastrana-Martínez, L.M.; Faria, J.L.; Manaia, C.M.; Fernandez-Ibanez, F.; Nunes, O.C.; Silva, A.M. Solar treatment (H_2O_2, TiO_2-P25 and GO-TiO_2 photocatalysis, photo-Fenton) of organic micropollutants, human pathogen indicators, antibiotic resistant bacteria and related genes in urban wastewater. *Water Res.* **2018**, *135*, 195–206. [CrossRef] [PubMed]
27. Mascolo, M.C. Synthesis of wide spectrum of mesoporous titania materials by forced co-hydrolysis of Zr–Ti alkoxides. Micropor. *Mesopor. Mater.* **2013**, *181*, 160–165. [CrossRef]
28. Petrella, A.; Cozzoli, P.D.; Curri, M.L.; Striccoli, M.; Cosma, P.; Agostiano, A. Photoelectrochemical study on photosynthetic pigments-sensitized nanocrystalline ZnO films. *Bioelectrochemistry* **2004**, *63*, 99–102. [CrossRef] [PubMed]
29. Camarillo, R.; Rincon, J. Photocatalytic discoloration of dyes: Relation between effect of operating parameters and dye structure. *Chem. Eng. Technol.* **2011**, *34*, 1675–1684. [CrossRef]
30. Yagub, M.T.; Sen, T.K.; Afroze, S.; Ang, H.M. Dye and its removal from aqueous solution by adsorption: A review. *Adv. Colloid Interface Sci.* **2014**, *209*, 172–184. [CrossRef]
31. Labidi, A.; Salaberria, A.M.; Fernandes, S.; Labidi, J.; Abderrabba, M. Functional chitosan derivative and chitin as decolorization materials for Methylene Blue and Methyl Orange from aqueous solution. *Materials* **2019**, *12*, 361. [CrossRef]
32. Tan, I.A.W.; Ahmad, A.L.; Hameed, B.H. Adsorption of basic dye using activated carbon prepared from oil palm shell: Batch and fixed bed studies. *Desalination* **2008**, *225*, 13–28. [CrossRef]
33. Srinivasan, A.; Viraraghavan, T. Decolorization of dye wastewaters by biosorbents: A review. *J. Environ. Manag.* **2010**, *91*, 1915–1929. [CrossRef] [PubMed]
34. Badr, Y.; El-Wahed, M.A.; Mahmoud, M.A. Photocatalytic degradation of methyl red dye by silica nanoparticles. *J. Hazard. Mater.* **2008**, *154*, 245–253. [CrossRef] [PubMed]
35. Waghmode, T.R.; Kurade, M.B.; Sapkal, R.T.; Bhosale, C.H.; Jeon, B.H.; Govindwar, S.P. Sequential photocatalysis and biological treatment for the enhanced degradation of the persistent azo dye methyl red. *J. Hazard. Mater.* **2019**, *371*, 115–122. [CrossRef] [PubMed]
36. Mittal, A.; Malviya, A.; Kaur, D.; Mittal, J.; Kurup, L. Studies on the adsorption kinetics and isotherms for the removal and recovery of methyl orange from wastewaters using waste materials. *J. Hazard. Mater.* **2007**, *148*, 229–240. [CrossRef]
37. Lellis, B.; Fávaro-Polonio, C.Z.; Pamphile, J.A.; Polonio, J.C. Effects of textile dyes on health and the environment and bioremediation potential of living organisms. *Biotechnol. Res. Innov.* **2019**, *3*, 275–290. [CrossRef]
38. Petrella, A.; Boghetich, G.; Petrella, M.; Mastrorilli, P.; Petruzzelli, V.; Petruzzelli, D. Photocatalytic degradation of azo dyes. Pilot plant investigation. *Ind. Eng. Chem. Res.* **2014**, *53*, 2566–2571. [CrossRef]
39. Petrella, A.; Mascolo, G.; Murgolo, S.; Petruzzelli, V.; Ranieri, E.; Spasiano, D.; Petruzzelli, D. Photocatalytic oxidation of organic micro-pollutants: Pilot plant investigation and mechanistic aspects of the degradation reaction. *Chem. Eng. Commun.* **2016**, *203*, 1298–1307. [CrossRef]
40. Petrella, A.; Spasiano, D.; Cosma, P.; Rizzi, V.; Race, M. Evaluation of the hydraulic and hydrodynamic parameters influencing photo-catalytic degradation of bio-persistent pollutants in a pilot plant. *Chem. Eng. Comm.* **2019**, *206*, 1286–1296. [CrossRef]
41. Konstantinou, T.; Albanis, A. TiO_2-assisted photocatalytic degradation of azo dyes in aqueous solution: Kinetic and mechanistic investigations. A review. *Appl. Catal. B Environ.* **2004**, *49*, 1–14. [CrossRef]
42. Kim, S.H.; Ngo, H.H.; Shon, H.K.; Vigneswaran, S. Adsorption and photocatalysis kinetics of herbicide onto titanium oxide and powdered activated carbon. *Sep. Purif. Technol.* **2008**, *58*, 335–342. [CrossRef]
43. Tang, W.Z.; An, H. Photocatalytic oxidation of commercial dyes in aqueous solutions. *Chemosphere* **1995**, *31*, 4157–4170. [CrossRef]
44. Giraldo, A.L.; Penuela, G.A.; Torres-Palma, R.A.; Pino, N.J.; Palominos, R.A.; Mansilla, H.D. Degradation of the antibiotic oxolinic acid by photocatalysis with TiO_2 in suspension. *Water Res.* **2010**, *44*, 5158–5167. [CrossRef] [PubMed]
45. Friedmann, D.; Mendice, C.; Bahnemann, D. TiO_2 for water treatment: Parameters affecting the kinetics and mechanisms of photocatalysis. *Appl. Catal. B* **2010**, *99*, 398–406. [CrossRef]

46. Comparelli, R.; Fanizza, E.; Curri, M.L.; Cozzoli, P.D.; Mascolo, G.; Passino, R.; Agostiano, A. Photocatalytic degradation of azo dyes by organic-capped anatase TiO_2 nanocrystals immobilized onto substrates. *Appl. Catal. B* **2005**, *55*, 81–91. [CrossRef]
47. Guillard, C.; Lachheb, H.; Houas, A.; Ksibi, M.; Elaloui, E.; Herrmann, J.M. Influence of chemical structure of dyes, of pH and of inorganic salts on their photocatalytic degradation by TiO_2 comparison of the efficiency of powder and supported TiO_2. *J. Photoch. Photobio. A* **2003**, *158*, 27–36. [CrossRef]
48. Franco, A.; Neves, M.C.; Ribeiro Carrott, M.M.L.; Mendonca, M.H.; Pereira, M.I.; Monteiro, O.C. Photocatalytic decolorization of methylene blue in the presence of TiO_2/ZnS nanocomposites. *J. Hazard. Mater.* **2009**, *161*, 545–550. [CrossRef]

MDPI
St. Alban-Anlage 66
4052 Basel
Switzerland
Tel. +41 61 683 77 34
Fax +41 61 302 89 18
www.mdpi.com

Processes Editorial Office
E-mail: processes@mdpi.com
www.mdpi.com/journal/processes

www.ingramcontent.com/pod-product-compliance
Lightning Source LLC
Chambersburg PA
CBHW041947260926
51850CB00018B/1860
9783036589459